Making Sense of Data through Statistics: An Introduction
By Dorit Nevo

Second Edition Copyright © 2017
First Edition Copyright © 2014

All rights reserved. No part of this publication shall be reproduced, distributed, or transmitted in any form or by any means, electronic or mechanical, including photocopying, recording, or by any information retrieval system without the prior written permission of the publisher, except in the case of brief quotations embodied in critical reviews and certain other noncommercial uses permitted by copyright law. For permission requests, email the publisher at: admin@ldpress.com.

Published by Legerity Digital Press, LLC

A catalog record for this book is available from the U.S. Library of Congress.

ISBN 978-0-9857955-8-0

Although every precaution has been taken in the preparation of this book, the publisher and author assume no responsibility for errors or omissions. Neither is any liability assumed for damages resulting from the use of this information contained herein.

Ordering information:
For all ordering inquiries, please visit www.ldpress.com, email sales@ldpress.com or call toll free at 855-855-9868. Special discounts are available on bulk purchases by academic institutions, associations, corporations, and others.

Printed in the United States of America.

Cover Illustration by Aaron Z. Williams

DEDICATION

To Lia, Liv and Saggi

TABLE OF CONTENTS

Chapter 1: Introduction to Data and Statistics..	1
Chapter 2: Data Presentation..	35
Chapter 3: Measures of Centrality and Variation...	88
Chapter 4: Probability..	128
Chapter 5: Discrete Probability Distributions...	162
Chapter 6: Continuous Probability Distributions..	203
Chapter 7: Introducing Hypothesis Testing..	234
Chapter 8: Additional Concepts in Hypothesis Testing..................................	286
Chapter 9: Hypothesis Testing for a Population Proportion and Variance.....	321
Chapter 10: Hypothesis Testing of Parameters from Two Populations.........	355
Chapter 11: Chi-Square Tests..	413
Chapter 12: Analysis of Variance...	438
Chapter 13: Regression Analysis..	493
Appendix A: Using the Microsoft Excel *Analysis ToolPak* for Testing Hypotheses for Parameters from Two Populations..	A-1
Excel Function Index...	EFI-1
Equation Index..	EI-1
Key Term Index...	KTI-1

CHAPTER 1: Introduction to Data and Statistics

Statistics is the study of data and how it can be collected, organized, analyzed and interpreted to obtain insights about people's opinions and behavior, about the success of experiments or about industry trends, to mention only a few examples. **Descriptive statistics** focuses on organizing and summarizing data so that it is better understood. For example, when you organize your monthly expenses in a table and calculate your average spending, you use descriptive statistics to better understand your budget. **Inferential statistics** leverages data from a small group to arrive at conclusions about the entire population of which the small group is a part. For example, you can study the difference in monthly expenses of male versus female students at your school and *infer* from your data conclusions about relationships between gender and spending for all college students.

Statistics is part of everyday life. When you read newspapers, talk to your friends or play sports, you hear statements such as: "70 percent of Americans believe that global warming is a real problem."; "The average commute time in New York City is 31 minutes."; or, "My softball batting average is 65 percent." In business, too, statistics plays an important role in making decisions. In fact, one of the most sought after job areas these days is **business analytics**, which refers to the application of statistics to obtain important insights from data available to organizations. For example, statistics is used in marketing to analyze market trends, compare among different groups of consumers and find out the success of various promotions. In finance, the standard deviation[1] of a stock's returns is an important indicator of its volatility and risk. In programming, we can reduce the cost of testing software code by scrutinizing a sample of code and drawing an *inference* about the total number of errors remaining in a program under development. In human resources, we can examine average salaries of different groups within the organization and compare these to industry standards. These are only a few of the applications of statistics in business and we will look at many more examples throughout this book.

In this chapter, we introduce key concepts in statistics, along with definitions and examples. Each unit provides a review of one topic, followed by a unit summary section reiterating the key lessons of the unit, and a unit exercise section ensuring concepts are clear before moving on to the next unit. At the end of the chapter, there is a chapter practice section covering all the topics previously introduced. This chapter covers the following topics:

- Variables and Data
- Data Types
- Scales of Measurement
- Populations and Samples
- Sampling Approaches
- Sample Size
- A Quick Look at Data Collection
- From Statistics to Analytics

---

[1] What is meant by 'standard deviation' will be covered later in this book.

UNIT 1

Variables and Data

At the heart of statistics are variables and data. A ***variable*** is a characteristic of individuals or objects. For example, students' grades, height and income are all variables of students. Variables have values they can attain. For example, a student's grade can have any value between 0 and 100. ***Data*** are the observed values of the variables. In this grades example, we can record data on the grades of all students in our class (e.g., 67, 74, 71, 83, 93, 55, 48 …). Consider, as another example, a laptop computer. Brand, RAM and Screen Size are examples of variables associated with laptop computers. The Brand variable can have values such as 'Dell', 'HP', 'Lenovo', etc. If we collect data on the laptops used by students in a particular class, we may find that three students have a Dell laptop, that five students have an HP laptop, and so on.

In and of themselves, data do not really tell us much about the world. However, when we classify, present, summarize and analyze data, we can draw important conclusions about phenomena of interest. For example, consider the data presented in Table 1-1. The numbers represent the percent of Internet users (over 18 years of age) who engage in different online activities, such as sending or reading email, getting news online or using social networking sites, based on the 2010 census data. To obtain these data, the researchers likely asked respondents to report on three variables: Internet Activity, Age and Gender. We can use these data to make inferences about patterns in overall Internet usage. For example, the data tell us that 62% of adult Internet users use email and 49% use search engines to find information. Take a few moments to consider this table and see what you can learn from it. Do there appear to be different use patterns for the different age groups? Are there some apparent use differences between genders?

Table 1-1
Typical Daily Internet Activities of Adult Internet Users[2]

Activity	Percent of Internet Users	Age				Sex	
		18 to 29	30 to 49	50 to 64	65 and over	Male	Female
Send or read email	62	62	67	60	55	61	63
Use a search engine to find information	49	55	54	42	34	49	49
Get news online	43	44	45	42	34	48	38
Check weather reports and forecasts online	34	38	37	27	27	37	31
Look for news or information about politics	19	18	22	17	19	23	16
Do any banking online	26	27	30	22	19	27	25
Watch a video on a video-sharing site	23	39	20	12	17	26	20
Use a social networking site	38	60	39	20	13	34	41
Send instant messages	15	24	15	9	4	13	17
Visit a local, state or federal government website	12	11	12	12	10	15	9
Get financial information online	12	9	14	13	14	19	6
Buy a product online	8	7	10	6	6	8	7
Look online for info. about a job	10	13	10	7	4	10	9
Use online classified ads or sites like Craig's List	11	14	13	6	5	13	8
Create or work on your own online journal or blog	4	6	4	4	4	4	5
Buy or make a reservation for travel	5	5	6	4	3	5	5
Rate a product, service or person	4	4	4	6	2	4	5
Participate in an online auction	4	5	5	4	2	6	3
Download a podcast so you can listen to or view it later	3	5	2	3	1	4	2
Make a donation to a charity online	1	2	1	1	1	1	1
Look for information on Wikipedia	17	29	15	11	4	21	14
Look for religious or spiritual information online	5	5	5	6	4	6	4

---

[2] 2010 US Census Table 1159. Source: Pew Internet & American Life Project Surveys.

From the data in Table 1-1, it seems that a higher percentage of female Internet users (41%) use social networking sites, such as *Facebook*, than do male Internet users (34%). At the outset, you may choose to believe or not to believe this claim, based on your own knowledge and experience. You could also use statistics to test this claim and reach a conclusion with some level of confidence. For example, you may design a study in which you compare the number of female and male *Facebook* users. In this case, your *research question* would be whether or not gender affects the use of social networking sites.

Data Types

As discussed in the previous section, a variable is a characteristic of an individual or an object, and data are the observed values of variables (it is what we collect, analyze and report). Data can be *qualitative* or *quantitative*. **Qualitative** data are categorical (e.g., marital status, gender, laptop brand name). **Quantitative** data are numerical (e.g., height, income, age). Quantitative data can be further characterized as either *continuous* or *discrete*. **Continuous** data are the result of a *measurement* process, while **discrete** data are the result of a *counting* process. To understand the difference between continuous and discrete data, consider an analog versus a digital clock. The analog clock shows continuous time; that is, the hands move over the full perimeter of the clock's face. However, the digital clock is limited to displaying the units in which the clock operates (seconds, milliseconds, etc.). Figure 1-1 further illustrates these data types, using examples of business-related variables.

Figure 1-1
Data Types

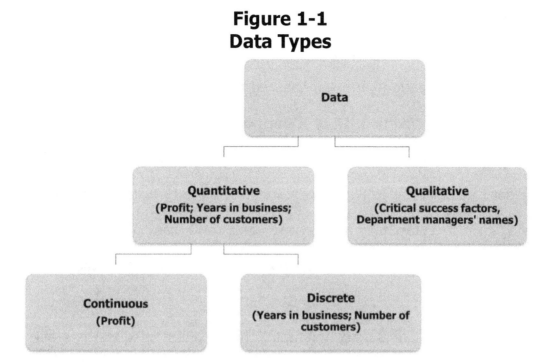

Scales of Measurement

Beyond the above classification of data as being qualitative or quantitative and as being continuous or discrete, it is also important to identify how we measure variables, or the **scales of measurement**. For example, distance is measured as miles, temperature is measured as degrees, and profits are measured as currency. There are four broad scales of measurement, namely *nominal, ordinal, interval* and *ratio*, that ultimately determine the type of analysis that can be conducted on a variable.

Assume you are asked to respond to a survey conducted by a market research company. The first set of questions ask: (1) "Where do you live?", (2) "What is your gender?", and (3) "What is your marital status?".

Let us say that your answers are (1) New York, (2) female and (3) single. Note that the answers to these questions are words that have no numerical interpretation. Therefore, these questions are all measured on a **nominal** scale.

Nominal variables can only be measured in terms of distinct *categories*. For example, gender can be measured as either male or female; and, departmental affiliation in an organization can be measured by where a person works (i.e., in an accounting, finance, marketing, IT or 'Other' department). There is no ordering to nominal data (e.g., we cannot say that the data value 'Accounting' is more or less than the data value 'Finance') and we cannot apply any arithmetic operations on these data, such as computing an average value. We can, however, study counts and proportions of nominal data (e.g., number of males, percentage of females, etc.).

Continuing with our survey example, the next set of questions ask you to state your level of education (i.e., high school, some college, bachelor degree, master's degree or higher degree) and how satisfied you are with your current cell phone provider ('not at all satisfied', 'somewhat satisfied' or 'extremely satisfied'). These two variables, education level and cell phone satisfaction, are measured on an **ordinal** scale. Beyond counts and proportions, ordinal scales allow us to rank order the items we measure. For example, we can say that 'somewhat satisfied' is better than 'not at all satisfied'. We cannot, however, quantify the difference between the different levels of satisfaction. In other words, we don't really know whether the difference between 'not at all satisfied' and 'somewhat satisfied' is the same as the difference between 'somewhat satisfied' and 'extremely satisfied'. To quantify such differences, we need to move into the quantitative realm.

Going back to our survey example, let us look at a third set of survey questions: (1) "What is your GMAT[3] score?", (2) "What is your annual income (in thousands of dollars)?" and (3) "What is your age?".

Consider first the question concerning your GMAT score. It is measured on an **interval** scale. The questions about your income and your age are measured on a **ratio** scale. Both interval and ratio scales are quantitative and allow us not only to rank order the items that are measured, but also to quantify and compare the differences between values. The difference between interval and ratio measurement scales is that while the ratio scale has a well-defined zero value, the interval scale does not. Consider, for example, the GMAT score question. The value of the GMAT score ranges from 200 to 800. Therefore, a GMAT score of 800 should not be interpreted as twice as high as a score of 400 (since the starting point of the measurement scale is 200 rather than zero). In other words, we cannot talk about ratios[4] with an interval scale. Another example for interval scale is temperature. When the weather outside is 80°F, we cannot say that it is *twice as hot* as 40°F. This is because a temperature of 0°F does not mean 'no temperature'; rather, it means 'a temperature of 0°F'.

Now consider a ratio scale example. When we consider income (a ratio variable), we can certainly say that a person who makes $80,000 per year makes exactly twice as much as a person who makes $40,000 per year. This is because an income of zero dollars indeed means 'no income'. Most quantitative variables you normally think of tend to fall under the ratio scale (income, height, distance,

---

[3] GMAT stands for Graduate Management Admissions Test, the standardized test generally taken when applying to an MBA program.
[4] By 'ratio' we mean the form: $\frac{x}{y}$.

age, length of time, debt amount, etc.). Interval variables are encountered less often, with common examples being IQ scales and temperature scales (Fahrenheit and Celsius).

Figure 1-2 summarizes the key attributes of the four scales of measurement.

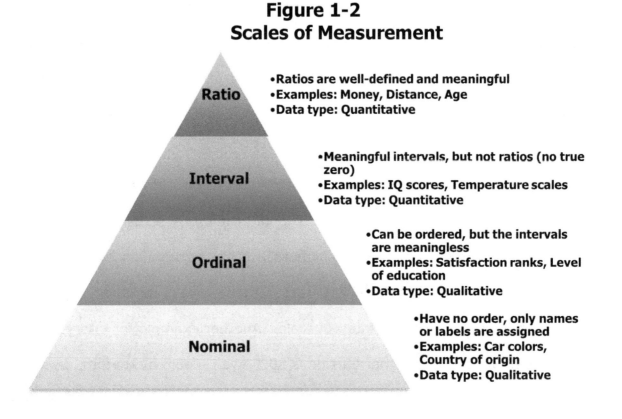

To make sure these concepts are clear, consider a final example relating to college course grades. Some schools grade students on a scale of 0 to 100, while other schools use a letter grade scale (A+, A ...). Although the variable itself remains the same (Course Grade), the variable's measurement differs: the numerical grade is measured on a ratio scale and the letter grade is measured on an ordinal scale. You can see the difference between the scales by trying to calculate your grade average. This is not a difficult task if your grades are: 95, 93, 88 and 79. But, what is your average if you have grades of: A+, A+, A, and B? As mentioned above, arithmetic operations cannot be carried out with ordinal-scaled data.

Unit 1 Summary

- ***Statistics*** is the study of data and how it can be collected, organized, analyzed and interpreted. ***Descriptive*** statistics focuses on organizing and summarizing data so that it is better understood, while ***inferential*** statistics leverages data from a small group to arrive at conclusions about the population as a whole. A popular application of statistics to the business world is ***business analytics,*** which focuses on analyzing business data for the purpose of obtaining business-relevant insights.

- A **variable** is a characteristic of individuals or objects. **Data** are the observed values of variables.

- Data can be **qualitative** or **quantitative**, **continuous** or **discrete**. Qualitative data are categorical, whereas quantitative data are numerical. Furthermore, quantitative data can be the result of a measurement process (continuous) or a counting process (discrete). The type of data you are working with and the measurement scale you use will determine the type of analysis you can conduct.

- Data can be collected using different **scales of measurement**, which define how we measure the values of a given variable. The four scales differ in terms of the arithmetic operations you can use with the data (e.g., can you compute an average or a ratio of two data points?).

- **Nominal** and **ordinal** scales are qualitative, thus arithmetic operations beyond counts and proportions are meaningless. The difference between the two is that ordinal scales provide insights on the rank order of items, allowing us to make statements such as 'A is greater than B'.

- **Ratio** and **interval** scales are quantitative. Since interval scales do not have a meaningful zero value, certain arithmetic operations (such as those involving ratios) are not allowed. Variables representing ratio scales are not restricted in the arithmetic operations they can support.

Unit 1 Exercises

1. The following survey was designed to assess the usage of email versus text messages by young professionals. For each question, please identify the proper measurement scale (nominal, ordinal, interval or ratio):

    a. How often do you use email at work? (a) Never; (b) Sometimes; (c) Often; (d) All the time

b. On average, how many text messages do you send on a given day?

   c. Do you read email on devices other than your personal computer? Yes/No

   d. How many email accounts do you have?

   e. Do you send and receive text messages while you are at work? Yes/No

   f. How concerned are you about your work/life balance? (a) Not at concerned; (b) Somewhat concerned; (c) Extremely concerned

   g. In what industry do you work?

   h. What is your level of education? (a) High school; (b) Some college; (c) Bachelor degree or professional certificate; (d) Master's degree or higher

   i. How many years of experience do you have?

   j. What is your age? (a) 18-24; (b) 25-34; (c) 35-44; (d) 45 or older

   k. What is your gender?

2. We discussed temperature and IQ scores as examples of interval scales. Another example is the calendar year (e.g., 2011). Explain why the calendar represents an interval measurement scale. What scale is the measurement scale for time (e.g., seconds, minutes, hours, etc.)?

3. Table 1-e1, below, was constructed to learn about the demographics of college professors in the United States. As data are collected, they would be recorded in the table below.

   a. What are the variables in this table?

   b. For each variable, state whether data collected would be qualitative or quantitative and whether it is discrete or continuous.

   c. Give an example of one data value that can be assigned for each variable.

Table 1-e1
Demographics

State of Residence	Age	Income	Marital Status	Number of Children	Highest Level of Education
...	...	...	...	...	...

4. Examine Table 1-1 in the opening section of this chapter. What is the measurement scale for the Age variable, as measured in this table?

5. You were hired to conduct a market study for a new soft drink company. Your task is to characterize the typical customer for the company's products on campus. Identify five variables of interest for this study and how each of these variables can be measured.

UNIT 2

Populations and Samples

A ***population*** is a collection of all objects or individuals of interest. It is the group from which we wish to collect data and about which we would like to learn. For example, if we were interested in studying the effect of gender on the amount of time spent using *Facebook*, our population of interest would be all *Facebook* users (both male and female).

To answer our research question of whether there are gender differences in the time spent on *Facebook*, we may decide to collect data about the gender of all *Facebook* users. This is called a *census*. A ***census*** collects data about each and every member of a population. A census is generally difficult to conduct and very resource intensive (just consider the fact that there are over one and a half billion active *Facebook* users!). A more efficient approach might be to select a *sample* out of the population of *Facebook* users.

A ***sample*** is a subset of the population being studied and, based on which, some knowledge about the population is obtained. For example, we can select a sample of 50 *Facebook* users to study a claim about gender differences. Figure 1-3 illustrates the notions of population and sample.

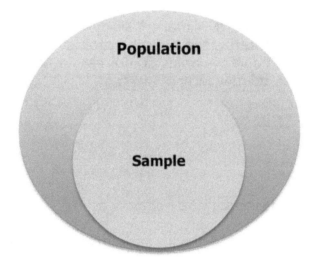

**Figure 1-3
Populations and Samples**

There are several approaches to ***sampling*** (i.e., selecting samples from populations) and several important decisions to make. We review these decisions in this unit.

Consider first the following possible samples to test a *Facebook* claim about gender differences:

1. A sample consisting of 100 of your classmates
2. A sample consisting of all 350 of your *Facebook* friends
3. A sample consisting of 50 randomly selected *Facebook* users

What is a strength and a weakness of each sample? The first two samples are *convenient* in the sense that you have easy access to these people and they are likely to give you the data you are looking for. These two samples' weakness, however, is that they most likely are not *representative* of the entire *Facebook* population. For example, if you are a male, you likely have more male than female friends. The results you would obtain will likely be *biased* toward any overrepresented characteristics of the population. The third sample may be more representative of the population, but consists of only 50 people, so we need to consider the *sample size* and whether it is appropriate for our study. The sections below detail some of the key decisions around statistical sampling.

Sampling Approaches

There are many factors that need to be taken into account when selecting samples for analysis. A key concept in sampling is *randomness*, and it has to do with the chances of each person, or object, in the population being selected for the study. **Random sampling** means that samples are selected so that every person, or object, in a population has an equal chance of being chosen. Random sampling is important in order to prevent **sampling bias**. If a sample has a group that is overrepresented or underrepresented, the sample is considered to be biased. Biased samples can cause the researcher to make inaccurate conclusions about a population.

There are several causes of sampling bias. One cause of sampling bias is the timeframe at which data are collected. For example, if you wanted to find out the number of cars that use the highway, you should collect data at different times of the day and different days of the week. This ensures that you do not obtain only rush hour data or only weekend data, which are not representative of all traffic patterns.

Another cause of sampling bias is when certain subjects are more likely to be selected as a sample than others. This could be due to biased selection by either the researchers or the subjects themselves. Researchers could cause bias in sample selection by not advertising widely for subjects or by selecting subjects only from a certain group. Examples of this type of sampling bias include the two convenience samples of your classmates and friends in the *Facebook* example discussed above (samples 1 and 2). The subjects themselves can also cause sampling bias. For example, often the people who take the time to respond to surveys are those who really care about the topic (and have extreme opinions, positive or negative). Students filling out evaluations of their professors on the 'www.ratemyprofessors.com' website are a good example of this bias.

To overcome potential bias, researchers must decide on the most suitable sampling method given their research objectives and the characteristics of the population. We now cover the *simple random*, *stratified* and *cluster* sampling methods.[5]

Simple random samples are selected from the population such that every member of the population has an equal chance of being included in the sample. Consider, for example, a population consisting of 100 people, from which you wish to select a simple random sample of ten people. You can assign each person in the population a number (from 1 to 100) and then use a random number generator (such as the =RAND() function in Excel) to select ten people from this population. This would be the same as putting 100 names in a hat, shaking the hat, and drawing out ten of the names to create your sample. A specific type of random

---

[5] For an interesting discussion of sampling methods in real life, see: "On Sampling Methods and Santa Rosa", *The New York Times*, March 29, 2011.

sampling is the *systematic* sample, in which you sample every kth person from the population (for example, every 10th person).

Stratified samples take into account different *layers* within the population. For example, assume we would like to survey the opinions of business students about the difficulty of the business curriculum. It is reasonable to expect that there would be differences in the difficulty of courses across the curriculum, so it would make sense to differentiate the responses of first-, second-, third- and fourth-year students. Using a stratified sampling approach, we would *randomly* select a sample of students from each of the four years. We ensure that our overall sample is representative of the population by keeping the same ratio across drawn sample sizes as the ratio across the population strata. For example, if there are more first-year students than second-year students, then we would take a larger sample out of the first-year population than out of the second-year population.

The stratified sampling approach is illustrated in Figure 1-4. The four strata (purple, green, orange and blue) together represent the population of interest, e.g., the population of business school students. Note in Figure 1-4 that the strata are not necessarily the same size. This depends on the actual size of the population you are sampling from. In this example, we are sampling from a population that has many members in stratum 1, fewer in stratum 2, and yet fewer in stratum 3 and stratum 4. As explained earlier, the respective sample sizes (inner rectangles) mimic the size of the population; hence sample 1 is the largest. Another example might be residents of a city, where the strata represent high-, medium- and low-income groups and we wish to study spending behavior within and across the three income groups. Within each stratum, we still use the simple random sampling approach to avoid sampling bias. Stratified sampling is beneficial to ensure that enough data from each category is collected to make meaningful inferences about a population, e.g., to draw conclusions about each of the four student groups or each of the three income groups.

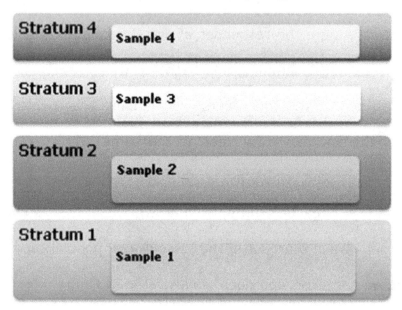

**Figure 1-4
Stratified Sampling**

Cluster samples focus on representative subsets within a population. For example, if we wish to survey people across the United States, we may first select a sample of states and then select a random sample of people within each state. Or, suppose that we wish to interview elementary school students in a city school district with eight elementary schools. Because it may be time consuming or costly to reach all students in all schools, we use cluster sampling by first selecting three of the eight schools, and then select a sample of students from each of these three schools. This is illustrated in Figure 1-5. Here, the city school district is the population (the rectangle), and we randomly select three of the elementary schools (the to-be-studied clusters) for our sample. Within each of these three elementary schools, we select a random sample of students for our study. Cluster sampling is used when it is difficult or impractical to randomly draw samples across an entire population. It is important that the clusters from which the samples are drawn are representative of the population.

**Figure 1-5
Cluster Sampling**

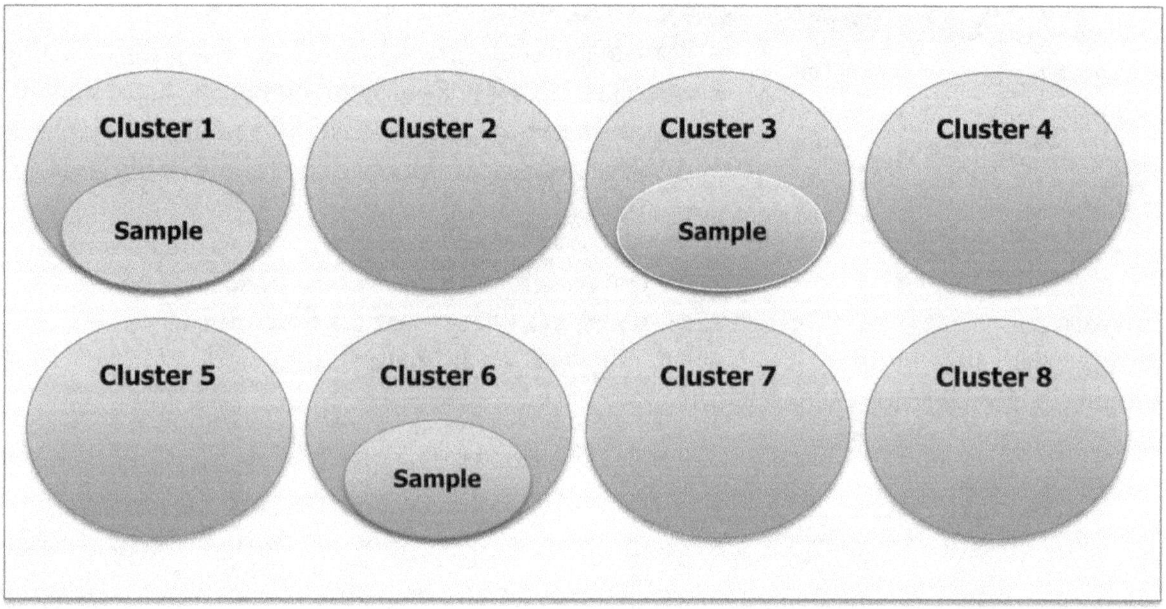

Sample Size

Consider the excerpt in Figure 1-6 on the next page, taken from the *New York Times*.[6] How would you know if the sample of 40 top M.&A.[7] bankers and lawyers, discussed in the story, is a large enough sample?

---

[6] "More Good Times Ahead for M.&A., Survey Finds", *The New York Times*, March 30, 2011.
[7] M.&A. stands for 'Mergers & Acquisitions'.

Figure 1-6
Sampling Example

MARCH 30, 2011, 6:57 AM MERGERS & ACQUISITIONS

More Good Times Ahead for M.&A., Survey Finds

BY MICHAEL J. DE LA MERCED

Article Tools

Recommend | E-mail This
Print | Share
1 Comment | Twitter
Recommend

Deal-makers are by their very nature an optimistic lot. And as many of the top legal specialists in the field head down to the annual Tulane Corporate Law Institute in New Orleans on Wednesday, it appears that their confidence is rising.

Of the roughly 40 top M.&A. bankers and lawyers surveyed by the Brunswick Group, a public relations firm, about 92 percent said they believed the rest of this year would bring continued strong growth in mergers and acquisitions.

The question of sample size is important and there are four criteria that can help you decide on a sample size: the *population size*, the *population variation*, the *resources* available to you, and the amount of *error* that you are willing to tolerate. Recall that we are sampling from a population of interest. Obviously, the larger this population is, the larger is the sample we can obtain. Therefore, a sample of 40 students out of a class of 200 might sound like a pretty reasonable sample size, but the same sample of 40 students out of the over one and half billion *Facebook* users is likely too small. In general, we can conclude that *the larger the population, the larger the sample should be.*

The *variation* in the population is also important. ***Variation*** refers to the extent to which members of the population differ from each other (this concept will be discussed further in later chapters). Say you want to study the income of all first-year students at your school. Assuming a pretty homogenous student population, we can expect a fairly similar income earned by students. A relatively small sample-to-population size ratio might be sufficient to represent the income of students. Now say you want to study the income of residents of Chicago. Obviously, there will be high variation in the income of Chicago residents. Therefore, you would have to obtain a much larger sample-to-population size ratio to adequately represent this population (and in this specific example, you would probably employ a stratified sampling approach, as discussed above). So, *the larger the variation in the population, the larger the sample (and the larger the sample-to-population size ratio) should be.*

The resources available to you would also determine your sample size. With enough resources, you can conduct a census and survey each member of the population. With very limited resources, you would be forced to work with a small sample. Thus, *the more resources available to you, the larger the sample can be*.

Finally, there is the notion of error. We will discuss error in more depth in future chapters, but for now you should know that statistical error is different than simply making a mistake. Let us consider a simple example of determining the average height of your classmates. If you measure each and every person in the class, you would obtain the *true* average height for the class. This true average is called the **population parameter**. If you take a sample of twenty students, you would obtain an *estimate* of the true average based on your sample. This is called the **sample statistic**. The difference between the population parameter and the sample statistic is called the **sampling error**. Obviously, surveying 100% of the population would result in a sampling error of zero (you would obtain the population parameter as explained above). However, as you reduce the sample size, you would, generally, be working with a less accurate representation of the population and increasing the sampling error. Now assume that you are dealing with a life and death question and would not be willing to tolerate any error whatsoever. It would make most sense for you to obtain the largest sample you are able to obtain. On the other hand, if you are just after a general estimate of the population parameter and would be happy with a ballpark figure, you can reduce your sample size (and save your resources). So, *the more tolerant you are of error, the smaller the sample can be*.

Getting back to the M.&A. sample size question at the beginning of this section, you can see why it is difficult to judge the size of this sample simply based on the information provided in this news article. In order to know whether this sample is sufficiently large, we would need to know the total number of top M.&A. bankers and lawyers (our population size), the extent their opinions vary (population variation), whether we can afford to survey more of these M.&A. bankers and lawyers (available resources), and how accurate we want or need to be in our conclusions (tolerable error).

Unit 2 Summary

- At the heart of any statistical study is **sampling**, that is, selecting samples from populations. Obtaining a good sample ensures that your data are representative of the population, that you avoid bias, and, ultimately, that your results can be trusted and used to draw conclusions about the population of interest.

- Make sure that you correctly identify the **population** of interest. Population refers to the collection of individuals, or objects, of interest to you. The scope of the population you focus on must be rooted in your research question.

- Study the population characteristics to identify the best sampling approach for the study. A **sample** is a subset of the population selected for your study. When you sample the whole population, you are conducting a **census.**

- Depending on the characteristics of the population and the resources available to you, you can choose between different sampling approaches. **Simple random** samples employ a probabilistic selection of members of the population; **stratified** samples mimic the structure of the population within the sample; and **cluster** samples focus on representative subsets of the population.

- Regardless of the overall sampling approach, **random sampling** must always be employed. Random sampling implies that every member of the population has an equal likelihood of being included in the sample. It ensures that you do not over- or under-represent a specific group, which would be considered as **sampling bias**.

- Consider the population size, variation, your available resources, and how tolerant you are of error in deciding what your sample size should be. These considerations are summarized in Figure 1-7. **Variation** refers to the extent to which members of the population differ from each other. **Sampling error** represents the difference between the true value of a variable (the **population parameter**) and the value computed from your sample data (the **sample statistic**).

Figure 1-7
Sample Size Determinants

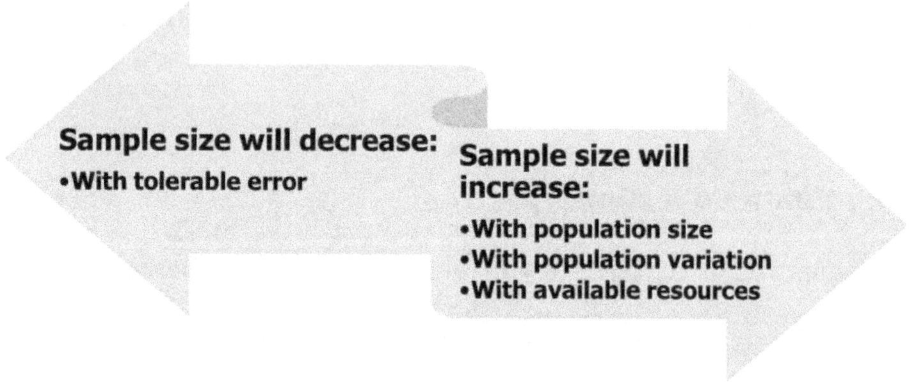

Sample size will decrease:
- With tolerable error

Sample size will increase:
- With population size
- With population variation
- With available resources

Unit 2 Exercises

For each of the research questions below, identify and describe (a) the population of interest, (b) the variables you would record data on, (c) the most suitable sampling approach, and (d) the general magnitude of the sample size you would require:

1. I would like to study the email usage habits of college students.

2. I would like to study the average growth of trees along the Appalachian Trail.

3. I would like to study the performance of technology funds in 2011.

4. I would like to study the sales of Honda cars over the last five years.

5. I would like to survey the political views of Americans between the ages of 18 and 25.

    a. Would your sample approach change if you removed the age restriction and decided to survey, instead, the political views of *all* Americans?

6. I would like to study the association between smoking and lung cancer.

7. I would like to compare unemployment rates in rural versus urban areas in the United States.

8. I would like to understand the success of a marketing promotion taking place in grocery stores.

9. I would like to learn about the brand preferences of young consumers.

UNIT 3

A Quick Look at Data Collection

Collecting data is a crucial step in any study, and you want to be sure that the data you analyze is of high quality and does not include any errors or biases. In this section, a brief overview of the different approaches to data collection is provided. Students interested in learning more are advised to attend a course on research methods.

We can directly collect data or we can use data collected by others (e.g., market research companies, financial institutions, etc.). The former is called **primary data** and the latter is called **secondary data**. To obtain data, we need to decide on the most suitable data collection method. Common choices are *surveys*, *experiments*, *observations* or *interviews*.

Surveys are questionnaires distributed to members of a sample, asking them to respond to a set of predefined questions. An example of a survey is shown in Question 1 in the Unit 1 Exercises section. Another example is the *Student Evaluation of Teaching* that students are often asked to complete at the end of courses.

Surveys can be paper-based, sent by mail, by phone, handed in person or delivered over the Internet. For example, you can be asked to complete a survey at the mall, at your university, through email or on various websites. Have you noticed that retail stores increasingly include a link at the bottom of sales receipts asking you to complete a customer satisfaction survey online or by phone?

Paper-based surveys are resource-intensive and require the person who administers the survey to type the data into a spreadsheet or database, where input errors are inevitable. More common today are web-based surveys that allow data to be captured automatically, rather than keyed-in by hand. Web-based surveys can also be easily distributed to a large, geographically-dispersed sample and are easier for respondents to complete since they include drop-down menus, selection buttons, etc. However, one of the downsides of web surveys is that they often have a lower *response rate* than paper-based surveys. **Response rate** refers to the ratio of the number of surveys completed to the number of surveys distributed. For example, if you mailed out 100 questionnaires and received 30 completed ones in return, your survey's response rate is 30%. Low response rates are not desired, as they increase the risk of *non-response* bias. **Non-response bias** occurs when respondents differ from non-respondents (for example, if more people or if fewer people from a specific ethnic group respond to a survey about racial profiling).

An interesting example of the trade-offs that arise in selecting a surveying approach is the *Student Evaluation of Teaching* survey completed at the end of most university courses. In recent years, many universities are shifting from paper-based surveys to online surveys. Reasons for the shift include: speed with which data can be collected and analyzed (no need for manual input or scanning); flexibility and ability to customize the questions; lower cost of data capture; and, more convenient input mechanism for students, who can type in their evaluations at their chosen time and place. On the other hand, concerns are raised about response rates (many students admit they only complete a survey when it is provided physically in class) and about maintaining respondents' anonymity. And, since the evaluations are often used in tenure and promotion decisions, it is crucial that non-response bias is eliminated from the surveys. Thus, many institutions think hard and long about making this shift from a paper-based survey to an online survey.

Surveys are considered to be **generalizable** if they allow us to draw conclusions about the wider population from our sample. However, when using surveys, we give up some degree of control over the environment of our study. For example, when a survey is distributed by mail (physically or electronically), it is difficult to know for certain when and where the questionnaire was completed or even who provided responses to the survey questions. Even when we give the survey in person (e.g., at the mall, on the street or even in people's houses), we have no control over external interference (e.g., other people walking by, noise level, etc.).

A more controlled approach to collecting data is through *experiments*. **Experiments** are commonly conducted in laboratory settings, where the researcher controls the conditions under which data are collected. The testing of drugs in the pharmaceutical industry provides a good illustration of data collection through experiments. An experiment may be designed in which an existing drug, a new drug and a placebo[8] are each given to three separate groups of patients, and then these patients' reactions are recorded. Having control over other factors likely to affect the patients' reactions, valid inferences are now possible about the effect of the tested drug. Experiments are, therefore, more suited to situations where you would like to collect data about a specific cause-and-effect relationship, holding all other factors constant.

Another method for collecting data is by **observations**. For example, you can sit on the side of a busy intersection and observe the number of cars that pass through the intersection at given times (perhaps to propose a change in the traffic light system for that intersection). Observations give you the opportunity to record various types of data. For example, you can observe a business meeting and count the number of conflicts that arise (counts are ratio data), you can record the different professional jargon words that are used (nominal data), and you can

---

[8] A 'placebo' refers to an innocuous treatment and is often used in medical studies.

record people's body language (e.g., their movement and gestures; also nominal data). You can also supplement your data later with a brief questionnaire, asking the business meeting participants to rate their satisfaction from the meeting (ordinal data). Observations are, of course, very resource-intensive and they also introduce a different type of bias called **observer bias**. The data collected while using observations are interpreted through the eyes of the observer. While it may not make a difference when you are counting cars, it is certainly something to consider when you are interpreting people's body language.

Finally, you can use **interviews** to collect data. We have previously discussed interviews as one way to obtain people's responses to a survey questionnaire. Interviews can also be used in a more open-ended format, with the interviewer having the freedom to ask questions not previously defined in a questionnaire. Interviews are conducted face-to-face or by telephone, which makes them much more conversational and, thus, they allow for the collection of more qualitative data such as stories, experiences, feelings, etc. Similar to observations, interviews are quite resource-intensive, as they require someone to personally interview each and every subject. They are also more susceptible to observer bias, since the researcher is the one interpreting the stories told by interviewees and drawing conclusions based on these stories.

Unit 3 Summary

- Data can be collected using a host of methods and the choice depends on factors such as the type of data you wish to collect, the amount of control you wish to have over the data collection process, the resources available to you, and your approach to various sources of bias.

- **Primary data** are collected by the researcher. **Secondary data** are collected by another party and available for use by the researcher (for example, U.S. Federal Government statistics data, financial markets information, etc.).

- Data can be collected using different methods. **Surveys** are questionnaires distributed online, in person or by mail to the respondents. Surveys are considered to be **generalizable**, meaning that the results can be used to draw conclusions in other settings and for other groups. An important notion in survey research is **response rate**, which refers to the percent of people who responded to the survey out of those that were asked to participate. The reason we look at response rate is to study whether there was a possibility of **non-response bias**, i.e., when the characteristics of respondents are significantly different than those of non-respondents. If we obtain a low response rate because only a small subset of the sample with special interest in the topic of our study has chosen to respond to the survey, then our findings may be biased.

- Another method for data collection is **experiments**, in which data are collected by measuring responses to specific conditions set up in a controlled environment (such as a laboratory). The main strength of experiments is that they allow researchers to control many of the factors affecting a study's results.

- **Observations** and **interviews** are qualitative approaches to data collection that are useful for obtaining rich data such as personal stories, feelings, behaviors, group dynamics, etc. Using observations, the researcher observes and records data, while taking into account **observer bias,** which reflects the fact that the data are collected through the eyes of the researcher. With interviews, data are gathered through conversations with research subjects.

Unit 3 Exercises

For each study below, decide whether you would use primary or secondary data to conduct the study and explain your decision. If you plan to use primary data, select what you consider to be the most appropriate data collection method (a survey, an experiment, observations or interviews), again explaining your decision:

1. I would like to study the email usage habits of college students.

2. I would like to study the average growth of trees along the Appalachian Trail.

3. I would like to study the performance of technology funds in 2011.

4. I would like to study the sales of Honda cars over the last five years.

5. I would like to survey the political views of Americans between the ages of 18 and 25.

6. I would like to study the association between smoking and lung cancer.

7. I would like to compare unemployment rates in rural versus urban areas in the United States.

8. I would like to understand the success of a marketing promotion taking place in grocery stores.

9. I would like to learn about the brand preferences of young consumers.

UNIT 4

From Statistics to Analytics

At the start of this chapter, we mentioned the area of business analytics, sometimes used interchangeably with the term *data science*.[9] Business analytics is an increasingly popular theme in business today, with a recent CIO survey by *Gartner* placing 'Business Intelligence and Analytics' as the top investment priority for managers in 2015.[10] As often becomes the case with successful innovations, a bandwagon effect has emerged, with nearly every organization today asking "How can we get into business analytics?". Well, one of the first steps always mentioned involves acquiring or developing the right talent and skills within the organization.

The path to becoming a business analytics expert begins with the basics, and there are three broad categories of foundational knowledge: knowledge of business analytics tools and technology; knowledge of data manipulation techniques; and, a good understanding of the data domain, or the application area of the data (e.g., healthcare, finance, retail, sports, etc.). This book offers an introduction to all three of these knowledge areas, with an emphasis on how they apply to the business domain. We will cover foundational concepts in statistics and data

---

[9] To understand the difference between a data science career and a business analytics career, please see this excellent infographic: http://www.kdnuggets.com/2015/10/infographic-data-scientist-business-analyst-difference.html
[10] Source: http://www.gartner.com/imagesrv/cio/pdf/cio_agenda_insights2015.pdf

manipulation using Excel, and we provide many examples illustrating how statistics can be applied within the business domain.

Speaking of data, you probably have heard or seen references to the phrase 'big data'. **Big data** refers to the application of business analytics to massive collections of data, commonly defined in terms of three Vs: volume, velocity, and variety.[11] **Volume** refers to the large amount of data that is being generated for potential analyses. Just think of transactions data generated daily at *Target* and *WalMart*, or the amount of data generated on social media sites such as *Facebook*, *Twitter*, and *Instagram*. Or, just think about the number of emails in your inbox or the number of notifications you receive daily from *Facebook*. In many cases, we are no longer talking about terabytes (1,024 Gigabytes) or even petabytes (1,024 Terabytes) of data, but much larger volumes. **Velocity** refers to the speed at which new data is generated and moves around. Think about messages going viral, or how fast your credit card company can identify fraud if someone uses your card to buy gas in another state. Finally, **variety** refers to breadth that exists with different types of data. While some of these data are *structured* (that is, they follow universal definitions and relationships to other data items, and are amenable to traditional data processing), the majority are unstructured (no universal definition and relationships; not amenable to traditional data processing) and are comprised of differing media types (e.g., text, images, videos, etc.).

Storage and processing problems with big data have fueled the development of many new technologies, such as *Hadoop* and *Spark* for data handling, *IBM Watson* for data analysis, and *Tableau* and *Spotfire* for data visualization. With all

---

[11] Some add two more Vs: *veracity* is the ability to verify or trust the data, and *value* is the ability to turn the data into value.

the hype surrounding big data and with new technologies developing at a fast pace, it is easy to overlook the importance of foundational knowledge areas, such as statistics. While traditional statistical techniques – such as the basic approaches to sampling – may not always fit with big data projects, statistics remains a core building block for anyone leaning toward a business analytics or a data science career.

First and foremost, statistics provides the ability to make sense of data, and data sense is a necessary condition for business analytics. The common estimate is that about 80% of data work focuses on putting together the data set and then cleaning and preparing it for analysis, a process known as **data wrangling**. While it may sound like a simple task, getting the data set needed for a business analytics project can be a time-consuming and challenging activity. It begins with a deep understanding of the business problem and the sought business objectives, that together drive the search for data to be analyzed. As previously mentioned, the identified data can come in different forms, and often in very large volumes, which adds to the challenges of constructing the data set for analysis. Once this data set is constructed, it must be properly cleaned and prepared for the chosen analysis techniques. Auditing the data to identify problems (errors, redundancies, desired level of detail, etc.) requires a strong understanding of descriptive statistics, frequency distributions, data visualization, outlier and missing-values analysis, etc. These skills are developed in Chapters 1 through 3 of this book, and re-iterated in later chapters, as well.

When the data set is ready for analysis, the analyst can choose from a range of **analytics techniques**, including dimension reduction techniques, variable association techniques, and prediction models. It is the collective use of these techniques that enables analysts to draw business insights from big data. Dimension reduction techniques, like Principal Component Analysis (PCA), allow the analyst to reduce the size of a data set, moving from tens or hundreds of variables to a much smaller variable set. Understanding PCA requires a strong understanding of the concepts of variance, covariance and correlation, explained in Chapter 3 of this book. Association techniques and prediction models, such as decision trees, association rules, cluster analysis, or logistic regression, to mention only a few, also build on the foundations introduced in this book, including Bayesian probabilities (introduced in Chapter 4), probability distributions like the chi-square and F distributions along with associated significance tests (Chapters 5 through 12), and regression analysis (Chapter 13). A strong understanding of statistical concepts and techniques, such as that developed through this book, thus serves as a core foundation of a business analytics or data science career.

Unit 4 Summary

- **Big data** is commonly defined in terms of three Vs. **Volume** refers to the large amount of data that is being generated. **Velocity** refers to the speed at which new data is generated and moves around. **Variety** refers to different types of data.

- **Data wrangling** refers to putting together a data set and cleaning it in preparation for analysis.

- **Analytics techniques** refer to a range of methods for dimension reduction, variable association and prediction, that together enable analysts to draw business insights from big data.

END-OF-CHAPTER PRACTICE

1. In this chapter we learned about variables and data. Remember that in order to work with data, we first need to understand it. The most effective way is to simply take some time to look at the numbers in front of you. If

you recall our opening Internet use example (Table 1-1), we took the time to look at the table and understand what the numbers mean and what story the data tell us. Similarly, consider the course performance data of ten students, shown in Table 1-e2, below. Answer each of the following questions:

a. How many variables are there in the table?

b. What is the scale of measurement used to measure each variable?

c. What is the population from which this sample was taken?

d. What is the sample size?

e. What are some of the questions we can answer with these data?

Table 1-e2
Course Performance Data

Professor	Final Grade	Letter Grade	GMAT score
Jones	88	A-	716
Jones	44	F	482
Jones	77	B	542
Smith	96	A+	792
Smith	85	A-	725
Smith	82	A-	621
Smith	70	B-	554
Taylor	91	A	795
Taylor	74	B-	555
Taylor	66	C+	528

2. For each research question below, identify and describe the population of interest.

   a. I am interested in studying the job satisfaction of women in IT positions.

   b. I am interested in studying the average life span of small businesses.

   c. I am interested in studying the effect of a new drug on reducing sugar levels in patients with Type II diabetes.

   d. I am interested in studying weather patterns throughout the year.

3. Identify the sampling approach that is best-suited for each of the following studies and explain why:

   a. A study of the time invested in extracurricular activities by undergraduate business students

b. An exit poll on Election Day

c. A study of product placement effectiveness in Wal-Mart stores

d. A study of consumers' trust in online vendors

e. A study of mutual funds' performance over time

4. Table 1-e3 displays data on fifteen sales contacts of an office supplies salesperson. Identify the variables in this table, the type of data captured for each variable, and the measurement scale used to capture each variable's values.

Table 1-e3
Sales Leads

Title	Industry	Annual Revenue (rounded for presentation)	Number of Employees	State
Sales Exec	Automotive	$2,000,000	200	AL
Sales Exec	High Technology	$58,423,292	1,600	CA
Sales Exec	Retail	$73,587,938	1,500	NY
Sales Exec	Energy	$134,246,523	2,000	NY
Sales Exec	Telecommunications	$156,993,493	900	CA
Sales Exec	High Technology	$164,575,816	2,000	AK
Sales Exec	Retail	$179,740,462	500	VA
Sales Exec	Energy	$187,322,785	1,100	VA
CFO	Manufacturing	$58,000,000	700	FL
CFO	Manufacturing	$66,779,867	1,100	FL
CFO	Financial Services	$81,944,513	2,000	CA
CFO	Pharmaceuticals	$97,109,159	1,300	IL
CFO	Financial Services	$135,020,775	1,600	IL
CFO	Pharmaceuticals	$150,185,421	300	CA

CHAPTER 2: Data Presentation

In the previous chapter, we discussed different types of data and the ways such data can be collected. In this chapter, we focus on presenting data that we have collected in a way that helps us deliver meaningful insights about these data. Effective data presentation techniques enable us to make sense of data that otherwise would be overwhelming in its raw form. Consider the example of a social network, which can be used to help identify relationships among people in your class. To collect this data, you ask people who their friends are and then record their answers. Your data tells you, for example, that Laura is friends with Mike and Shannon; Shannon is friends with Dan, Laura and Jessica; Jessica is friends with Shannon, Mike, Iesha, Sean and Lauren; and, so on. You can see that, quite quickly, it becomes difficult to keep track of these relationships and especially difficult to visualize the class's social network.

Now consider the graph in Figure 2-1 (a social network map), in which the ovals represent the members of a social group and the lines represent relationships among these members. From this graph, it should be much easier to infer the friendship relationships in your class; that is, who is friends with whom. You can see that Jessica, for example, plays an important role in bridging two groups of people, and that Ryan, Iesha and Dan are only linked to one other person. Presenting data in such graphic form is called **data visualization**.

Figure 2-1
Social Network Graph

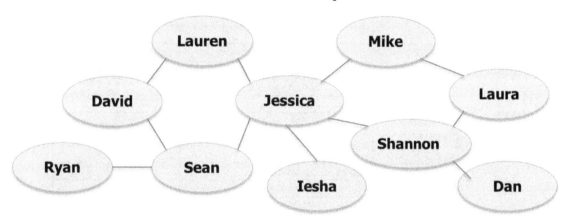

We are not limited to using graphs to present data in more understandable forms. To see this, consider Table 2-1a and Table 2-1b, which display Jane Doe's credit card spending in 2015. Table 2-1a provides a subset of Jane's detailed spending data. Each row in this table is a transaction placed on Jane's credit card. The complete file contains over 300 transactions. You can see that it is quite difficult to understand Jane's spending habits using Table 2-1a. For example, can you tell what percent of Jane's spending was on gas? What percent did she spend at restaurants? We can use Table 2-1b, which offers a different view of Jane's credit card transactions during 2015, to better grasp Jane's spending. In Table 2-1b, Jane's spending transactions have been grouped into meaningful categories. For example, all merchandise purchases were placed in the 'Merchandise & Supplies' category, all gas purchases were placed under the 'Transportation' category, and so on.

Table 2-1a
Partial Spending Data: Detailed Transactions

Date	Transaction	Amount
5/19/15	Macy's	$108.47
9/19/15	Bistro Bar	$50.00
1/11/15	Apple Store	$17.84
6/27/15	Banana Republic	$72.78
12/13/15	Fast Fix Jewelry	$32.35
3/16/15	Foot Locker	$36.40
8/16/15	Sporting Goods	$175.58
12/9/15	Amazon.com	$163.27
5/4/15	Apple Store	$16.19
5/27/15	Unisex Hair Place	$11.00
12/23/15	Mexican Café	$25.27
11/24/15	Sunoco	$36.59
11/6/15	Staples	$283.09
3/14/15	Sunoco	$44.30
5/22/15	Price Chopper	$65.72
8/23/15	Sears Roebuck	$75.59
10/22/15	Price Chopper	$7.97
2/26/15	Royal Theater	$206.55
10/24/15	Royal Theater	$68.00
7/20/15	Gap	$62.85
9/10/15	Book House	$20.43
11/14/15	Hollister.com	$50.00
12/1/15	Lindt Chocolate	$20.77
9/26/15	Hess	$48.59
5/7/15	Verizon	$108.00
...	...	...

Table 2-1b
Summarized Spending Data

Category	Q1 (Jan-Mar)	Q2 (Apr-Jun)	Q3 (Jul-Sep)	Q4 (Oct-Dec)	Total
Communications	$164.01	$164.01	$164.01	$164.01	**$656.04**
Entertainment	$152.47	$691.00	$122.70	$213.25	**$1,179.42**
Merchandise & Supplies	$995.36	$904.62	$699.31	$231.57	**$2,830.85**
Restaurant	$800.50	$881.16	$527.22	$ -	**$2,208.88**
Transportation	$858.92	$852.76	$709.69	$281.88	**$2,703.25**
Total	**$2,971.26**	**$3,493.55**	**$2,222.93**	**$890.71**	**$9,578.44**

Grouping the detailed transactions by category (rows) and by time period (columns) makes it easier to understand Jane's spending patterns. For example, you can see from the bottom row of Table 2-1b that Jane spent the most money in the second quarter of the year ($3,493.55) and the least money in the fourth quarter ($890.71). You can also see that she spent the same amount of money on 'Communications' in each quarter, and that 'Merchandise & Supplies' and 'Transportation' are her two highest spending categories.

These two highest spending categories are also quite easily seen from the *pie chart* below, in which each section of the pie represents a spending category. The green and light blue sections correspond to the 'Merchandise & Supplies' and 'Transportation' categories, respectively.

Figure 2-2
Pie Chart Presentation of Spending

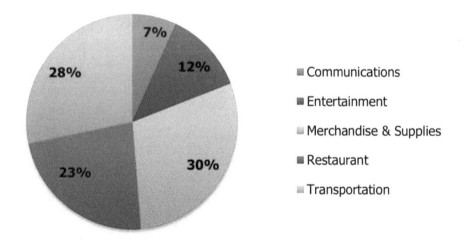

Now imagine that you are working for a credit card company and you wish to understand the transactions of not one but millions of customers. Are there any patterns in customer spending that might be useful for marketing complementary services? Do certain types of transactions suggest the potential of card misuse? Being able to summarize and present data either in tabular or graphical form would be extremely helpful in this case. Moreover, using tables and charts, we can make more sense of our data and we can communicate the meaning of our data to others more easily. Therefore, this second chapter focuses on data presentation, beginning with the very important notions of *frequencies* and *frequency tables*. We will also begin using the Microsoft Excel spreadsheet software tool. Excel is integrated into the chapter in two ways. First, instructions and screen shots for conducting analysis are provided, as needed, throughout the chapter. Second, an Excel data file, corresponding to the examples and exercises being used in this chapter, is available as an accompanying file to the text. You are invited to follow our examples using this accompanying Excel file and encouraged to practice these Excel skills by doing the Unit Exercises and the End-of-Chapter Practice exercises. The following topics are covered in this chapter:

- Frequency Tables
- Frequency Tables for Qualitative Data
- Frequency Tables for Quantitative Data
- Column Chart
- Histogram
- Stem-and-Leaf Plot
- Relative Frequency
- Cumulative Relative Frequency
- Comparing Nominal Data
- Describing the Relationship between Two Quantitative Variables
- Describing Time Series Data

UNIT 1

Frequency Tables

In statistics, ***frequency*** refers to the number of times a value occurs in a data set. For example, in the small data set below, the frequency of the color 'White' is five and the frequency of the color 'Red' is three:

Table 2-2
Frequency: Colors Example Data Set

Blue	Red	Green	Yellow	White
Blue	Red	White	Yellow	Purple
White	White	Red	White	Purple

Table 2-3 presents data of students' grades on a statistics quiz (out of 100 possible points). In this data set, the frequency of a grade in the 80s is three and the frequency of a failing grade (lower than 50) is nine.

Table 2-3
Frequency: Grades Example, Data Set 1

39	59	94	60	72	75	50	63	61	57
65	49	59	72	70	53	66	79	63	41
53	72	80	54	51	56	68	76	26	67
67	72	65	35	75	53	43	78	80	40
71	76	52	63	39	56	57	34	84	75

Notice that there are slight differences between the two examples shown as Tables 2-2 and 2-3. The first data set includes qualitative, discrete data (names of colors) and we simply *count* occurrences of specific colors. The second data set includes quantitative data (grades on a percentage scale), which is still discrete, but with a wider range of values (0 to 100). Therefore, we examined the frequency of values within groups, or categories, of this second data set (e.g., a failing grade or a grade between 80 and 89). Let us explore such differences between qualitative and quantitative data further.

Frequency Tables for Qualitative Data

Recall from Chapter 1 that qualitative data are those measured on the nominal and ordinal scales, and the only operation we can perform with such data are counts and proportions. To understand how qualitative data are categorized into frequency tables, consider the question: *"Which country dominated the 2016 Summer Olympics?"*

One way to answer this question is to study the medals won during the *Olympic Games* and see which country obtained the most gold medals. Suppose you decide to take this route. You create a data set by listing all gold medals won and recording the country that won them. Your data might look something like:

'USA, 'China', 'USA', 'Italy', 'China', 'USA', 'Great Britain', …

indicating the first gold medal was won by the United States, then a gold medal by China, then another gold medal won by the USA, and so on. Based on your data, you construct the **frequency table** shown as Table 2-4, which lists the true medal counts for the top ten countries in the 2016 *Summer Olympics* in Rio, Brazil.[12] Each row in Table 2-4 provides a value (i.e., a country) alongside this value's observed *frequency* (i.e., the number of gold medals won by the country).

---

[12] Source: http://www.nbcolympics.com/medals

Table 2-4
Frequency Table: Gold Medal Count, 2016 *Summer Olympics*

Country	Gold Medals Won
United States	46
Great Britain	27
China	26
Russia	19
Germany	17
Japan	12
France	10
South Korea	9
Italy	8
Australia	8
Total	**182**

While interpreting the table, we note that the frequency of the data value 'China' is 26, indicating that the country China appears 26 times in our data set. Hence, 26 gold medals were won by Chinese athletes. The frequency of the data value 'United States' is 46, indicating that 46 gold medals were won by the United States team. The total number of medals is 182 and it is calculated by adding up the frequencies listed in the table. A useful self-check to avoid errors is to verify that your data set indeed includes 182 observations. (By the way, in the 2008 *Summer Olympics*, which took place in China, China won 51 gold medals. Brazil won only three gold medals in each of the past two *Olympic Games*, but seven medals in 2016. Is there a link between hosting the *Olympic Games* and a nation's performance? In future chapters, you will learn how to use statistics to answer such questions.)

Let us consider another example to understand how a frequency table is constructed. You are about to buy a new mobile phone and would like to find out the best provider with whom to contract. You quickly survey 30 classmates and obtain data (shown in Table 2-5) on their mobile providers.

Table 2-5
Frequency: Mobile Providers Example, Data Set 1

Verizon	Sprint	Tracfone	Sprint	AT&T	AT&T
Verizon	T-Mobile	T-Mobile	Sprint	Verizon	Verizon
Verizon	T-Mobile	Sprint	Tracfone	T-Mobile	AT&T
AT&T	Other	Verizon	AT&T	Other	AT&T
AT&T	Verizon	Verizon	AT&T	Verizon	Verizon

Looking at Table 2-5, you see that there are six different *values* in the data set: 'Verizon', 'AT&T', 'Sprint', 'T-Mobile', 'Tracfone' and 'Other'. To construct your frequency table, you would first list these values in the left-most column of the frequency table (as shown in Table 2-6a). Next, you simply count how many times each data value appears in Table 2-5. For example, you can see that 'Verizon' appears ten times, 'AT&T' appears eight times, 'Sprint' and 'T-Mobile' each appear four times, and 'Tracfone' and 'Other' each appear twice. These are the values that are listed under the 'Frequency' column in Table 2-6b. Again, it is a good idea to make sure that the total is indeed 30, since you asked 30 classmates.

Table 2-6
Frequency Table: Mobile Providers Example

Table 2-6a

Provider	Frequency
Verizon	
AT&T	
Sprint	
T-Mobile	
Tracfone	
Other	
Total	

Table 2-6b

Provider	Frequency
Verizon	10
AT&T	8
Sprint	4
T-Mobile	4
Tracfone	2
Other	2
Total	**30**

Using Excel

Excel provides several approaches to constructing frequency tables. With qualitative data, a very easy way is to use the =COUNTIF() function. Open this chapter's Excel file and look at the first worksheet, titled 'Mobile Provider Example'. Figure 2-3 illustrates what you should see. The worksheet lists the 30 responses obtained from your classmates and also displays the same frequency table shown as Table 2-6b. If you click on any of the calculated values in this table (cells B11 through B16), you will see that we have used the =COUNTIF() function to compute these values. The =COUNTIF() function uses two arguments.[13] The first argument is *range* and the second is *criteria*: =COUNTIF(range,criteria). In this function, *range* refers to the spreadsheet cells containing your data values. In our example, the data on the 30 mobile providers is shown in cells A3 through F7. This is indicated in Excel as: A3:F7. Next, *criteria* refers to the value you wish to count in the data set. For example, if we wish to count how many times the data value 'Verizon' appears in our data set, we can type: =COUNTIF(A3:F7,"Verizon"). We use quotation marks to indicate to Excel that we are looking for the specific word. Alternatively, since we have already inserted the data value 'Verizon' into cell A11, we can also type: =COUNTIF(A3:F7,A11). This second approach to specifying the *criteria* argument is shown in Figure 2-3.

---

[13] *Arguments* refer to the values placed within a function's parentheses.

Figure 2-3
Screen Shot: Mobile Providers Example Frequency Table in Excel[14]

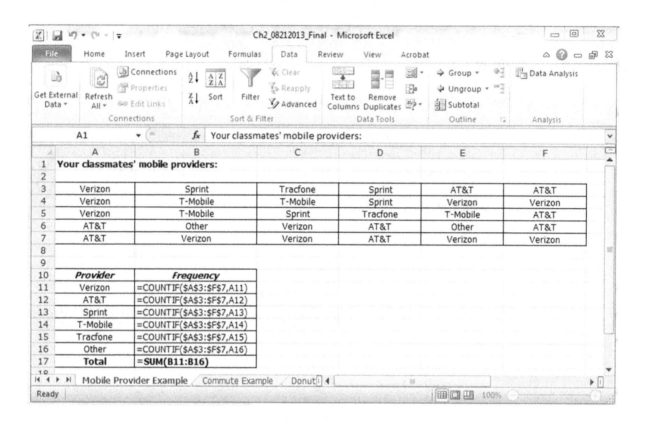

Thus, =COUNTIF($A$3:$F$7,A11)[15] places the number 10 in cell B11, as it counts the number of times the data value 'Verizon' appears in cells A3 through F7. The rest of the values in the table are computed in the same way. For example, =COUNTIF($A$3:$F$7,A12) places the number 8 in cell B12, since there are eight occurrences of the data value 'AT&T' in the data set.

As always, we also make sure that the sum total of our frequencies equals the number of observations in our data set. In cell B17, we have used the =SUM() function to compute the total frequencies in our table. The =SUM() function takes

---

[14] All of the screen shots in this book were taken in the 2010 version of Excel.
[15] Excel note: We have used the $ sign in this equation. When placed before the row number, as in our example above, the $ sign tells Excel to keep referring to this specific row (i.e., Row 3) even when you copy your equation up or down. When placed before the column letter, the $ sign tells Excel to keep referring to this specific column, even when you copy your equation sideways.

in a set of numbers and adds them up. There is a single argument for the =SUM() function: the range of the data values to be added. Typing =SUM(B11:B16) into cell B17 results in adding the values in cells B11 through B17 of the table and displaying this sum in cell B17.

Frequency Tables for Quantitative Data

We now focus on constructing frequency tables for quantitative data. Quantitative data are numerical and can be either discrete (countable) or continuous (generally uncountable, as listing all the values present in a data set would result in much too large a frequency table). Consider the grades, measured as percentages, shown in Table 2-7. Even through these data are discrete, it does not make sense to construct a frequency table with 101 rows (with each row representing one possible grade from zero through 100). We would have many empty cell counts (for example, no student received a grade of 98) and numerous cells with counts of 1 (for example, the frequency for a grade of 60 is exactly one student). Therefore, with quantitative data we commonly group values into *classes* prior to constructing a frequency table. In the grades example below, we can create a frequency table using intervals of ten and record the number of students in each interval. This is shown in Table 2-8.

Table 2-7
Frequency: Grades Example, Data Set 2

26	40	51	54	59	63	67	72	75	79
34	41	52	56	59	63	67	72	75	80
35	43	53	56	60	65	68	72	76	80
39	49	53	57	61	65	70	72	76	84
39	50	53	57	63	66	71	75	78	94

Table 2-8
Frequency Table: Grades Example 2

Class	Interval	Frequency
1	0 to 9	0
2	10 to 19	0
3	20 to 29	1
4	30 to 39	4
5	40 to 49	4
6	50 to 59	13
7	60 to 69	11
8	70 to 79	13
9	80 to 89	3
10	90 to 100	1
	Total	50

A very important observation to make about Table 2-8 is that these classes are *mutually exclusive* and *exhaustive*. **Mutually exclusive** means that each data value falls into a single class; thus, the first class goes from 0 to 9, the second from 10 to 19, then 20 to 29, and so on. **Exhaustive** means that all of the data values are represented in the table. In particular, note that the final class is 90 to 100 rather than 90 to 99. This is done to ensure that the value of 100 is represented in the frequency table. As before, we again make sure that the frequencies in our table add up to 50, which is the total number of observations (students) in our data set.

There are three more important points to make with respect to the example shown in Figure 2-8. First, we still have a relatively large number of classes (in this case ten) for a relatively small number of observations (50 observations). We may wish to consider a different way of grouping the data into classes. For example, we could use the letter grade scale and create five classes representing the grades A, B, C, D and F. Second, our classes are defined in terms of integers and, therefore, we are not capable of handling a grade of, say, 79.5 (Would you place a grade of

79.5 in 'Class 8' or in 'Class 9'?). This is not a problem with the data shown in Table 2-7, as there are no half-grades, but this is something to keep in mind. Third, we defined our classes arbitrarily as intervals of ten and some of these classes have very low frequencies. For example, the first two classes each have a frequency of zero and the next two ('Class 3' and 'Class 4') a frequency of one. It is always best to carefully consider the data and your research requirements before defining classes.

Two important decisions are involved in making more informed decisions about constructing frequency tables. The first of these decisions involves determining the number of classes to use, and the second involves determining the width of these classes.

Decision 1: Number of Classes

There are no strict rules for determining the *right* number of classes to use, and the decision is generally guided by the number of observations in the data set, the researcher's knowledge of what is being described by the data, and other research requirements. As a general rule of thumb, any number between five and twenty classes is acceptable, with the specific number of classes used being situationally dependent.

Alternatively, with no clear research requirements to guide our decision, we can use **Sturges' formula,** which offers a way to calculate the number of classes to use for a data set of size n:

$$\text{Number of classes} = 1 + 3.3 * \log(n)$$

For example, in a data set of 50 observations we would have $1+3.3*\log(50)=6.6$, or seven, classes, and in a data set of 200 observations we would have

1+3.3*log(200)=8.59, or nine, classes. Note that in both examples, we have rounded the number of classes up, from 6.6 to seven and from 8.59 to nine. You cannot have partial classes (i.e., the number of classes being used must be an integer) and it is best to always round up the number of classes to ensure that you do not leave out any values that are in the data set (that is, you do not run out of classes before you get to your highest data value).

Decision 2: Class Width Intervals

A second decision concerns the width of each class. Consider another example, this time of the commute times of 200 students at State University. These commute times (in minutes) are provided in Table 2-9.

Table 2-9
Class Width Intervals: Commute Time Example, Data Set 1

15	22	30	38	45	56	62	69	76	85
15	23	30	38	45	56	63	69	76	85
15	23	31	38	45	57	63	70	77	85
15	23	31	38	45	57	63	70	77	86
15	23	31	39	47	58	64	70	78	86
16	24	32	39	48	58	64	71	78	86
17	25	32	39	48	58	65	71	79	86
17	25	32	39	48	58	65	71	79	86
18	25	33	40	48	59	66	72	79	87
18	25	34	40	49	59	66	74	79	88
18	26	34	40	49	59	66	74	80	88
19	26	34	41	50	60	66	74	82	88
20	27	35	41	50	60	66	74	82	89
20	27	35	41	51	61	67	75	83	89
20	27	35	41	51	61	67	75	83	89
20	27	35	42	52	61	67	75	83	89
20	28	36	42	54	61	67	75	83	90
21	28	37	43	54	61	68	76	84	90
21	29	37	44	55	62	68	76	84	90
22	29	38	45	55	62	69	76	84	90

There are 200 observations in the above data set and we have no obvious research requirements to help us determine the right number of classes. Thus, we use Sturges' formula to calculate the number of classes as nine (1+3.3*log(200)=8.59~9). Once the number of classes is known, we need to decide on the **class width interval**, or the **lower** and **upper bounds** for the data values in each class. In other words, what are the lowest and highest data values to be counted as part of a class? We usually aim to create classes with more or less equal intervals so that none of the intervals are over- or under-represented by being larger or smaller than the others. For this reason, we calculate the class width interval by dividing the *range* of the data by the *number of classes*, thereby creating equal class widths.

The **range** of the data is calculated as the largest minus the smallest value in our data set. Therefore, in our commute time example, we can calculate the range as: 90 minutes (the longest commute) minus 15 minutes (the shortest commute) equals 75 minutes. (Note that a nice way to ease our calculations is to first sort the values in Table 2-9 from smallest to largest; this is very easy to do if your data is placed into an Excel spreadsheet.) Now, we calculate the class width interval by dividing the range (75) by the number of classes (nine):[16]

$$\text{Classwidth} = \frac{90-15}{9} = 8.34$$

The resulting classes for our commute time example are shown in Table 2-10. Note that in our specific example, there are no decimals in the data (our data values are given as minutes). This implies that it may be better to use rounding again, for a class width of nine minutes. For clarity of our explanation, however,

---

[16] A simple way to think about this is: we have 75 possible commute time values in the data set and we need to divide them into nine equal groups (classes). Therefore, we divide 75 by nine to find out how many values to place in each group and this is our class width.

we will continue with the exact value of 8.34 in creating the frequency table for this example.

Table 2-10
Frequency Table: Commute Time Example

Class	Lower Bound	Upper Bound	Frequency
1	15	23.34	25
2	Greater than 23.34	31.68	20
3	Greater than 31.68	40.02	26
4	Greater than 40.02	48.36	18
5	Greater than 48.36	56.7	13
6	Greater than 56.7	65.04	26
7	Greater than 65.04	73.38	21
8	Greater than 73.38	81.72	22
9	Greater than 81.72	90.06	29
		Total	200

To create the frequency table, we begin with the lowest value in our data set, which, in this example, is 15 minutes. This is the *lower bound* of our first class. To find the *upper bound*, we add 8.34 to obtain 23.34. Therefore, our first class is 15 to 23.34. The second class will not include the value of 23.34 as its lower bound, giving us a lower bound of >23.34. We use the greater than (>) symbol to ensure that the classes remain mutually exclusive. Adding 8.34 give us an upper bound of 31.68. Thus, the second class is >23.34 to 31.68. We continue creating all classes in this manner. All nine classes, and their frequencies, are shown in Table 2-10.

Using Excel

The contents of Table 2-10 are provided in this chapter's Excel file, in the worksheet titled 'Commute Example'. Column A contains the 200 data values shown in Table 2-9. Columns C through F contain the frequency table shown above as Table 2-10. Figure 2-4 provides a screen shot of the top of this worksheet. To illustrate how a frequency table can be constructed, we placed in cell D1 the value of the class width (8.34 in our example). You can change this value to nine and see what happens to the numbers in the table (simply type 9 in cell D1 and press 'Enter'). What happens when you change it to 20? How about 200?

Figure 2-4
Screen Shot 1: Commute Time Example Frequency Table in Excel

We previously used the =COUNTIF() function to create the frequency table for qualitative data. However, as explained earlier, when a data set has many differing data values, we no longer count the frequencies of individual values but rather use class intervals to construct our frequency table. In Excel, this implies the use of a different function than the =COUNTIF() function. Here, we use the =FREQUENCY() function, that returns a set of frequency counts for class intervals as its result. The function involves two arguments, respectively termed (in Excel)

the *data-array* and the *bins-array*. The *data-array* is the range of your data set. The *bins-array* is the range of class upper bounds. Using this function, Excel will automatically count the number of observations that fall into each class. Let us go through the steps of using the =FREQUENCY() function.

In the 'Commute Example' worksheet, the frequency table is located in cells C3 through F13. To use the =FREQUENCY() function, first clear the contents of cells F4 through F12. Now, follow these steps *exactly*:

- Click the mouse on cell F4 and drag it down to cell F12 in order to select the range F4:F12, which is where you would like to place the frequency counts (this function's results).

- Cells F4:F12 will now be highlighted. Type '=FREQUENCY(A2:A201,E4:E12)' as shown in Figure 2-5. Do *not* type the single quote marks! Because of your selected range, this will automatically show in cell F4.

- When you are done typing, press and hold (simultaneously) the 'Ctrl' and 'Shift' and 'Enter' keys (on your keyboard) to obtain the results. This specific key combination is used to trigger the calculations that produce the frequency counts.

You should see a table similar to Table 2-10 with cell F13 showing a value of 200, which is the sum of all frequencies in your table. To obtain this value, we have used the =SUM() function, typing =SUM(F4:F12) into cell F13.

Figure 2-5
Screen Shot 2: Commute Time Example Frequency Table in Excel

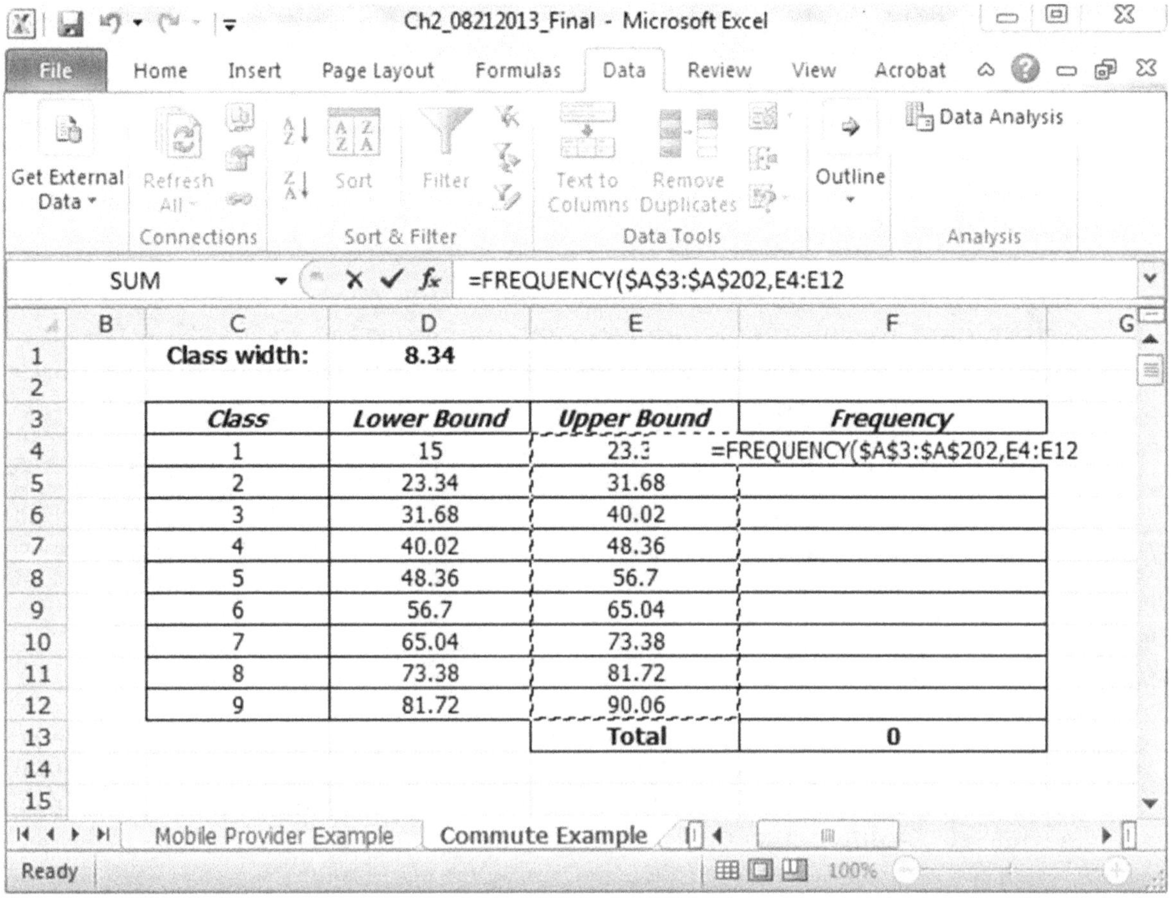

Class width:	8.34		
Class	Lower Bound	Upper Bound	Frequency
1	15	23.3	=FREQUENCY($A$3:$A$202,E4:E12)
2	23.34	31.68	
3	31.68	40.02	
4	40.02	48.36	
5	48.36	56.7	
6	56.7	65.04	
7	65.04	73.38	
8	73.38	81.72	
9	81.72	90.06	
		Total	0

Unit 1 Summary

- **Data visualization** refers to presenting data sets in ways that are easier to read and comprehend. One such way involves summarizing data using **frequencies**, defined as counts of occurrence. With any data set that you have, it is a good tactic to begin by looking at the frequencies of data classes of interest in order to better understand your data. A **frequency table** lists data categories, or classes, along with their respective frequencies.

- Constructing frequency tables involves a great deal of common sense. You should take the time to fully understand: what the data are about, the unit of measurement, whether there are natural groupings of data values, etc.

- Working with qualitative data, the values can be used directly in the frequency table and simple counts of the number of occurrences of each value are used to compute frequencies.

- When working with quantitative and discrete (countable) data, you can decide whether to list individual values or group them into classes. The decision should be made based on the number of discrete values in the dataset. If you have a data set with a relatively small number of discrete data values, you may still list and count the frequency of each value. Use judgment to decide which approach would result in a more meaningful frequency table.

- When working with quantitative and continuous (uncountable) data or with discrete data having many different values, you need to create a frequency table that is based on *classes*, or groups of data values. Three important steps (shown in Figure 2-6 below) are involved in constructing classes.

    o Step 1: Identify the desired number of classes. You can decide on the number of classes either based on your domain knowledge[17] or using **Sturges' formula**, which calculates the number of classes based on the sample size:

    $$\text{Number of classes} = 1 + 3.3 * \log(n)$$

    o Step 2: Compute the **class width interval**, which defines the **lower** and **upper bounds** of each class. Class width is computed by dividing the **range** of the data (highest minus lowest value) by the number of classes obtained in Step 1.

    o Step 3: Ensure that your resulting classes are **exhaustive** (that is, covering the whole data set) and **mutually exclusive** (that is, each data value falls in one and only one class).

Figure 2-6
Constructing a Frequency Table

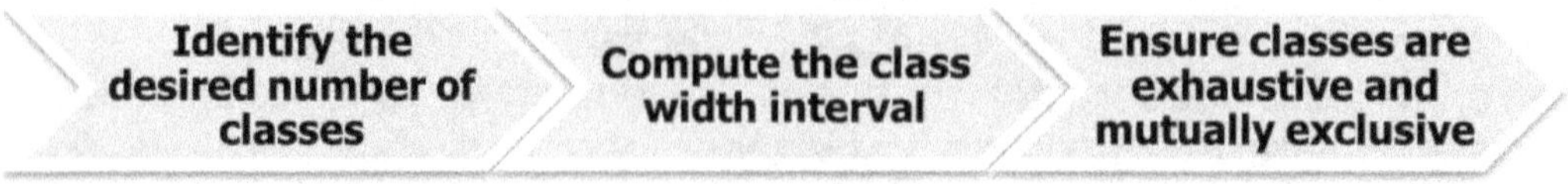

Identify the desired number of classes → Compute the class width interval → Ensure classes are exhaustive and mutually exclusive

- Finally, when rounding the number of classes or upper/lower class boundaries, make sure you are staying true to the original scale (integer, decimals, etc.) and that no omissions of data values inadvertently occur.

---

[17] *Domain knowledge* refers to your existing understanding of the subject matter about which you have collected data.

Unit 1 Exercises

1. Construct a frequency table for each of the data sets in Tables 2-e1a, 2-e1b and 2-e1c.

Table 2-e1a
Car Brands Data Set

Toyota	Hyundai	Toyota	Toyota	Honda	BMW	Honda
Kia	Hyundai	Hyundai	Hyundai	Toyota	Toyota	

Table 2-e1b
Wait Time (in minutes) Data Set

8	9	3	2	5	5	4	7	8	6	1	6	1	2	9	2	8	2	7	6

Table 2-e1c
Daily Interest Rates (percent) Data Set

0.513	0.49	0.377	0.474	0.251	0.623	0.185	0.586	0.513	0.459
0.486	0.354	0.369	0.396	0.424	0.947	0.109	0.009	0.088	0.346

2. In a data set with 100 observations, use Sturges' formula to determine the number of classes to use in a frequency table.

3. In a data set with 500 observations, in which the lowest value is 50 and the highest is 160, what would be your class width?

4. A professor collected data on students' grades on the letter grade scale. How should she construct the frequency table for her data?

5. Table 2-e2 lists each of the doughnuts sold at *Donut Heaven's* drive-through window on a Saturday morning. For planning purposes, the manager would like to find out how many of each kind were sold. Create a frequency table based on these data. (Note: the data for this question are also provided in the chapter's Excel file in the worksheet titled 'Donut Heaven'. You can practice your Excel skills for building the frequency table).

Table 2-e2
Donut Heaven Items Sold Data Set

Apple	Chocolate	Vanilla	Vanilla
Apple	Glazed	Glazed	Vanilla
Glazed	Apple	Cream-filled	Glazed
Glazed	Chocolate	Cream-filled	Glazed
Glazed	Cream-filled	Cream-filled	Chocolate

6. Which is the most popular phone app among college students? You collected responses from a sample of 50 students regarding which phone app each student thought was the best. The responses included the seven apps listed in Table 2-e3a. To make your data analysis easier, you coded the responses, meaning that you assigned a numerical value to each app (shown in the second column of Table 2-e3a). Table 2-e3b shows the 50 values collected in your sample, according to the assigned codes (i.e., a value of 1 means the answer was 'Snapchat', a value of 2 means the answer was 'Instagram', and so on). Based on these data, create a frequency table for the most popular apps. These data are available in the Excel file in the worksheet 'Popular Apps'.

Table 2-e3a
Apps and Associated Codes

App	*Code*
Snapchat	1
Instagram	2
Uber	3
Venmo	4
Spotify	5
Tinder	6
2048	7

Table 2-e3b
Popular Apps Data Set

2	1	2	5	4	7
1	6	6	6	2	2
6	5	5	3	3	2
3	7	3	4	1	4
2	5	2	7	5	3
5	6	2	4	7	2
3	4	4	6	5	6
7	6	3	5	3	7
4	5	2	4	3	1
5	1	1	2	1	6

7. A bank branch manager tracks her employees' performance by measuring the time it takes each employee to serve a customer. The times it took to serve each of 50 customers at the branch are listed in Table 2-e4 (and in this chapter's Excel file in the worksheet titled 'Service Time'). Create a frequency table for the service times at the branch.

Table 2-e4
Service Time (in minutes) Data Set

2	4	2	2	9	11	2	5	10	8
3	2	11	7	2	6	7	3	6	3
11	10	6	11	10	9	5	5	8	7
8	7	7	9	3	2	10	11	6	4
6	8	5	3	5	5	3	9	7	5

8. You collected data about the financial performance of 1,000 organizations. As part of understanding your data, you would like to find out more about the sizes of the organizations that participated in your survey (e.g., are they small, medium, large, etc.). The collected size data are provided in this chapter's Excel spreadsheet in the worksheet titled 'Organization Size'.

   a. Determine the number of classes to use and class width. Why did you choose these classes?

   b. Construct a frequency table for organization size using the =FREQUENCY() function in Excel.

   c. What can you say about the sizes of the organizations in your data set?

9. The data in the worksheet labeled 'Tax Rates' within this chapter's Excel spreadsheet contains the sales tax rates for each of the 50 states in the United States. Construct a frequency table for these data. Explain your decisions in creating this table.

10. The cost of a restaurant meal for two people can range from $10 to $1,000. Data were collected on average meal costs from 1,000 restaurants and summarized in the frequency table shown in Table 2-e5. Identify the errors that exist within this table. Correct those errors you are able to correct without having access to the actual data set.

Table 2-e5
Meal Costs Frequency Table

Class ($)	Frequency
0 to 50	241
50 to 100	111
100 to 150	24
150 to 200	66
200 to 250	155
250 to 1000	403
Total	995

UNIT 2

Column Chart

The frequency tables constructed in Unit 1 can be presented graphically for easier visualization of the data. Qualitative data are presented using a **column chart**, in which a column is created for each category with the height of each column reflecting the number of items in that category. Figure 2-7 displays a column chart for the frequency table used in the Mobile Providers Example shown in Unit 1 as Table 2-6b.

Figure 2-7
Column Chart: Mobile Providers Example

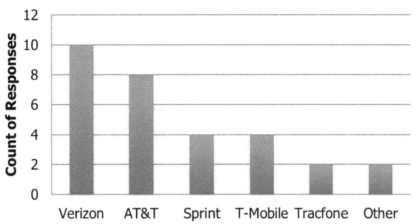

Using Excel

The above graph was produced using Excel. To create the graph, follow these steps:

- In the worksheet titled 'Mobile Provider Example', highlight cells A10 through B16.

- Click on the 'Insert' tab to insert a Chart.

- Select the Column chart type and then select your preferred column style from the options provided. Right-click on the graph to adjust any options, as needed.

Histogram

You may note that the column chart shown in Figure 2-7, above, has gaps between each two columns. This type of chart, where the individual columns represent specific data values, works well here since the data are qualitative and the frequency categories are distinct from each other.

In the case of continuous data, however, we would prefer that the columns touch in order to represent that the classes are, in fact, continuous. A graph with no gaps between the columns is called a *histogram*. A **histogram** displays classes on the x-axis and frequency counts on the y-axis. To illustrate a histogram, we use the second grades example from Unit 1 (shown again as Table 2-11). Figure 2-8 provides a graphical presentation of these data using a histogram.

Table 2-11
Frequency Table: Grades Example 2

Class	Interval	Frequency
1	0 to 9	0
2	10 to 19	0
3	20 to 29	1
4	30 to 39	4
5	40 to 49	4
6	50 to 59	13
7	60 to 69	11
8	70 to 79	13
9	80 to 89	3
10	90 to 100	1
	Total	**50**

Figure 2-8
Histogram: Grades Example 2

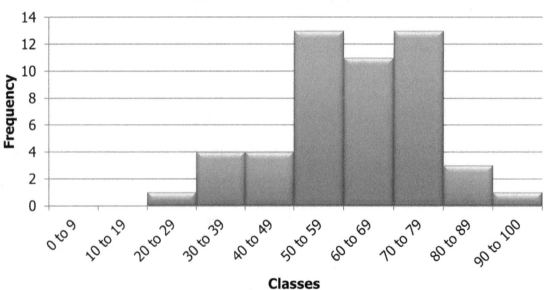

You can see that the graph has no gaps between the columns, representing the fact that grades are continuous (in this case on the integer scale). For example, there is no gap between a grade of 79 and a grade of 80. Each column represents the count of the grades within a specific data class. For example, from the histogram we can see that there are no grades below 20 (the first two columns have a height of 0), that there are four students with a grade between 30 and 39, and that the majority of grades are between 50 and 79 (represented by the three tallest columns). When combined, these three tallest columns account for 13+11+13=37 of the 50 students in our data set.

Histograms are a powerful tool for presenting data, as this type of graph can tell us many things about the *distribution* of values in our data set. By **distribution** we mean the spread of the data values. Frequency tables are commonly referred to as **frequency distribution tables** since they depict how data are spread (or distributed) *across* the range of data values. When looking at histograms, we look for things such as symmetry and shape. Consider the three graphs shown in Figure 2-9. The histogram on the left (Figure 2-9a) is **symmetrical**, meaning that the left side is a mirror image of the right side. The histogram in the middle (Figure 2-9b) is also symmetrical. The one on the right (Figure 2-9c), however, is not. The histogram on the right is *negatively* (or *left-*) *skewed*. The **skewness** of data tells us the extent to which more values fall on one side or the other. When we say that a data set is *negatively skewed*, we mean that there are more values on the right (higher data values) side of the scale than on the left (lower data values) side. Data which are *positively skewed* contain more values on the left side of the scale than on the right side. Grades on an easy test are an example for data which would likely be negatively skewed, with many students receiving high grades. Income is often used as an example for positively skewed data, with the majority of incomes being relatively low and only a few incomes having extremely high values. A good illustration here would be a company where most employees earn a relatively low salary, but a few executives earn very high salaries. We will discuss distributions, symmetry and skewness in more depth in later chapters of this book.

Figure 2-9
Three Histogram Examples

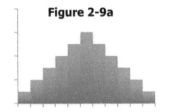
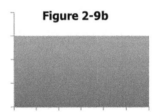
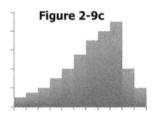

Figure 2-9a Figure 2-9b Figure 2-9c

Using Excel

You can create histograms using Excel in two ways. First, if you have already created a frequency table, you can use the column chart (as explained earlier in this Unit). Once you create the chart, right-click anywhere on the chart and select 'Format data series …'. Then, click on the 'Options' tab and change the value for 'Gap width' to 0, eliminating the gaps between columns. This last step is shown as the screen shot in Figure 2-10.

Figure 2-10
Screen Shot: Histogram Construction in Excel

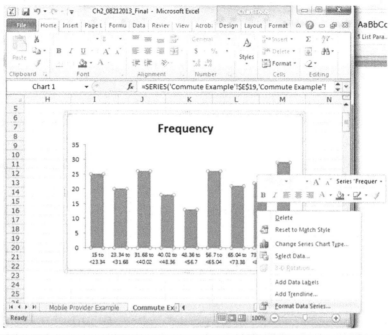

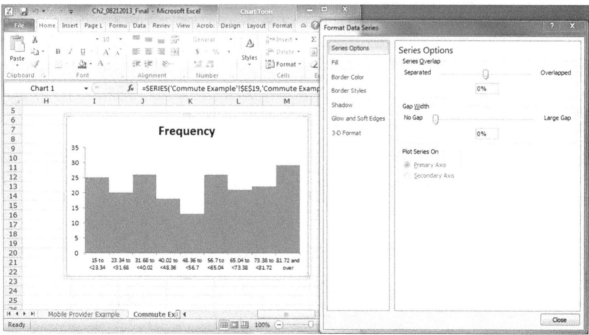

Second, there is a statistical add-in that you can use in Excel, called the *Analysis ToolPak*.[18] Once you have installed the *Analysis ToolPak*, you can select among a variety of statistical analysis options. One of these options is for creating histograms. To construct the histogram using the *Analysis ToolPak*, use the grades example data (provided in this chapter's Excel file in the worksheet titled 'Grades Histogram') and follow these steps:

- Under the 'Data' tab, click on 'Data Analysis'.
- Select 'Histogram' from the list and click 'OK'.
- In the window that opens (see Figure 2-11), select the data values in Column A as the 'Input Range' and the bins in Column D as the 'Bin Range'. The bins are the upper bounds of the classes.
- Select your desired 'Output Range' (where the output will be placed).
- Check the 'Chart Output' checkbox to generate a chart and click 'OK'.
    - You will need to format this chart to eliminate the gaps, as explained above.

---

[18] Here is the information on how to install the *Analysis ToolPak*:
http://office.microsoft.com/en-us/excel-help/load-the-analysis-toolpak-HP010021569.aspx.
Mac users should refer to: https://support.microsoft.com/en-us/kb/2431349

Figure 2-11
Screen Shot: Histogram Construction in Excel Using the *Analysis ToolPak*

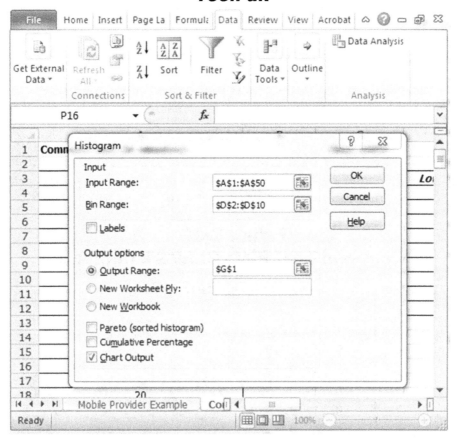

Stem-and-Leaf Plot

Another way to graphically present the frequency of quantitative data is with the *stem-and-leaf* plot, shown in Figure 2-12 (Figure 2-12 uses data from Grades Example 2 in Unit 1). A **stem-and-leaf** plot displays data broken into the stem, which is on the left-hand side of the line and represents the grouping level of the data, and the leaves, shown on the right-hand side of the line. In the example shown in Figure 2-12, we divided each grade into two parts: the *tens digits* became the stem and the *units digits* became the leaves. Hence, each row in the stem-and-leaf plot shows the grades for a given range of data values. We can reconstruct our data by putting the stem and the leaves back together. For example, the first seven numbers in the data set can be reconstructed as:

26, 34, 35, 36, 36, 40, 41

Because the stem-and-leaf plot uses the original data, it helps us to visualize the shape and distribution of the data in a manner similar to a histogram. With the stem-and-leaf plot, however, we do not lose potentially useful information about the data itself (an advantage of the stem-and-leaf plot over the histogram). Note that we can still see that the majority of grades are in the range of 50 to 79.

Figure 2-12
Stem-and-Leaf Plot: Grades Example 2

Stem	Leaf
2	6
3	4 5 6 6
4	0 1 3 9
5	0 1 2 3 3 3 4 6 6 7 7 9 9
6	0 1 3 3 3 5 5 6 7 7 8
7	0 1 2 2 2 2 5 5 5 6 6 8 9
8	0 0 4
9	4

Unit 2 Summary

- Graphs can be used to visualize the distribution of data. **Distribution** refers to how data are spread within their range of data values. Frequency tables are commonly referred to as **frequency distribution tables** since they depict how data are spread (or distributed) by reducing large data sets into a smaller number of classes. Graphical presentations help us to visualize the frequency distribution table and to obtain insights regarding the shape and spread of a data set.

- **Column charts** are used for discrete data. Each column represents one row in the frequency table and the height of each column is determined according to that row's frequency.

- **Histograms** are used to visualize frequencies of continuous data. Each column represents one class in the frequency table and its height is determined by the frequency of that class. Columns touch each other to represent the continuity of the data (there are no gaps between classes).

- The shape of histograms teaches us about the distribution of the data. We look for **symmetry**, with one side of the distribution mirroring the other, and we look for **skewness**, which is a measure of asymmetry.

- **Stem-and-leaf** plots are also used to visualize distributions. When constructing stem-and-leaf plots, we split each data value into the stem and the leaf. Stem-and-leaf plots overcome one limitation of histograms by maintaining a visible link with the original data. By putting together the stems and leaves, we can reconstruct the data set.

Unit 2 Exercises

1. Give an example of left-skewed data and an example of right-skewed data.

2. Refer to the data in Question 6 from the Unit 1 Exercises (provided in this chapter's Excel worksheet titled 'Popular Apps'). Draw the appropriate graph to represent a frequency distribution for these data.

3. Refer to the data in Question 7 from the Unit 1 Exercises (provided in this chapter's Excel worksheet titled 'Service Time'). Draw the appropriate graph to represent a frequency distribution for these data.

4. Refer to the data in Question 8 from the Unit 1 Exercises (provided in this chapter's Excel worksheet titled 'Organization Size'). Draw the appropriate graph to represent a frequency distribution for these data.

5. Draw a stem-and-leaf plot for the data in Question 9 from the Unit 1 Exercises (provided in this chapter's Excel worksheet labeled 'Tax Rates').

6. Use the stem-and-leaf plot shown in Figure 2-e1 to reconstruct the original data set. List these data values. What can you say about the distribution of these data based on the stem-and-leaf plot?

Figure 2-e1
Stem-and-Leaf Reconstruction

Stem	Leaf
12	7 7 8 9 9
13	0 0 2 2 3 3 4 4 5 5 6 7 8
14	0 1 2 3 4 5 5 5 6 7 7
15	0

7. Looking at the histogram shown in Figure 2-e2, what can be said about this distribution's characteristics?

Figure 2-e2
Histogram Shape

8. Create a frequency distribution for the data set displayed in Table 2-e6.

   a. Draw a histogram for these data.

   b. Draw a stem-and-leaf plot for these data.

   c. Compare the two plots you have created and discuss the differences and similarities between them.

Table 2-e6
Histogram and Stem-and-Leaf Plot Comparison Data Set

0.01	0.20	0.31	0.48	0.67	0.85
0.07	0.22	0.33	0.51	0.69	0.86
0.07	0.26	0.34	0.55	0.76	0.89
0.10	0.28	0.35	0.55	0.77	0.91
0.11	0.29	0.41	0.56	0.79	0.92
0.11	0.29	0.42	0.58	0.79	0.93
0.12	0.29	0.45	0.59	0.80	0.93
0.15	0.29	0.47	0.59	0.82	0.96
0.16	0.30	0.47	0.62	0.82	0.97
0.19	0.30	0.47	0.65	0.85	0.98

UNIT 3

Relative Frequency

Often, we wish to compare two data sets. For example, we might wish to compare: the incomes of different populations, the grades of students in different sections, the Internet usages of different age groups, the financial returns of different mutual funds, and so on. As noted earlier, an important first step in any interaction with data is to simply look at the data – its range, values and frequency distribution. We have already discussed the notion of frequencies and learned how to graph a frequency distribution table. In this section, we will introduce two new concepts: *relative frequency* and *cumulative relative frequency*, as well as two new graphs: the *pie chart* and the *ogive*. Let us consider a new example as we discuss these concepts.

University students sometimes take speed-reading courses to improve their reading skills. A student considering taking such a course collected data on the reading speeds (measured as words per minute or wpm) of 800 students who did not take the course and 300 students who took the course. The complete data set is provided in the chapter's Excel file in the worksheet titled 'Speed Reading'.

As a first step, we construct a *frequency distribution table* for each of the two data sets, i.e., reading speed for the 800 students who did not take the course and for the 300 students who did. As these data are *quantitative,* we will use *classes* to create our frequency table. We could use Sturges' formula to compute the number of classes and our class interval width, but, based on our knowledge of the speed-reading domain, we know that class widths of 100 words per minute (wpm) are representative of different reading skills. Therefore, we will use class intervals of 100 wpm. Looking at the data in the Excel worksheet, you will see that the lowest data value is 200 wpm and the majority of values are between 200 wpm and 600 wpm, with a few values higher than 600 wpm. Consequently, we create a frequency distribution table with five classes, using class widths of 100 wpm. This is shown in Table 2-12. To make sure you understand, use the data in the 'Speed Reading' worksheet to create this table on your own.

Table 2-12
Frequency Table: Speed Reading Example

Class (wpm)	Frequency (f_i) – 'No Course' Data	Frequency (f_i) – 'Course' Data
200-299	160	45
300-399	200	56
400-499	210	86
500-599	220	99
600 and above	10	14
Total	**800**	**300**

The frequencies described in the second column of Table 2-12 refer to the reading speeds of those who *did not* take the course and the frequencies in the third column refer to the reading speeds of those *who did take the course*. For example, 160 students who did not take the course tested at a reading speed of

200 wpm to 299 wpm. Note that we use a new notation in this table: we use the notation f_i to represent the frequency (f) of class i. So f_3 would equal 210 for the 'No Course' group and 86 for the 'Course' group.

Looking at the data in Table 2-12, what can you say about the effectiveness of the speed reading course? Is it effective (meaning that students taking the course read faster than those who do not)? How can you compare these two data sets? It is difficult to draw a conclusion by looking at Table 2-12 because the group sizes are different (there are more students in almost all of the classes for the 'No Course' group). We need to consider some other comparison approaches.

As a first step, we can look at the histograms of the two frequency distributions, shown in Figure 2-13. Review your understanding of histograms by creating the two histograms depicted in Figure 2-13 on your own in Excel.

Figure 2-13
Histograms: Speed Reading Example

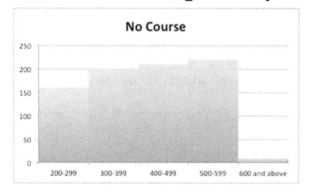
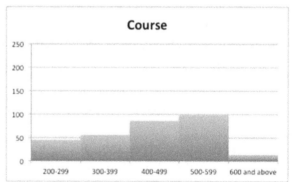

Comparing these two histograms shows a relatively similar pattern in the data, with both graphs displaying an increasing number of students up to the 599wpm speed. Does this mean that the course has no impact on reading speed? It thus is difficult (and somewhat misleading) to compare the two histograms directly because there are fewer people in total for the 'Course' group (300 versus

800). What we need is some measure of the number of people who fall in each class, *relative* to the size of the group. The **relative frequency** is the measure we are looking for, as it describes *the frequency of occurrence of specific values as a percentage of the total number of observations*. To create the relative frequency (rf_i) distribution table, we divide the frequency of each class (f_i) by the total number of observations (N):

$$rf_i = \frac{f_i}{N}$$

This is shown, for the speed-reading example, in Table 2-13. The relative frequency (rf_i) values for the 'No Course' group were calculated by dividing the frequency of each class by 800, which is the number of students surveyed. Thus, the relative frequency for the first class is 160 divided by 800, or 20% (0.20). Similarly, for the group of students who took the course, relative frequencies were calculated by dividing the frequency of each class by 300 (the number of students surveyed in this group). Note that in both cases the sum of all relative frequencies adds up to one (or 100%). (In Table 2-13, the percentages in column 5 add up to 1.0001 due to rounding of the presented numbers.) You can continue working in Excel to add the relative frequency columns to your frequency table.

Table 2-13
Relative Frequency Distribution: Speed Reading Example

Class	No Course		Course	
	Frequency (f_i)	Relative Frequency (rf_i)	Frequency (f_i)	Relative Frequency (rf_i)
200-299	160	160/800=0.20	45	45/300=0.15
300-399	200	200/800=0.25	56	56/300=0.1867
400-499	210	0.2625	86	0.2867
500-599	220	0.275	99	0.33
600 and above	10	0.0125	14	0.0467
Total	800	1	300	1

Look at the second to last class in Table 2-13 (the class for 500 wpm to 599 wpm) and compare the 'No Course' and 'Course' groups. While there were 220 students reading at this level who did not take the course, compared with only 99 students who did take the course, this number of students constitutes a lower percentage of the overall number of students in the 'No Course' group (27.5%) than in the 'Course' group (33%). The relative frequency provides a more accurate view of the share of each class (out of the total number of observations) in a data set. You can obtain a similar view of the data by using a *pie chart*, as shown in Figure 2-14.

Figure 2-14
Pie Charts: Speed Reading Example

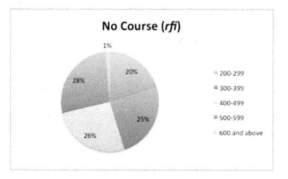
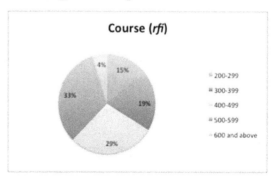

A ***pie chart*** displays how the total data set (the *whole pie*, or 100%) is divided among the five classes: the larger the slice, the larger the relative frequency of that class. Comparing the two purple slices, which represent the class of 500 wpm to 599 wpm, we can see that this constitutes a larger part of the pie on the right, which is the pie chart for the 'Course' group. Further, compare the area captured by the two lower classes (red and blue) and you can see that it is smaller in the 'Course' group than in the 'No Course' group.

Using Excel

To obtain these pie charts in Excel, highlight the column in your frequency table that contains the relative frequencies. Then, click on the 'Insert' tab to insert a chart and choose the pie chart from the available options.

Cumulative Relative Frequency

To follow up on the relative share of the two lowest reading groups, another type of analysis you can do is add together the relative frequencies of these two classes (shown in Figure 2-14 as the dark blue (200 wpm to 299 wpm) and red (300 wpm to 399 wpm) slices), and compare these combined shares for both groups. Note that nearly half of the students (45%) in the 'No Course' group read at a slower pace (399 wpm or lower), compared with about a third of the students in the 'Course' group (33.67%). The above calculation provides us with what is called the *cumulative relative frequency*. The **cumulative relative frequency** is obtained by adding up the relative frequencies of classes as you move down a frequency table. Table 2-14 provides an example for the 'Course' group in the speed reading example.

Table 2-14
Cumulative Relative Frequency: Speed Reading Example ('Course' Group)

Class	Frequency (f_i)	Relative Frequency (rf_i)	Cumulative Relative Frequency (crf_i)
200-299	45	0.15	0.15
300-399	56	0.1867	0.3367
400-499	86	0.2867	0.6234
500-599	99	0.33	0.9534
600 and above	14	0.0467	1
Total	300	1	

The first number in the column titled 'Cumulative Relative Frequency' is the same as the relative frequency of the first class (0.15). The second number was obtained by adding the relative frequency of the second class (0.1867) to the cumulative relative frequency of the first class (0.15). To obtain the cumulative relative frequency of the third class, we computed: 0.3367+0.2867=0.6234. The cumulative relative frequency of the third class represents *the percent of observations in the data set that fall below the upper bound of that class*. In other words, approximately 62% of the students in the 'Course' group read at a pace that is up to, and includes, 499 wpm. This interpretation holds for any class in the table. For example, 34% of the students read at or below 399 wpm. You will note that the cumulative relative frequency of the final class is one, meaning that 100% of the students read at one of the reading speeds in the table.

Another way of presenting the above information is to use an **ogive**, which is a graph representing the cumulative relative frequency. An ogive for the speed reading example is shown in Figure 2-15. The cumulative relative frequencies are shown on the *y-axis* and the reading speeds are shown on the *x-axis*. Note that the points identified on the x-axis are the upper bound of each class. We have also added the value of 800, the highest value in the data set (hence, it is the upper bound of the final class).

Figure 2-15
Ogive: Speed Reading Example ('Course' Group)

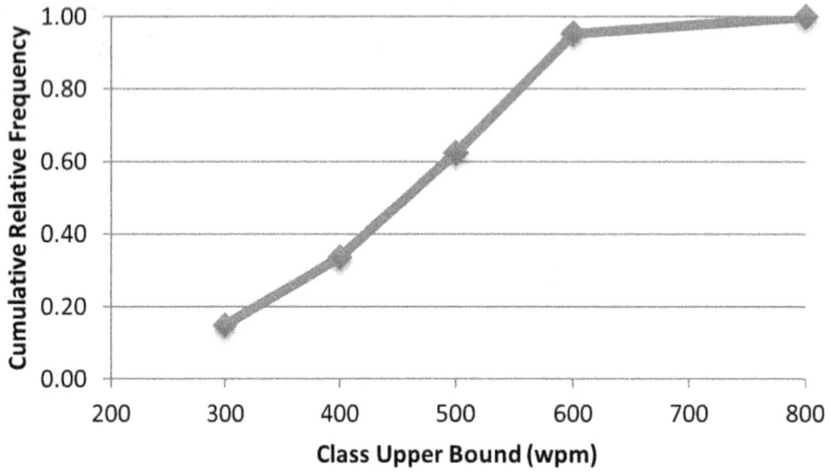

Using Excel

The above graph was created as an *XY scatter* graph in Excel, with x values representing class upper bounds and y values representing cumulative relative frequencies. To create this graph, first compute the column of cumulative relative frequencies for your data. Next, highlight this column, as well as the column containing the upper class bounds (i.e., your bins). Note that the bins column needs to be on the left of the cumulative frequency column, so that Excel will identify the two columns as x and y, respectively. In order to highlight two columns that are *not* adjacent to each other, highlight the first column (the bins column) and then press and hold the 'Ctrl' key as you highlight the second column (the cumulative frequency column). Now, click the 'Insert' tab to insert a new chart and select the scatter chart type.

Another way to create the cumulative relative frequency table and graph in Excel is by using the *Analysis Toolpak* and corresponding 'Data Analysis' option from the 'Data' menu. Follow the same steps as you did to create the histogram (as described earlier in Unit 2), but this time check the box for 'Cumulative Percentage' (shown in the screen shot in Figure 2-16). As you click 'OK', the output will provide the cumulative relative frequency in the table, as well as both the histogram and the ogive.

Figure 2-16
Screen Shot: Cumulative Relative Frequency Table Construction in Excel Using the *Analysis ToolPak*

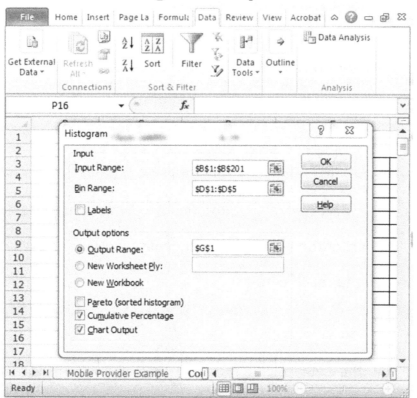

Unit 3 Summary

- ***Relative frequency*** takes into account the size of the data set and is used to compare the frequencies of a phenomenon of interest between groups whose size varies. The relative frequency of a class is calculated by dividing the frequency of that class by the total number of observations in the data set.

- ***Cumulative relative frequency*** allows us to look beyond class boundaries and make more general observations about our data. For example, it can help us identify the percentage of the data that lies below a certain value or the percentage of the data that lies between two values. Cumulative relative frequency is calculated by adding up the relative frequencies as you move down the table.

- The ***pie chart*** can be used to graphically represent relative frequency and the ***ogive*** is the graphical representation of the cumulative relative frequency.

Unit 3 Exercises

Using the Excel data file for Chapter 2, conduct the following calculations for the data in the worksheets labeled 'Donut Heaven', 'Popular Apps', 'Service Time', 'Organization Size' and 'Tax Rates':

1. Create a frequency distribution table for each data set.

2. Compute the relative frequency column in each table.

3. Compute the cumulative relative frequency column in each table.

4. Draw a pie chart of the relative frequency distribution for each data set.

5. Draw the ogive for each data set.

UNIT 4

If you review the chart menu in Excel, you will notice that many types of graphs are available to choose from. In this final unit of the chapter, we will briefly introduce some of these graph types and indicate when you might use each.

Comparing Nominal Data

Table 2-15 displays a frequency table listing the number of gold, silver and bronze medals won by the ten leading countries in the 2016 *Summer Olympics*. Here, we will use the *column chart,* introduced in Unit 2, to compare these data. (We could just as well use the **bar chart**, which is the horizontal equivalent of a column chart.)

Table 2-15
Frequency Table: Gold, Silver and Bronze Medal Counts, 2016 *Summer Olympics* Example

Country	Gold Medals Won	Silver Medals Won	Bronze Medals Won
United States	46	37	38
Great Britain	27	23	17
China	26	18	26
Russia	19	18	19
Germany	17	10	15
Japan	12	8	21
France	10	18	14
South Korea	9	3	9
Italy	8	12	8
Australia	8	11	10
Total	**182**	**158**	**177**

Figure 2-17 displays two column charts created from the data shown in Table 2-15. Note that the data in Table 2-15 vary along two dimensions – by country and by medal type. This means that we can choose to present these data to emphasize either of these two dimensions. Accordingly, Figure 2-17a shows, *for each country*, the medal count by type of medal. In contrast, Figure 2-17b shows, *for each type of medal*, the medal count by country. The choice of presentation depends on the conclusions we are interested in reaching. Figure 2-17b, for example, makes it easier to see that while Great Britain won more Gold and Silver medals than China, China surpassed Great Britain in the Bronze medal count. We can also see that

France won a high number of Silver medals and Japan won a high number of Bronze medals, compared with other countries. Thus, Figure 2-17b is more suitable for making comparisons between countries. Figure 2-17a, on the other hand, makes it easier to see that Great Britain, for example, won more gold medals than silver or bronze medals, and that France won more silver medals than gold or bronze. Thus, Figure 2-17a is more suitable for making comparisons between medal types within a given country.

Figure 2-17
Column Charts: Gold, Silver and Bronze Medal Counts, 2016
Summer Olympics Example

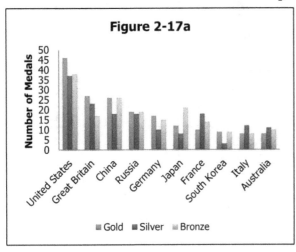
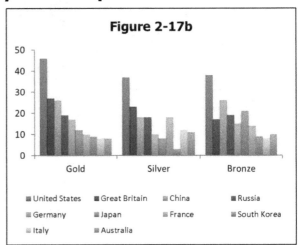

Describing the Relationship between Two Quantitative Variables

It is frequently desired to graphically describe the *relationship* between two quantitative variables. We introduced an example about speed reading in Unit 3. Let us consider whether taking a speed reading course can improve your overall grade point average (GPA). (Please note that we are using *made-up* data. This is not an endorsement for speed reading courses.) We look at two variables for a sample of 100 students: Reading Rate (wpm) and GPA. To observe whether there is a relationship between these two variables, we can use a *scatter plot* as shown in Figure 2-18. A **scatter plot** is a graphical presentation of the relationship between two variables in which data points are charted within a two-dimensional plane. Thus, each data point has an (x,y) value. Note that the data points in Figure 2-18a are randomly distributed across the graph, displaying no systematic pattern. Such a visual pattern suggests that there is no observed relationship between the two variables. On the other hand, looking at the data points in Figure 2-18b, there seems to be a positive, fairly linear (i.e., straight line) relationship. With a positive relationship, the two variables move in the same direction; that is, when the value of one variable increases (i.e., higher words-per-minute Reading Rate), the value of the other variable (i.e., GPA) increases as well.

**Figure 2-18
Scatter Plot: Reading Rate and GPA Example**

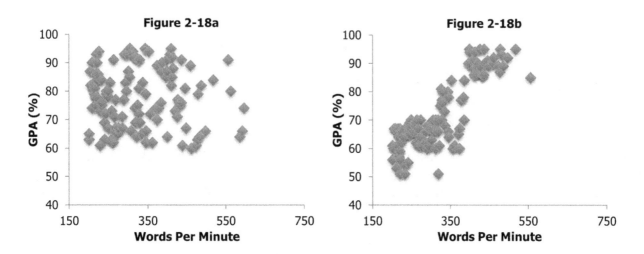

Other possible relationships we might observe using scatter plots are a negative linear relationship, where the variables move in opposite directions (e.g., the lower the price of a product, the higher the demand for the product), and non-linear relationships. An example of a non-linear, or curvilinear, relationship from the field of economics is that of diminishing marginal productivity (as illustrated in Figure 2-19): adding additional units of labor contributes less and less to the total quantity of goods being produced.

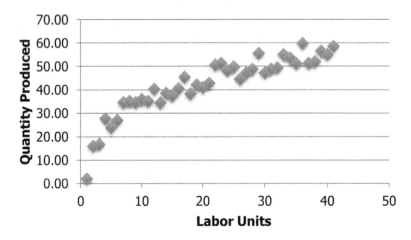

**Figure 2-19
Scatter Plot: Diminishing Marginal Productivity Example**

Describing Time Series Data

The final type of graph introduced is a **line graph**, commonly used to describe *time series data*. **Time series data** represent the values of a given variable over time, where time can be measured in days, quarters, years or any other temporal unit of measurement. Time series data are useful for observing trends in a data set over time. For example, Figure 2-20 shows a graph of the *Nasdaq Composite Index* from 2007 until 2012.[19] You can clearly see the trends in the stock market over time from this line graph. For example, the stock market crash of 2008-2009 is shown as the large dip in the center of the line (just before January of 2009).

---

[19] Source: Yahoo Finance.

Figure 2-20
Time Series Graph: *Nasdaq* Example

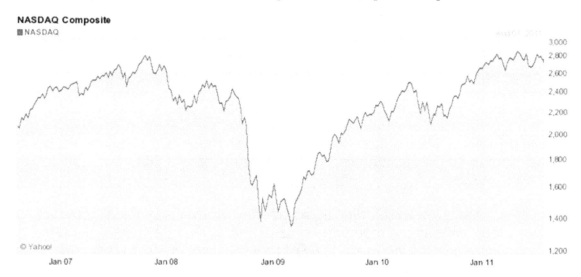

Using Excel

To create a line graph in Excel, you list your time periods in the first column, and the corresponding data values in the adjoining column. Then highlight both columns and, from the 'Insert' menu, select a 'Line' graph. In the above graph, the time period is shown in months on the x-axis, and the data values shown on the y-axis are the *Nasdaq Index*.

Unit 4 Summary

The main idea to take away from Unit 4 is that different graphs are suitable for representing different types of data.

- Use bar charts, column charts and pie charts for nominal data. **Bar charts** and column charts (covered in Unit 2) show counts of occurrence of given values. Pie charts (covered in Unit 3) show the relative share of given categories in the data set.

- Use scatter plots to show relationships between two quantitative variables. **Scatter plots** map each (x,y) pair of data points onto a two-dimensional space and allow us to examine the shape of the relationship (i.e., positive or negative relationship, linear or curvilinear relationship).

- Use line graphs to show trends within **time series data**, which represent the values of a given variable over time. A **line graph** charts a variable against time. The x-axis of the line graph is the time period and the y-axis represents the values of the variable of interest.

Unit 4 Exercises

1. A realtor wants to display how housing prices have changed over the past ten years. Which type of graph should he use?

2. A restaurant owner wants to study whether there is a relationship between the number of soups ordered and the temperature outside. Which graph should she use?

3. In this chapter's Excel file, look at the worksheet titled 'Sales'. This worksheet contains a company's sales data over 25 time periods, each period representing quarterly sales in specific years. Graph this data using the most suitable chart type.

The worksheet titled 'Medal Count' displays the full medal count of the 2016 *Summer Olympic Games*. The four questions that follow refer to these data:

4. Would it make sense to use a bar chart for the data as currently formatted in the worksheet? Why or why not?

5. Create the best graph for displaying which continent won the most medals in the 2016 *Summer Olympics*. Now, see if you can provide a better answer by creating more than one graph to answer the question. What is the difference between the graphs you created?

6. Is there a relationship between winning gold and silver medals? What does this relationship look like?

7. When you created the graph to answer Question 6, was there a problem with how the graph was displayed? If so, what is the nature of this problem and what might be done to resolve the problem?

END-OF-CHAPTER PRACTICE

For questions 1 through 4 below, use the data in this chapter's Excel worksheet titled '*IMDb*'.[20] This worksheet provides a data set listing: the 250 best movies as voted by *IMDb* members, the year in which the movie was released, the *IMDb* rating for the movie, and the number of votes used to create the rating.

---

[20] Source: Internet Movie Database, http://www.imdb.com/

1. Create a frequency table to show the number of best movies for each decade, beginning with the earliest movie that is included in the data set.

2. Use an appropriate graph to represent your frequency table.

3. Calculate the relative frequencies and cumulative relative frequencies based on the frequency table produced while answering Question 2 and draw the associated ogive. What is the percentage of the best movies that are pre-1990?

4. One critic argues that the more voters there are, the lower are a movie's ratings. Is there a relationship between the variables Rating and Votes that justifies such a claim?

5. The worksheet titled 'APPL' within the Chapter 2 Excel file shows data on the *Apple Inc.* stock price in 2011.[21] Use a graph to show the performance of this stock over time. What can you say about this stock's performance?

6. The worksheet labeled 'MSFT' within the Chapter 2 Excel file provides similar data for *Microsoft* stock in 2011. Is there an observed relationship between the stock prices of *Apple* (APPL) and *Microsoft* (MSFT)?

7. The Hours-of-Service regulations put limits on when and how long commercial motor vehicle drivers (i.e., truck drivers) may drive. These regulations are in place to ensure that drivers get the necessary rest to avoid accidents. To learn more about the relationship between driving hours and accidents, data were collected from twenty truck drivers on their daily hours of service and the number of accidents these drivers were involved in over the past year. The data are displayed in the worksheet titled 'Hours of Service' within the Chapter 2 Excel file. Create a graph to show the relationship between these two variables.

8. A statistics professor needs to create a histogram to display 700 observations. How many classes should she use?

9. 100 people were asked about their search engine usage and their responses are shown in Table 2-e7. Create a frequency distribution for these data.

---

[21] Source: Yahoo Finance (http://finance.yahoo.com/q/hp?s=AAPL+Historical+Prices).

Table 2-e7
Search Engine Usage Data Set

Google	Google	Google	Google	Google	Google	Google	Google	Bing	Yahoo
Google	Google	Google	Google	Google	Google	Google	Google	Bing	Yahoo
Google	Google	Google	Google	Google	Google	Google	Google	Bing	Yahoo
Google	Google	Google	Google	Google	Google	Google	Google	Bing	Yahoo
Google	Google	Google	Google	Google	Google	Google	Google	Bing	Yahoo
Google	Google	Google	Google	Google	Google	Google	Google	Bing	Yahoo
Google	Google	Google	Google	Google	Google	Google	Google	Bing	Ask
Google	Google	Google	Google	Google	Google	Google	Google	Bing	Ask
Google	Google	Google	Google	Google	Google	Google	Google	Yahoo	AOL
Google	Google	Google	Google	Google	Google	Google	Google	Yahoo	AOL

10. Draw a graph showing the market share (i.e., the percentage of a market that has been captured) of each search engine for which data was collected, as presented in Question 9.

11. Create a histogram for the data in the worksheet titled 'Histogram' within the Chapter 2 Excel file.

12. A researcher collected data on the performance of 30 software development teams. Data were collected for each team on each of six performance indicators, using a rating scale of 1 (low performance) to 5 (high performance). Table 2-e8 provides the performance of the 30 teams based on the indicator. For example, the first team scored 4 (out of five) on efficiency, the second team scored 3 (out of five), and so on. Create the frequency distribution for the efficiency scores of the 30 teams.

Table 2-e8
Software Development Team Efficiency Indicator Data Set

4	4	5	5	4	3
3	4	5	4	4	3
4	2	2	4	3	5
5	3	2	3	5	5
4	3	4	4	3	4

13. The organization in which the software development team performance data were collected would like to know what percent of teams perform below the acceptable threshold of 3 (out of five) on the efficiency indicator. Using the data provided in Table 2-e8, can you provide this information to the organization? Use either a table or a graph to obtain your answer.

CHAPTER 3: Measures of Centrality and Variation

You have just graduated from college and have been offered a job that pays $40,000 per year. You are very happy with the offer, and then you come across the following piece of information on the U.S. Department of Education website: the median annual earnings of workers (ages 25 to 34) with a Bachelor's degree, in 2013 was $51,940 for males and $44,620 for females.[22] As you read this, several questions may come to mind: "What does median mean?", "Why is my salary below it?", "Why do they report the median salary and not the average salary?", "Am I still happy about my offer?", and "Why is there a difference between male and female salaries?" Well, we won't be able to answer all of these questions here, but we will certainly be able to explain what median means, as well as related concepts such as mean, mode, variance and standard deviation.

Fundamental to any interpretation of data in statistics, as well as in other fields, is the *context* in which these data lie. **Context** refers to the circumstances or conditions in which an event occurs. For example, the number 21 used as a data value may mean many different things, such as a stock's selling price, the legal drinking age in the United States, or a grade on a statistics test. Data without context are impossible to interpret and understand. In statistics, a commonly used context is the *distribution* from which specific data values are drawn. For example, when you see your course grade (a single data value), you often wish to find out how your grade compares to the grades obtained by the other students in your class (the distribution). You may wish to find out if your grade is above or below

---

[22] Source: U.S. Department of Education, http://nces.ed.gov/fastfacts/display.asp?id=77

the class average, if you are in the top 5% of students in the class (or the bottom 5%), or how many students received a grade better than yours.

In order to make such comparisons, we need to know more about the frequency distribution of the data than what was explored in Chapter 2. Specifically, we need to understand two types of measures – those that describe the center of the frequency distribution, called **measures of centrality**, and those that describe the dispersion of data around the center of the distribution, called **measures of variation**. This is the objective of Chapter 3.

Before we delve into this chapter's concepts, it is useful to make a quick comment about the notation to be used from here on. In Chapter 1, we distinguished between a *population* and a *sample*, where a sample represented a subgroup of the population. Throughout this chapter and the ones that follow, we use different notations to represent values derived from samples (such as the sample's mean) versus the same values when derived from populations. Specifically, we generally use *lower case letters* for sample values (e.g., n represents the size of a sample) and either *upper case letters or Greek notation* for population values (e.g., N represents the size of the population and µ represents the population mean). Table 3-1 provides examples of some of the different notations used in this chapter.

Table 3-1
Examples of Notations Used in this Chapter

	Population	*Sample*
Size	N	n
Mean	μ	$\bar{X}$
Standard Deviation	σ	s
Variance	σ^2	s^2

This chapter covers the following topics:

- Measures of Centrality
- The Mean
- The Median
- The Mode
- Measures of Variation
- Variance
- Standard Deviation
- Coefficient of Variation
- Percentiles: Measures of Relative Standing
- Interquartile Range and Box Plots

UNIT 1

Measures of Centrality

The *central tendency* of a frequency distribution refers to where its center lies. This definition is somewhat elusive since there are different ways to statistically define a distribution's *center*. We can say that the center of the distribution is the most common class (graphically, this would be the peak of an associated histogram). Such a measure is called the *mode*. Alternatively, we can say that the center of the distribution is the value that splits the distribution in half – meaning that 50% of the values are below it. This value is called the *median*. And most commonly, we refer to the **mean** as the center of the distribution. Let us explore these measures further.

The Mean

The **_arithmetic mean_**, or the average, of a set of numbers is calculated by summing up all values and dividing this sum by the number of observations. For example, the mean of the following set of ten SAT scores

1410, 1371, 1369, 1441, 1381, 1363, 1328, 1342, 1463, 1370

is 1383.8. Thus, the arithmetic mean was obtained by summing up all scores and dividing by 10, or:

$$\frac{1410 + 1371 + 1369 + 1441 + 1381 + 1363 + 1328 + 1342 + 1463 + 1370}{10} = 1383.8$$

Now let us consider a different example. Assume that you invested $1,000 for five years with the following yearly returns: 5%, 7%, -2%, 12%, and 8%. At the end of Year 1, you would have $1,000*1.05=$1,050 (the original $1,000 plus the 5% return for Year 1). At the end of Year 2, you would have $1,050*1.07=$1,123.50 (the $1,050 from Year 1 plus the 7% return for Year 2), and so on. Below is a summary of your investment in each of the five years:

- End of Year 1: $1,000*1.05=$1,050
- End of Year 2: $1,050*1.07=$1,123.50
- End of Year 3: $1,123.50*0.98=$1,101.03 (you lost 2% this year)
- End of Year 4: $1,101.03*1.12=$1,233.15
- End of Year 5: $1,233.15*1.08=$1,331.81

Therefore, at the end of the five years you will have an amount of $1,331.81. Now let us calculate the _average return_ on your investment. Using the arithmetic mean formula, you can sum up your yearly returns and then divide them by the five years of investment. Your return in Year 1 was $50 ($1,050-$1,000), in Year 2 it was $73.50 ($1,123.50-$1,050) and so on, as shown in Table 3-2.

Table 3-2
Return on Investment Example

Year	Rate	End of year	Yearly return
1	5%	$1,050.00	$50.00
2	7%	$1,123.50	$73.50
3	-2%	$1,101.03	$(22.47)
4	12%	$1,233.15	$132.12
5	8%	$1,331.81	$98.65

Your average return is therefore:

$$\frac{50 + 73.50 - 22.47 + 132.12 + 98.65}{5} = 66.36$$

$66.36 is the yearly return amount (in dollars) you would receive, on average, from investing $1,000 for five years at the above return rates. If you multiply this number by 5 ($66.36*5=$331.8), you would obtain the overall return amount on the five-year investment, yielding the same amount of $1,331.81 that we obtained by calculating the yearly returns above.

What if we did not average *the returns amounts* (the values in Column 4 of Table 3-2), but instead we averaged the actual *return rates* (the percentages in Column 2 of Table 3-2)? Would we obtain the same result? Here is the computed average rate of return:

$$\frac{5\% + 7\% - 2\% + 12\% + 8\%}{5} = 6\%$$

Averaging the return rates thus yields an average return of 6%. By using this number as our yearly return (shown in Table 3-3), we would obtain $1,338.23 at the end of the five years. However, comparing this result with that obtained using the initial return rates (shown in Table 3-2), you can see that the numbers do not match. 6% is an inaccurate estimate of the average rate of return. Why is this so?

Table 3-3
Average Return Rate Example

Year	Rate	End of year	Yearly return
1	6%	$1,060.00	$60.00
2	6%	$1,123.60	$63.60
3	6%	$1,191.02	$67.42
4	6%	$1,262.48	$71.46
5	6%	$1,338.23	$75.75

When the progression of numbers is not linear – such as in the case of rates of return or growth rates - we need to use a **geometric mean** to account for the fact that the base amount changes from year to year (for example, at the start of Year 2 you compute your interest from $1,050 and not $1,000). The geometric mean for this example can be calculated as

$$\left(\prod_{t=1}^{n}(1+r_t)\right)^{1/n} - 1$$

where n is the number of time periods and r_t is the return rate in each time period. For our investment example, the geometric average rate of return is, therefore, 5.898%:

$$\left(\prod_{t=1}^{5}(1+r_t)\right)^{1/5} - 1 = (1.05 * 1.07 * 0.98 * 1.12 * 1.08)^{\frac{1}{5}} - 1 = 0.05898$$

Note that the formula for the geometric mean uses the nth root of the expression in parentheses. This means that we cannot multiply any negative values. Since we had at least one negative return rate (-2% in Year 3), we added 1 to the interest rate to ensure that all numbers are positive (and, then, subtracted 1 from the resulting calculation).

So far we have discussed two ways to compute the mean of a set of numbers, the arithmetic and the geometric means. A third approach to computing the mean that is frequently used in statistics is the **weighted mean**, in which we allow some data to have greater weight in the computation of the mean. A common example is calculating your GPA. As course credits vary, we need to account for the number of credits assigned to each course. For example, if you took three three-credit courses and your grades (on a percentage scale) were: 80, 75 and 85, the simple average of these three grades would be 80 (add them all and divide by three). If you also took two four-credit courses, with grades of 90 and 95, then your simple average for the five courses would be 85 (summing up all grades and dividing by five). However, calculating the simple average treats all grades equally and does not account for the fact that two of the courses were four-credit, instead of three-credit, courses. To properly account for the credits of each course, we use the *weighted average* formula shown below

$$\text{Weighted Average} = \frac{\sum_{i=1}^{n} w_i x_i}{\sum_{i=1}^{n} w_i}$$

where n is the number of observations (courses), w_i is the weight assigned to each observation (course credits), and x_i is the value of each observation (course grade). Hence, for our GPA example, the weighted average grade would be:

$$\frac{\sum_{i=1}^{n} w_i x_i}{\sum_{i=1}^{n} w_i} = \frac{3*80 + 3*75 + 3*85 + 4*90 + 4*95}{3+3+3+4+4} = 85.8824$$

Figure 3-1 summarizes these three types of means and their calculation (using the sample notations).

Figure 3-1
Three Ways to Calculate the Mean

$$\bar{x} = \frac{\sum_{i=1}^{n} x_i}{n}$$

Arithmetic: the average of a set of numerical values

$$\left(\prod_{t=1}^{n}(1+r_t)\right)^{1/n} - 1$$

Geometric: the central number in a geometric progression

$$\frac{\sum_{i=1}^{n} w_i x_i}{\sum_{i=1}^{n} w_i}$$

Weighted: the average of a set of numerical values with varying weights

Using Excel

The following functions are used in Excel to calculate the three types of means described above:

- Arithmetic mean: =AVERAGE(value1,value2,value3,...)

- Geometric mean: =GEOMEAN(value1,value2,value3,...)

- Weighted mean: There is no direct function to calculate the weighted mean, but you can use the function =SUMPRODUCT(array1,array2) to multiply your values by their respective weights and obtain the numerator for the weighted mean formula. Note that the =SUMPRODUCT() function refers to *arrays* in its arguments. In Excel, an array is a range of cells (for example, A1:A10 is an array containing the values placed in cells A1 through A10).

The Chapter 3 Excel file provides three examples for calculating means for different types of data. The worksheet titled 'Arithmetic Mean' contains the annual cost of study at 50 top undergraduate business programs in the United States.[23] To calculate the arithmetic mean for these data, type: =AVERAGE(B4:B53) in cell B54 and hit 'Enter'.

The 'Geometric Mean' worksheet in the Chapter 3 Excel file contains the GDP growth data for the United States.[24] Looking over Column B of the 'Geometric Mean' worksheet, you can see that some years have negative growth values (for example, 1974), which will result in an error message if we directly employ Excel's =GEOMEAN( ) function on these data. To overcome this problem, you can create another column (Column C in the worksheet) calculated as 100 plus the growth rates in Column B. Now, we can calculate the geometric mean for these data. To do so, type: =GEOMEAN(C4:C43) in cell C44 and hit 'Enter'. Don't forget to take away the 100 that we added in Column C, to conclude that the average GDP growth in the U.S. is 2.89%.

Lastly, the worksheet titled 'Weighted Mean' contains four component course grades for each of 24 students. The final course grade for each student should be calculated as the weighted mean of these component grades, with the weight of each component grade reflecting the component's contribution to the final course grade. To calculate the weighted mean for Student1, we need to compute:

$$\frac{\sum_{i=1}^{n} w_i x_i}{\sum_{i=1}^{n} w_i} = \frac{20\% * 80 + 10\% * 76 + 20\% * 85 + 50\% * 85}{20\% + 10\% + 20\% + 50\%}$$

---

[23] Source: These numbers were taken from past rankings published by *BusinessWeek*. Most recent rankings are presented on: http://www.bloomberg.com/features/2016-best-undergrad-business-schools/. The Bloomberg ranking page includes very nice visualizations of these data.
[24] Source: The World Bank, http://data.worldbank.org

Since we are dealing with percentages, we can rewrite the above as:

$$\text{Weighted mean} = 0.2 * 80 + 0.1 * 76 + 0.2 * 85 + 0.5 * 85$$

To calculate the weighted average for Student1, type =SUMPRODUCT(B$5:E$5,B6:E6) in cell F6 (Excel recognizes that the numbers in Row 5, i.e., the *weights* in our weighted mean formula, are shown as percentages) to obtain a weighted average of 83.1. Now calculate the weighted course mean for the other students in the course using the =SUMPRODUCT() function. When you have calculated all the students' individual grades, use the arithmetic mean to find out the average course grade for this class. Your answer should be 74.59.

The Median

The *median* is the value that divides the data so that half the values are lower than it and half are higher. For example, in the SAT scores example we used earlier in this section, the median of the ten scores (1410, 1371, 1369, 1441, 1381, 1363, 1328, 1342, 1463 and 1370) would be 1370.5. To see this, let us first put these values in order, from lowest to highest:

$$1328, 1342, 1363, 1369, 1370, 1371, 1381, 1410, 1441, 1463$$

Now consider the value 1370.5: it divides the data so that exactly five observations are below it (i.e., 1328, 1342, 1363, 1369 and 1370) and exactly five observations are above it (i.e., 1371, 1381, 1410, 1441 and 1463).

More generally, the median is the observation in location L of the sorted data set, such that:

$$L=0.5*(n+1)$$

In the SAT example above, n=10 and, therefore, the median is the value located at the 0.5*11=5.5 place. In our data set, this location is the midpoint between the fifth value (1370) and the sixth value (1371) and, hence, the median is 1370.5.

Although the mean is the more commonly used measure of centrality, the median is often viewed as being more accurate. This is because the mean suffers from one key limitation, which is the fact that it can be affected by extremely high, or extremely low, data values. For example, if you study the salaries of employees in an organization, the mean salary would probably be pulled upwards by the salary of executives (which are likely to be very high relative to the salaries of all other employees). Similarly, if most students in a specific course section received grades around 70 percent on a recent quiz but two students who missed the quiz received a grade of zero, then the mean would be pulled down by these two extreme values. The median is less sensitive to extreme data values relative to the mean. Thus, in the presence of extreme values, the median can provide a more accurate view of the center of a data distribution.

Using Excel

Open the 'Arithmetic Mean' worksheet within the Chapter 3 Excel file. There are 50 observations in this data set and we are looking for the median. To find the median, first sort the data from smallest to largest. (In Excel: Highlight the data table, select the 'Data' menu and then select 'Sort...'. A window should open, allowing you to choose the column you wish to sort by. From the drop down menu, select the column 'Annual Cost' and the option to sort from 'smallest to largest'.) Next, find the location of the median by calculating: L=0.5*(50+1)=25.5. The median would be the value at location 25.5. Finally, calculate the value of the median as the average of the values in locations 25 (the value 36,602) and 26 (the value 36,792). Hence the median is (36,602+36,792)/2=36,697.

You can also use the =MEDIAN() function in Excel to calculate the median. Typing =MEDIAN(B4:B53) in cell B54 of the worksheet should return the same answer as above: 36,697.

The Mode

The final centrality measure we will discuss is the *mode*. Strictly speaking, the **mode** is defined as the most common value in a data set. For example, in the data set

$$1, 2, 2, 3, 4, 5$$

the mode is the value 2, which appears twice. In the data set

$$1, 2, 2, 3, 4, 4, 5$$

there are two modes: the value 2 and the value 4. In the SAT scores example data set

$$1410, 1371, 1369, 1441, 1381, 1363, 1328, 1342, 1463, 1370$$

there is no mode since each value appears exactly once.

Using Excel

Computing the mode for small, discrete data sets is straightforward. However, for populations and for large samples, as well as for continuous data, we report the *modal class*, which is the class with the highest frequency, rather than a specific mode value. (You can think about this graphically: with a bar chart, there is a specific value associated with the highest bar; with a histogram, there is a specific class, or range of values, associated with the highest bar). Let us use an example to demonstrate this difference as well as explain how the mode is computed in Excel.

In the worksheet titled 'Mode' within the Chapter 3 Excel file, you will see, again, the annual cost at each of the top 50 undergraduate business programs in the United States. Using the Excel function =MODE.SNGL(B4:B53)[25], we obtain the value $12,188 as the mode of the data set. You can review the numbers in Column B to see that, indeed, this value is the only one that appears twice in the data set (in Rows 19 and 20).

Now, let us create a frequency distribution starting from the value of $4,000 (since the lowest value in our data set is $4,420) and using class intervals of $5,000. The graph of this distribution is shown in Figure 3-2.

**Figure 3-2
Histogram: Undergraduate Business Program Cost Example**

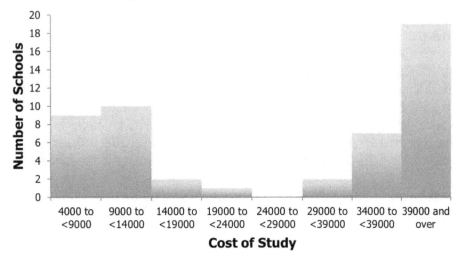

The **modal class** of a distribution is the class with the highest frequency. In the above example, the modal class is the last class, counting schools with an

---

[25] Another Excel function is =MODE.MULT(). This function can be used if it is thought there are multiple values for the mode in a data set. In this chapter, we will use the =MODE.SNGL() function only.

annual cost greater than $39,000. Although the value $12,188 is the only one that appears twice in this data set, it would be misleading to claim that it represents the most common value in this distribution. The majority of schools fall in the '$39,000 and over' class. For this reason, we more often create a histogram first and then identify the modal class (the peak of the histogram) rather than compute a specific mode value. A distribution with a single modal class (one peak) is called **unimodal**, and a distribution with two modal classes (two peaks) is called **bimodal**.

Unit 1 Summary

- **Measures of centrality** and **measures of variation** are two important types of **context** information used in statistics to help us understand data better. The former refers to where the center of a data set lies, and the latter refers to how the data are spread around this center.

- The center of a distribution can be represented using three measures: the mean, the median and the mode.

- The **mean** is the average value, and perhaps the most commonly used measure of centrality. The mean is sensitive to extreme observations, which can pull it up or down. We reviewed three types of means: the **arithmetic mean** is the commonly used average, computed by summing all values and dividing by the sample size (n). When the progression of value is compounded (such as with interest rates and growth rates) we use the **geometric mean**. Finally, when we would like some values to weigh more than others in the computation of the mean, we use the **weighted mean**.

- The **median** is the value that divides the sorted data set in half. The median is also a popular measure of centrality since it is not as sensitive as the mean to extreme values.

- The **mode** is the most common value (or most common class) in the data. We commonly use the **modal class** to learn about the shape and peaks of a distribution. **Unimodal** distributions have a single peak (mode) whereas **bimodal** distributions have two peaks.

Unit 1 Exercises

1. Find the arithmetic mean, the median and the mode for each of the data sets in Tables 3-e1 through 3-e5.

Table 3-e1

19	13	14	14	17	17	11	10	14	20
17	19	18	16	10	18	19	14	13	13

Table 3-e2

511	692	756	209	208	284	216	116	43	261
943	437	51	113	390	301	913	860	17	742

Table 3-e3

5	2	5	2	5	2	0	0	1	2
0	0	3	4	1	195	441	254	281	130

Table 3-e4

40	44	34	40	33	39	43	49	31	38
36	39	41	50	35	39	36	47	37	39

Table 3-e5

357	357	355	357	359	350	350	350	360	350
353	356	351	355	355	353	360	351	350	350

In the Chapter 3 Excel file, open the worksheet 'School Ranking Data'. This worksheet lists the top 50 undergraduate business programs in the U.S. (as created in 2012 by *Business Week* magazine), and displays data on six variables, in addition to the schools' rank and name: the state in which the school is located, the annual cost of study at the school, full-time enrollment, median starting salary of graduates, student faculty ratio, and average SAT scores of students. Using this data set, answer and explain Questions 2 through 7:

2. Calculate the average annual cost of study for university students in California, Massachusetts, New York and Pennsylvania.

3. Calculate the median full-time enrollment for the 50 schools, with and without Excel.

4. Use Excel to find the individual mode value for the variable Median Starting Salary. Then, create a frequency distribution for this variable and identify the modal class. Is there a difference between the two values?

5. Can you calculate the mean, median or mode for the variable State?

6. Can you use Excel to find the mode for the variable Full-Time Enrollment? Can you identify a modal class?

7. Calculate the mean and median for the variables Annual Cost, Student Faculty Ratio and Average SAT Scores. For which variable is the difference between the two measures the greatest?

A high school is raising funds for its Spring Dance by selling seeds purchased from the *Fedco* seed catalog. You can buy the seeds in bulk (prices are listed in Table 3-e6), repackage them into small consumer-size packets, and sell them. Table 3-e6 lists: the purchase price for a single bulk order of each seed type (cost per order), the selling price (per packet), and the expected sales (number of packets). Use this data set to answer Questions 8 through 11.

Table 3-e6
Spring Dance Fundraiser Data Set

Seed Type	Cost (per order)	Selling Price (per packet)	Expected Number of Sold Packets
Masai Green Beans	$11.00	$1.28	16
Kentucky Wonder Pole Beans	$7.00	$1.12	8
Sugarsnap Snap Pea	$11.00	$1.60	10
Scarlet Nantes carrot	$4.00	$0.64	8
Black Seeded Simpson Lettuce	$3.00	$0.80	7
Sweet Basil	$15.00	$1.20	28
Cilantro	$5.00	$0.80	28
Sensation Mix Cosmos	$7.00	$0.80	40
Blue Morning Glory	$7.00	$0.72	16
Autumn Beauty Mix Sunflower	$6.00	$0.80	28
Mammoth Grey Stripe Sunflower	$10.00	$0.96	16
Summer Sensation Sunflower	$8.00	$0.88	15

8. What is the average cost per order? What is the average selling price?

9. What are the median cost per order and median selling price?

10. What is your average revenue per packet sold? (Revenue is calculated as price*quantity sold). How is your calculation in this question different than that in Question 8?

11. Suppose that at the end of Month 1 of the fundraiser, you reassess your expected sales based on your recently gained knowledge of seed sales. Focusing on your sales of the Sugarsnap Snap Pea seeds, you believe that sales will increase by 10% in Month 2, by 10% in Month 3 and will then go down by 5% in Month 4. What is your average expected sales growth rate of Sugarsnap Snap Peas over the next three months?

UNIT 2

Measures of Variation

One of the most important measures you will learn about in statistics is the *variance*. For example, researchers are often interested in studying *expected* versus *observed* variance or in comparing the variance of two data sets under different scenarios. It is important to differentiate between the concepts of variation among data and the measure of the variance of a data set. *Variation* means that not all people or things are created equal. For example, the test scores of ten students are shown in Table 3-4:

Table 3-4
Student Test Scores Data Set

Test Scores
71
72
88
69
77
63
91
81
83
75
Mean=77

You can see that although the mean test score is 77 (the last row in the table), only one student received this test score. All other students received test scores that are either below or above the average. This is illustrated in Figure 3-3:

**Figure 3-3
Plot of Student Test Scores around the Mean Test Score**

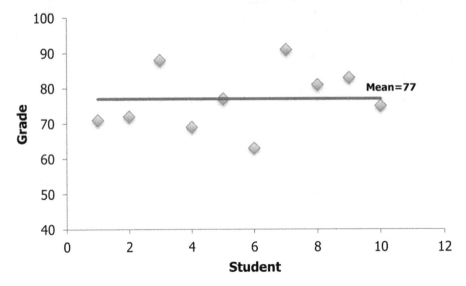

The distance between each student's grade and the mean is called a *deviation*. These deviations represent the fact that the grades in this data set are not identical. For any data set that you are likely to work with, deviations such as these will exist. They may be large in absolute size (for example, if you measure people's income) or they may be small (for example, if you measure the accuracy of a radiation machine). The extent of these deviations tells you the extent of variation that exists in your data. This is why we calculate the *standard deviation* for a given data set. To do so, however, we first need to calculate the *variance*, as explained below.

Variance

The **variance** is calculated as the *average of the squared deviations* of a data set. We will walk through an example to understand this calculation before providing the formula for it. We already discussed the notion of *deviation*, which is the distance between each observation and the mean. The deviations for the ten test scores in Figure 3-3 are shown in Table 3-5.

Table 3-5
Student Test Score Deviations

Test Score	Deviation from the Mean
71	71-77=-6
72	72-77=-5
88	88-77=11
69	-8
77	0
63	-14
91	14
81	4
83	6
75	-2
Mean=77	Sum of Deviations=0

Because of how the mean is computed, the positive deviations cancel out the negative deviations and, therefore, the sum of all deviations from the mean in a given data set is always zero. To avoid this problem in our computation of the variance, we use the squared deviations instead, shown in Table 3-6.

Table 3-6
Student Test Score Deviations (Using Squared Deviations)

Test Score	Deviation	Squared Deviation
71	71-77=-6	$(-6)^2=36$
72	72-77=-5	$(-5)^2=25$
88	88-77=11	$11^2=121$
69	-8	64
77	0	0
63	-14	196
91	14	196
81	4	16
83	6	36
75	-2	4
Mean=77	**Sum=0**	**Sum of Squared Deviations=694**

We explained that the variance is calculated as the average squared deviation, which suggests that the logical next step would be to divide the sum of squared deviation by the total number of observations. There is an important distinction, however, in the calculation of *sample variance* versus *population variance*. For the population variance, we indeed divide the sum of squared deviations by the population size (N), as shown in Figure 3-4a. For sample variance, we divide the sum of squared deviation by the sample size minus 1, or (n-1), shown in Figure 3-4b.

Figure 3-4
Variance Formulas

Figure 3-4a: Population

$$\sigma^2 = \frac{\sum_{i=1}^{N}(x_i - \mu)^2}{N}$$

σ^2 is the population variance
μ is the population mean
N is the population size

Figure 3-4b: Sample

$$s^2 = \frac{\sum_{i=1}^{n}(x_i - \overline{X})^2}{n-1}$$

s^2 is the sample variance
$\overline{X}$ is the sample mean
n is the sample size

Why do we use (n-1) and not simply n in the formula for calculating sample variance? Recall that our objective in statistics is to provide the best estimate for population values based on sample values. It has been shown that estimates obtained by dividing sample deviations by n are consistently smaller than the true population variances from which the sample was taken. To correct for this bias, we divide by (n-1), obtaining a more accurate estimate of the population variance.

Before we discuss the standard deviation, note that the units of measurement of the variance are the original units of the data, *squared*. For example, if our original data measured length in inches, then the variance is measured as inches-squared. If we measured income in dollars, then the variance is measured as dollars-squared, and so on. Since we square the deviations in the variance formula, we must also square the unit of measurement we are using.

Standard Deviation

The **standard deviation** is calculated, very simply, as the *square root* of the variance, as shown in Figure 3-5.

Figure 3-5
Standard Deviation Formula

Figure 3-5a: Population

$$\sigma = \sqrt{\sigma^2}$$

σ is the population standard deviation
σ^2 is the population variance

Figure 3-5b: Sample

$$s = \sqrt{s^2}$$

s is the sample standard deviation
s^2 is the sample variance

Because we are taking the square root of the variance, the unit of measurement of the standard deviation is the same as that of the original data (e.g., inches, dollars, etc.).

The standard deviation can be used to compare the variability of several distributions and to make statements about the general shape of a distribution. The higher the standard deviation, the wider the distribution is; meaning that the data values in this distribution are spread across a larger range. A smaller standard deviation means that the data values are clustered more closely around the mean. Going back to the grades example that opened this Unit, Figure 3-6 shows plots of student test scores, this time taking samples from two different sections of the course. The test scores for the section on the left (Section 1) seem to be clustered more closely around the mean than are the test scores for the section on the right (Section 2). We would, therefore, expect the standard deviation (of student test scores) to be higher for Section 2 than for Section 1. Let us calculate the two standard deviations and see whether this is the case.

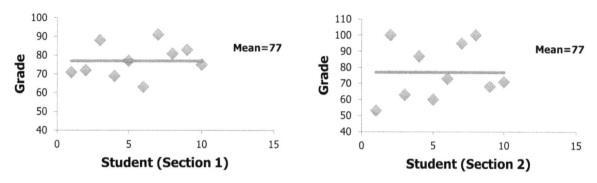

Figure 3-6
Plots of Student Test Scores for Two Course Sections

To calculate the standard deviation, we first calculate the variance and then take the square root of the value we obtain. Aside from the formula for calculating variance shown in Figure 3-4, there is another shortcut formula we can use, shown in Figure 3-7.

Figure 3-7
Shortcut Formula for Calculating the Standard Deviation

$$s^2 = \frac{1}{n-1}\left[\sum_{i=1}^{n} x_i^2 - \frac{\left(\sum_{i=1}^{n} x_i\right)^2}{n}\right]$$

We refer to it as a *shortcut formula* because it is easier to compute by hand. Table 3-7 shows the calculation of the standard deviation for Section 1 of the course. x_i are the actual test scores recorded in our data set. x_i^2 is a calculated value, computed by squaring each value of x_i. At the bottom of the table are the sums of x_i and x_i^2. These are all the values needed for our shortcut variance formula.

Table 3-7
Computing Student Test Score Standard Deviation for Section 1

x_i	x_i^2
71	5041
72	5184
88	7744
69	4761
77	5929
63	3969
91	8281
81	6561
83	6889
75	5625
$\sum_{i=1}^{n} x_i = 770$	$\sum_{i=1}^{n} x_i^2 = 59,984$

$$s^2 = \frac{1}{10-1}\left[59,984 - \frac{770^2}{10}\right] = 77.111$$

$$s = \sqrt{s^2} = 8.7813$$

Now, you try calculating the standard deviation for Section 2's test scores, provided in Table 3-8. Your work should reveal that the standard deviation for Section 2 (17.2434) is indeed larger than that of Section 1 (8.7813), as we expected from observing the two graphs in Figure 3-6.

Table 3-8
Student Test Score Data Set for Section 2

x_i
100
100
95
87
71
73
63
60
68
53

Using Excel

Of course, you can also use Excel to calculate variance and standard deviation. The Excel functions to use when you are working with *sampled* data are =VAR.S(value1,value2,value3,...) for the variance and =STDEV.S(value1,value2, value3,...) for the standard deviation (of course, you can always highlight a range of cells to apply the formula to rather than list specific values). For *population* data, use =VAR.P(value1,value2,value3,...) for the variance and =STDEV.P(value1,value2,value3,...) for the standard deviation. We will practice the use of these functions later in this chapter. As a final note, most calculators have statistical capabilities and can provide calculations of both variance and standard deviations. You should check the user manual of your calculator to learn how to use the statistical features.

The Empirical Rule

Another use for the standard deviation, aside from comparing distributions, is to gain information about the location of a specific data value within a distribution. We can say that a specific value lays, for example, two standard deviations above the mean, or one standard deviation below the mean, and so on. This information enables us to compute the probability of specific values within the distribution, as we will see in later chapters. One special case is that of the *normal distribution* (also known as the bell curve), discussed more thoroughly in Chapter 6. When data are normally distributed, the mean, median and mode are identical and the distribution is symmetrical around these values. For normal distributions, the **empirical rule** states that approximately 68% (roughly two-thirds) of all the values in the data set lie within one standard deviation from the mean. For example, if the grades of 100 students are normally distributed with a mean of 73 and a standard deviation of 8, then the empirical rule tells us that approximately 68 students have grades between 65 (that is, 73-8) and 81 (that is, 73+8). Further, approximately 95% of all values lie within two standard deviations from the mean, and nearly all values (99.7%) lie within three standard deviations from the mean.

The Coefficient of Variation

When two distributions have exactly the same mean (as was the case with the test scores example pictured in Figure 3-6), we can directly compare their standard deviations to draw conclusions about these distributions' relative variation. However, when the means of two distributions being compared differ, comparing the relative variation of the two distributions is not straightforward. Consider the example of comparing an established technology company's stock with an average price of $26.05 and a standard deviation of $7.00 against a penny stock with an average price of $0.77 and a standard deviation of $1.00. Which stock is more volatile (that is, which exhibits more variation over time in its stock price)?

To compare the two stocks, we use the **coefficient of variation**, which simply divides the standard deviation by the mean. We multiply the resulting number by 100 to obtain a percentage value rather than a fraction:

$$CV = \frac{\sigma}{\mu} * 100$$

Using the example of the two stocks, we can compare:

$$CV_{Tech\ Stock} = \frac{7}{26.05} * 100 = 26.87\%$$

$$CV_{Penny\ Stock} = \frac{1}{0.77} * 100 = 129.87\%$$

The penny stock is much more volatile. In this way, then, the coefficient of variation allows us to compare the standard deviations of different variables.

Unit 2 Summary

- Data values will almost always differ from each other. These differences represent variation in the data, which we measure by looking at how values are spread around the mean of the distribution. We looked at two measures that are related to each other: the variance and the standard deviation.

- The **variance** is defined as the average of the squared deviations from the mean, which are the squared distances between each data point and the mean.

- The variance is calculated differently for population data versus sample data, with the former using the population size (N) as the divisor and the latter the sample size minus 1 (n-1) as the divisor.

- The **standard deviation** is computed as the square root of the variance. The standard deviation is presented in the same units as the data. We use the standard deviation to compare the spread of two distributions and also to locate a specific value within a distribution based on its distance from the mean.

- The **empirical rule** tells us about how data are spread around the mean of the normal distribution. Approximately 68% of data are within plus or minus one standard deviation from the mean. 95% are within plus or minus two standard deviations from the mean. And nearly all data values (99.7%) lie within plus or minus three standard deviations from the mean.

- The **coefficient of variation** is another measure of the dispersion of data points around the mean, computed as the ratio between the standard deviation and the mean. The coefficient of variation allows us to compare the standard deviation of distributions with different mean values.

Unit 2 Exercises

1. Calculate the variance and standard deviation for each of the data sets below (assume these are population values, rather than sample values):

Table 3-e7

19	13	14	14	17	17	11	10	14	20
17	19	18	16	10	18	19	14	13	13

Table 3-e8

511	692	756	209	208	284	216	116	43	261
742	943	437	51	113	390	301	913	860	17

Table 3-e9

5	2	5	2	5	2	0	0	1	2
0	4	1	195	441	254	281	130	0	3

Table 3-e10

40	44	34	40	33	39	43	49	31	38
39	41	50	35	39	36	47	37	39	36

Table 3-e11

357	357	355	357	359	350	350	350	360	350
353	356	351	355	355	353	360	351	350	350

2. Using the data in the 'Weighted Mean' worksheet within the Chapter 3 Excel file:

   a) Calculate manually the variance and standard deviation of students' grades on the exam component of this course (in Column E).
   b) Are you treating these data as a sample or as the population? Why?
   c) Would you say that the standard deviation is high? Explain your answer.

3. Within the Chapter 3 Excel file, open the worksheet 'School Ranking Data' again. Use the Excel functions =VAR.S() and =STDEV.S() to calculate the variance and standard deviation for each of the five quantitative variables. Now use =VAR.P() and =STDEV.P() to compute these two values. Why are the results different from your previous numbers?

4. In the 'School Ranking Data' worksheet in the Chapter 3 Excel file, assume that the variable Student Faculty Ratio is normally distributed. You can treat this data set as *sampled* data, as it is based on a sample of 50 schools (albeit not a random sample).

   a. Find the mean and standard deviation of this variable.

   b. Based on our assumption of a normal distribution and the empirical rule explained in this Unit, use your answers to Question 4.a to determine approximately how many schools have a student faculty ratio of between eleven and 24 students. Check your answer against the actual data.

c. Of the five variables in the 'School Ranking Data' worksheet, which has the highest coefficient of variation and which has the lowest? What does this information tell you about these two variables?

5. Going back to the Spring Dance fundraiser example from the Unit 1 Exercises (these data are shown again in Table 3-e12), calculate the variance and standard deviation of the selling price of seeds and of the costs of seeds.

Table 3-e12
Spring Dance Fundraiser Data Set

Seed Type	Cost (per order)	Selling Price (per packet)	Expected Number of Packets Sold
Masai Green Beans	$11.00	$1.28	16
Kentucky Wonder Pole Beans	$7.00	$1.12	8
Sugarsnap Snap Pea	$11.00	$1.60	10
Scarlet Nantes carrot	$4.00	$0.64	8
Black Seeded Simpson Lettuce	$3.00	$0.80	7
Sweet Basil	$15.00	$1.20	28
Cilantro	$5.00	$0.80	28
Sensation Mix Cosmos	$7.00	$0.80	40
Blue Morning Glory	$7.00	$0.72	16
Autumn Beauty Mix Sunflower	$6.00	$0.80	28
Mammoth Grey Stripe Sunflower	$10.00	$0.96	16
Summer Sensation Sunflower	$8.00	$0.88	15

UNIT 3

Percentiles: Measures of Relative Standing

"Mensa, the high IQ society, provides a forum for intellectual exchange among its members. ... Membership of Mensa is open to persons who have attained a score within the upper two percent of the general population on an approved intelligence test that has been properly administered and supervised."[26]

Mensa's members are at the upper two percent of the population of those taking intelligence tests. This means that 98 percent of the people who take these tests score lower than Mensa members. This location, the upper two percent, is an example for what we term a **measure of relative standing**, or *percentile*. We already encountered one such measure, which is the median. The median is the

---

[26] Source: http://www.mensa.org/

50th percentile, as it divides the data so that 50% of the values are below (or above) it. More generally, the **Pth percentile** is the value that divides the data such that *P percent* of the values are below it. For example, if your SAT score indicates that you are at the 60th percentile, it means that 60% of those who took the test received scores lower than yours. In other words, it positions your score relative to the other scores.

In Unit 1 we provided the formula to find the location of the median and we can now extend this formula to calculate the location of any percentile. The location L of a percentile P, in a *sorted* data set, is calculated as

$$L_p = (n+1) * \frac{P}{100}$$

where n is the size of the data set. For example, consider the data set of ten SAT scores used earlier in this chapter:

1328, 1342, 1363, 1369, 1370, 1371, 1381, 1410, 1441, 1463

To find the 20th percentile, we need to first find its location in the data set. This can be computed as

$$L_{20} = (10+1) * \frac{20}{100} = 2.2$$

where n=10 and P=20. Similarly, the 60th percentile is the value at location:

$$L_{60} = (10+1) * \frac{60}{100} = 6.6$$

The interpretation of the above location means that after you sort the ten observations, from smallest to largest, the value that is located in place 6.6 is the 60th percentile. 60% of the observations would be below this value.

To calculate the actual value of the 60th percentile, rather than its location in the data set, we can use the weighted average. The value in location 6.6 is located between the sixth and seventh values, which in our SAT data set are 1371 and 1381, respectively. More specifically, we are looking for a value that is located 60% of the way between 1371 and 1381. This value is closer to 1381 and we, therefore, assign a larger weight to 1381 than to 1371. Using the 60% to 40% ratio for our weights, we calculate: 0.4*1371+0.6*1381=1377. This calculation is illustrated in Figure 3-8.

Figure 3-8
Finding a Percentile Value as a Weighted Average

Location										
6	6.1	6.2	6.3	6.4	6.5	6.6	6.7	6.8	6.9	7
1371	1372	1373	1374	1375	1376	1377	1378	1379	1380	1381
Value										

You can check the original data set to see that, indeed, six of the ten values are lower than 1,377 and four are higher. Another way to compute the value in location 6.6 is to take the value in location 6, which is 1371, and add to it 0.6*(1381-1371)=6, or 60% of the distance to the value in location 7. The result would be the same as before: 1377.

Interquartile Range and Box Plots

There are three measures of relative location that are of particular interest to statisticians. These are the 25th, 50th and 75th percentiles, also termed **quartiles**. The **interquartile range** (**IQR**), calculated as the difference between the first (Q1) and third (Q3) quartiles (i.e., the 25th and 75th percentiles), is another measure of the variability of a data set. The interquartile range gives us a view of how the middle 50% of the data are spread. Higher values of the interquartile range imply a larger spread of the data. A graphic representation of this interquartile range is called a **box plot**.

Consider the data in the worksheet 'Interquartile Range', found within the Chapter 3 Excel file. These data include two of the top 50 business schools' ranking variables: Annual Cost and Full-Time Enrollment. Both data sets have been sorted from smallest to largest. There are 50 observations in each data set. We can calculate the locations of the 25th and 75th percentiles as follows:

$$L_{25} = (50 + 1) * \frac{25}{100} = 12.75$$

$$L_{75} = (50 + 1) * \frac{75}{100} = 38.25$$

For the Annual Cost variable, the value of the 25th percentile is the value of the twelfth value (10,228) plus three-fourths of the distance between the twelfth and thirteenth values (0.75*(10,326-10,228)), or 10,228+73.5=10,301.5. Similarly, the value of the 75th percentile is 39,771 (that is, 39,768+0.25*(39,780-39,768)). Thus, the interquartile range of the Annual Cost variable is:

$$IQR = 39{,}771 - 10{,}301.5 = 29{,}469.5$$

The boxplot in Figure 3-9 uses the three quartiles to provide a graphical representation of the distribution of the Annual Cost data values. There are five values of interest shown in the boxplot: the minimum data value in the data set, the maximum data value in the data set, and the three quartiles (25th, 50th and 75th percentiles). (Note that in Figure 3-9 the 50th percentile is referred to as the median. Can you explain why?) We have marked these values on the boxplot in Figure 3-9 to help you interpret it.

Figure 3-9
Box Plot for the Annual Cost Data Values

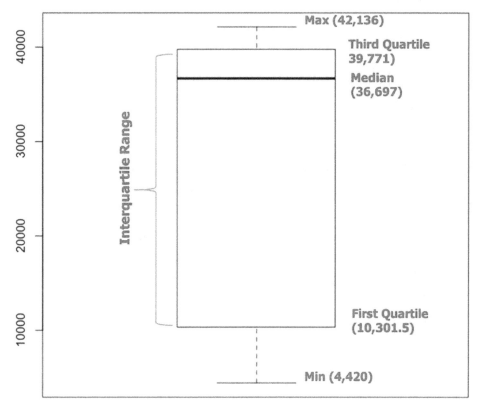

Looking at the above box plot, you can see that the distance between the first and third quartile is large, indicating a high spread of these data. You can also see that the median is closer to the third quartile, indicating the lower values are more spread than the higher values (i.e., the distribution is negatively skewed). The two lines extending out at the top and bottom of the box are called the 'whiskers'. In Figure 3-9, the whiskers extend to the maximum and minimum values in the data set. In some cases, as we will see below, the whiskers do not capture extremely high or extremely low values. Generally, the whiskers extend to the *smaller* of (a) the highest/lowest observations or (b) 1.5 times the interquartile range.

Now let us look at the interquartile range and box plot for the second variable in this worksheet: Full-Time Enrollment. You can calculate the first and third quartiles to see that they are 692.5 and 2,417.75, respectively. The interquartile range is therefore 1,725.25 (that is, 2417.75-692.5). Let us examine the box plot for these data, shown in Figure 3-10.

Figure 3-10
Box Plot for Full-Time Enrollment Data Values

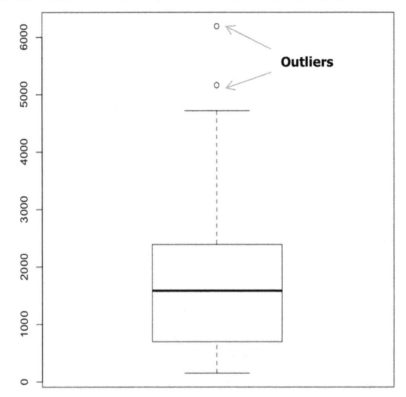

As you compare this box plot to the previous one, you can see that the interquartile range is smaller with these data values and that the data are more symmetrically spread around the median. Indeed, if you look back at your work computing the standard deviations for these variables (Unit 2 Exercises, Question 3), you will see a lower standard deviation for the Full-Time Enrollment variable than for the Annual Cost variable - hence, the smaller interquartile range. You will also see a longer tail at the upper end of the graph, indicating a larger spread among high values. Finally, there are two dots at the top of the graph. These represent **outliers**, or extreme observations, which in our case are unusually high (in other examples, outliers can also be unusually low values). In a box plot, outliers are values more than *1.5 times the interquartile range* from the box. To compute this threshold, find the interquartile range (IQR) and multiply it by 1.5. Then, add this value to the value of Q3 for the upper limit, and subtract it from the value of Q1 for the lower limit. If we decide to analyze these data further, we may wish to exclude such outliers, as they represent extreme and potentially unusual cases and may bias our findings.

Using Excel

There are several functions to compute percentiles in Excel. The one we will use is =PERCENTILE.EXC(), which provides results consistent with our percentile formula. You can read Excel's help files to learn about the differences between Excel's different functions for calculating percentiles.

Unit 3 Summary

- **Measures of relative standing** provide information on where a specific value lies with respect to the other values in the distribution.

- The **Pth percentile** value divides the data such that P percent of the values are below it.

- **Quartiles** are special cases of percentiles that divide the data into four parts. The first quartile (Q1) is the 25th percentile. The second quartile (Q2) is the median (or the 50th percentile), and the third quartile (Q3) is the 75th percentile. The difference between the first and third quartiles is called the **Interquartile Range** (IQR) and is another measure of the spread of data.

- **Box plots** are used to graphically display the interquartile range and provide graphical insights on the shape and spread of the distribution.

- **Outliers** are extreme values that may exist in the data set. Outliers are identified in the box plot as dots outside of the graph's whiskers.

Unit 3 Exercises

1. In the 'School Ranking Data' worksheet found within the Chapter 3 Excel file, find the 20th, 40th, 70th and 90th percentiles for the variable SAT Scores.

2. In the 'School Ranking Data' worksheet found within the Chapter 3 Excel file, calculate the interquartile range for the variable Student Faculty Ratio and draw the box plot for it. What can you say about this distribution? Are there any outliers?

3. Find the first and third quartiles for each of the data sets below:

Table 3-e13

46	33	81	8	56	66	61

Table 3-e14

49	52	10	97	34	94	24	87	19	27	92

Table 3-e15

45	97	87	8	65	53	97	65	84

Table 3-e16

81	45	31	97	60	36	28	96	60	7

4. A bank branch manager tracks her employees' performance by measuring the time it takes them to serve a customer. The time it took to serve each of 30 customers at the branch are listed in Table 3-e17. Draw a box plot for these data.

Table 3-e17

2	4	4	5	5	5	5	8	8	8	8	8	10	10	10
10	10	10	10	10	15	15	15	15	20	20	20	20	20	30

END-OF-CHAPTER PRACTICE

1. Provide a definition for a data set of size n=5 with a mean of 7 and a standard deviation of 0.

The next two questions are directed at the data sets provided in Tables 3-e18 through 3-e22.

2. For each of the data sets, compute the five box plot values (minimum, maximum, first quartile, median and third quartile).

3. For each of the data sets, identify any outliers, if they exist (outliers are values beyond 1.5 times the interquartile range from either the first or third quartiles). What problems can outliers cause during data analysis?

Table 3-e18

2	4	6	6	7	7	7	8	8	12

Table 3-e19

21	23	26	26	28	29	34	36	38	40	42	42	44	44	44	44	49	55	58	70

Table 3-e20

100	102	104	104	105	106	107	107	107	107	110	110	110	115	116

Table 3-e21

1	5	10	15	15	16	24	26	26	39	39	41	42

Table 3-e22

87	289	323	558	647	686	724	749	791	975

4. Calculate the average (i.e., mean) return for each of the following investments:

   a. A $1,000 investment that returns 5% in Year 1, 6% in Year 2, and 7% in Year 3.

   b. A $1,000 investment that returns 5% each year for five years.

   c. A $1,000 investment that returns $50 each year for five years.

   d. The yearly income from a portfolio consisting of an annuity providing a fixed income of $100/year and three $1,000 bonds, each with a 6% annual return.

5. An investor divided her funds between three different investment options. She invested 20% of her funds in Investment A, 30% in Investment B, and 50% in Investment C. The rate of return for Investment A is 5%, for Investment B it is 10%, and for Investment C it is 2%. What is the average return on her portfolio as a whole?

6. Create a data set in which it would be better to use the median, rather than the mean, as the measure of central tendency. Explain why.

7. If the age distribution of *iTunes* customers is normally distributed with a mean of 25 years and a standard deviation equal to five years, what percent of customers are older than 30 years?

8. 95% of all calls to a customer support center are between two and six minutes. If calls are assumed to be normally distributed, what is the average call time?

9. Calculate the variance and standard deviation for the following three data sets (assume these are samples and not populations):

Table 3-e23

2	4	6	6	7	7	7	8	8	12

Table 3-e24

1	5	10	15	15	16	24	26	26	39	39	41	42

Table 3-e25

87	289	323	558	647	686	724	749	791	975

10. Looking at the data in Tables 3-e23 through 3-e25, calculate and then use the coefficient of variation to comment on which of the three data sets has the higher variation.

The notion of comparing one value to the distribution from which it is drawn is fundamental to statistical analysis. Because the concepts discussed in this chapter are at the heart of statistical analysis, you should ensure that these are clear before moving on to other chapters. Throughout this chapter we have worked with the 'School Ranking Data' worksheet within the Chapter 3 Excel file and have conducted various analyses on the variables in this worksheet. It is time to bring it all together and try to tell the story that these data reveal.

11. In Table 3-26, determine the summary statistics for each cell in the table.

Table 3-e26
Summary Statistics for the 'School Ranking Data' Data Set

Summary Statistics	Annual Cost	Full-Time Enrollment	Median Starting Salary	Student Faculty Ratio	Average SAT Score
Sample Size					
Mean					
Median					
Standard Deviation					
Coefficient of Variation					

12. Now, create a frequency distribution chart (i.e., a histogram) for each of the five variables listed in Table 3-e26.

13. Looking at the completed Table 3-e26 and the associated histograms, what do we know about these data? For example, to what extent do the schools vary regarding annual costs or full-time enrollments? Are the distributions symmetrical or skewed? Are there outliers in these data sets? How similar are the schools' students in terms of Average SAT Scores and Median Starting Salary? What other observations can you make?

14. The final worksheet within the Chapter 3 Excel file is titled 'Team Social Ties'. The worksheet contains data on the responses of 481 professionals, each of whom responded to a survey concerning the number of team members with whom they work closely. Using what you have learned in this chapter, describe these data as best as you can: identify the mean, median and standard deviation; draw a box plot and discuss the distribution of the data set; and, compute any other value(s) that you feel might be helpful in understanding these data.

CHAPTER 4: Probability

In this chapter, we take a break from descriptive statistics to establish another important foundation for the chapters to come. Specifically, this chapter focuses on probability theory. Probability theory is the branch of mathematics that focuses on the likelihood of random events to occur. Probability plays an important role in business decisions by shaping our expectations about decision outcomes. For example, people often estimate the chances of getting a job or of obtaining funding for a project *prior* to investing the effort to apply for the job or to put together a project proposal. Similarly, predictions about market conditions are used to make stock purchase decisions and past project performance outcomes are used in estimating the probability of success in similar new projects.

In the context of this book, we will combine what we learn about probability in this chapter with what we have already learned about variables and their distributions in previous chapters in order to discuss *probability distributions* in Chapters 5 and 6. Then, we will apply what we know about probability distributions to test claims being made about variables.

For now, however, let us focus on obtaining a foundational knowledge of probability definitions, rules and relationships. This chapter covers the following topics:

- The Vocabulary of Probability Theory
- Assigning Probability
- Relationships Between Events
- Contingency Tables

- The Addition Rule
- Conditional Probabilities

- The Multiplication Rule

- Bayes' Theorem

UNIT 1

The Vocabulary of Probability Theory

Probability refers to the chance that an event will occur. *Impossible* events are assigned a probability of zero (0) and *certain*, or *sure*, events occur at a probability of 100% or one (1). Anything in between these two values (0 and 1) is probable. **Events** are outcomes of probability-producing *experiments*. For example, rolling a die represents an experiment that produces different events with corresponding probabilities. An **elementary event** is used to delineate the most basic outcome of an experiment (e.g., rolling a 2 on the die). Rolling an even number (i.e., 2, 4 or 6) on the die is an event that consists of three elementary events, namely: the possible outcomes of rolling a 2, 4 or 6. Similarly, tossing a coin is an experiment that produces two elementary events: the possible outcomes of the coin landing on Heads (H) or Tails (T).

The collection of all elementary events that can possibly come out of a specific experiment is called the **sample space** and it is denoted using curly brackets: {sample space}. For example, the sample space for rolling the die is {1, 2, 3, 4, 5, 6}, and the sample space for tossing a coin is {H, T}. The sample space for tossing *two* coins would be {HH, HT, TH, TT}, which enumerates all possible outcomes of the experiment.

Let us consider another example to demonstrate these concepts in a more real-life setting. *Kickstarter* is a crowd-funding platform aimed at helping filmmakers, musicians, artists and designers raise funds for projects. Project creators set a funding goal and a deadline. Visitors of the *Kickstarter* website can then pledge money to make the project happen. Funding on *Kickstarter* is all-or-nothing, meaning that projects must reach their funding goals to be deemed successful and receive any money. According to *Kickstarter*, 36% of the projects have reached their funding goals.[27] Among the projects that were not successful (that is, received no funding), various pledge levels (in terms of percentage of the funding goal) were still obtained, as shown in Table 4-1 below.

Table 4-1
Pledge Levels of Unsuccessful Projects

Percent Pledged	Number of Projects
0%	43,376
1% to 20%	118,521
21% to 40%	18,608
41% to 60%	7,213
61% to 80%	2,651
81% to 99%	1,484

Let us rephrase the above paragraph using our new probability language. Creating a new project on *Kickstarter* is an *experiment* which produces two possible *events*: (1) the project is successful (36% of projects are successful, hence the probability of success is 36%) and (2) the project is not successful (the remaining 64% of projects). The event 'unsuccessful project' can be further broken down into the six *elementary events* listed above in Table 4-1. We can further define the *sample space* for the *Kickstarter* funding experiment as: {0%, 1% to 20%, 21% to

---

[27] Source (data obtained in July, 2016): http://www.kickstarter.com/help/stats?ref=footer

40%, 41% to 60%, 61% to 80%, 81% to 99%, 100%}. Note that this sample space includes all the possible outcomes of the experiment. The next step in our analysis involves assigning probabilities to these outcomes.

Assigning Probability

There are three ways to determine the probability of an event. ***Classical probabilities*** are derived mathematically by dividing the number of successful outcomes from the experiment (i.e., the outcomes we are interested in) by the number of all possible outcomes (i.e., the sample space). We will use the single die example to demonstrate this concept. The sample space for this experiment is {1, 2, 3, 4, 5, 6}, which enumerates all possible outcomes of a single roll. The probability of rolling a 2 in this experiment is, therefore, 1/6 because a 2 represents one of the six possible outcomes. The probability of rolling an even number is 3/6 because three possible outcomes (2, 4, and 6) would be considered a success in this experiment. Classical probability assigns equal probabilities to each of the elementary events within the sample space.

Many classical probability examples can be found in Las Vegas gaming casinos. For example, with a roulette wheel having 36 pockets, the sample space will consist of 36 elementary events, each with a success probability of 1/36. Of course, we can also play different games using the wheel, like betting on colors or on various number combinations. Such bets involve *events* (again, as opposed to *elementary events*) and we can compute their probabilities from the elementary events that comprise these events, as will be explained in the next section. Other classical probability experiments include drawing cards out of a deck, tossing a coin or drawing lottery numbers.

A second approach to defining probabilities uses **relative frequency**. In Chapter 2 we learned how to create frequency tables and compute relative frequency, as well as cumulative relative frequency. We can use these concepts to find the probability of events based on past occurrence. Let us return to the *Kickstarter* example to demonstrate this approach. Recall that, according to *Kickstarter*, 36% of projects are successful, whereas 64% are not. We can use this information to infer future success probabilities. Specifically, we can expect a 36% chance of a future project being successful.

Assume, however, that I believe that my project is very unique and would appeal to a wide audience. I also believe that my project involves one of today's hottest topics. Based on these beliefs, I form a *subjective probability* of the success for my project, reflecting a belief that my project has a 60% chance of being funded. **Subjective probability** is the third approach used to determine the probability of an event and is simply based on my own beliefs and my interpretation of the world around me. Other examples in which we often use subjective probabilities include stock performance, baseball game outcomes or an expected grade on a test.

Regardless of how we defined the probabilities of elementary events (i.e., through classical probabilities, relative frequencies or subjective probabilities), the probability (P) of an event (E) is equal to the sum of the probabilities assigned to the elementary events (E_n) comprising the event

$$P(E) = P(E_1) + P(E_2) + P(E_3) + \ldots + P(E_n)$$

where n is the number of elementary events comprising event E.

In the die example, let E be the event of rolling an even number:

$$P \text{ (Rolling an even number)} = P(2) + P(4) + P(6) = \frac{1}{6} + \frac{1}{6} + \frac{1}{6} = \frac{1}{2}$$

Similarly, when tossing two coins the probability of one Heads and one Tails is:

$$P \text{ (one H and one T)} = P(HT) + P(TH) = \frac{1}{4} + \frac{1}{4} = \frac{1}{2}$$

In order to better see this, recall that the sample space for the experiment of tossing the two coins was: {HH, HT, TH, TT}. Out of the four possible outcomes, two of the outcomes include one Heads and one Tails.

Finally, the probability that an experimental outcome will include events from the sample space is always one. This is because, by definition, the sample space covers every possible outcome of the experiment.

Unit 1 Summary

- **Probability** describes the chances of an event to occur. Probability values range between zero (0) and one (1), with higher numbers representing higher probability.

- **Elementary events** are individual probabilistic outcomes of some experiment of interest. **Events** are composites of elementary events. An experiment is any process that yields different outcomes with corresponding probabilities, be it rolling dice, flipping a coin, choosing an investment or applying for a job.

- The **sample space** enumerates all possible outcomes of a given experiment.

- In **classical probability**, we assign equal probability to each elementary event in the sample space. We can also determine probability based on the **relative frequency** of occurrence of events in the past. Finally, we may use **subjective probability** to estimate the chances of an event to occur.

- The probability of an event is determined as the sum of the probabilities of the elementary events that comprise it. The sum of all probabilities in the sample space is one.

Unit 1 Exercises

Identify the sample space, experiment and probability approach (classical, relative frequency or subjective) for each of the following scenarios.

1. You are applying for two scholarships. For each scholarship you can either be selected to receive it in full or be rejected. You believe that you have a 60% chance of obtaining the first scholarship and about a 40% chance of obtaining the second one.

2. You are investing in a mutual fund. The fund can outperform the market, display average performance or perform below average. Based on past data you collected, you know that: the fund has outperformed the market 60% of the days, showed average performance 20% of the days, and performed below average the remaining 20% of the days.

3. You draw straws to select a leader for a group project. There are seven straws, one of which is the shorter one.

4. You answer a multiple choice question on a quiz by guessing. There are four possible answers to the question.

5. You answer a multiple choice question on a quiz based on what you have learned. Your course average so far is 80%.

6. You answer a multiple choice question on a quiz. You believe that you are well-prepared for the test and expect a grade of 90%.

Table 4-e1 displays unsuccessful projects on *Kickstarter* (a total number of 191,853) and the level of funding that they have obtained. Column 1 shows the pledge level obtained (the *classes* in our frequency distribution), Column 2 shows the number of projects in each class (the *frequency*) and Column 3 shows the percent of each class out of the total number of projects (the *relative frequency*). Answer Questions 7 - 11 based on this table:

Table 4-e1
Unsuccessful Projects on *Kickstarter*

Pledge Level Percentage	Number of Projects	Percent of Total Projects
0%	43,376	23%
1% to 20%	118,521	62%
21% to 40%	18,608	10%
41% to 60%	7,213	4%
61% to 80%	2,651	1%
81% to 99%	1,484	1%
Total	191,853	100%

7. Define the sample space for the unsuccessful projects based on the information provided in the table.

8. What is the probability of an unsuccessful project receiving pledges between 21% and 40%?

9. What is the probability of an unsuccessful project receiving pledges of 80% or less?

10. What is the probability of an unsuccessful project receiving pledges between 1% and 60%?

11. What is the probability of an unsuccessful project receiving less than 100% pledges?

UNIT 2

Relationships Between Events

We have so far introduced the basic concepts and definitions of probability theory. In this unit, we review probability rules and relationships that allow us to compute more complex probabilities. Imagine that our sample space represents the outcome of the experiment 'Selecting a person out of a group of executives'. Assume further that we define event A as 'the person is a female'. We can use a graphical tool to describe the sample space and event A within it. The **Venn diagram** that is shown in Figure 4-1 is commonly used to describe sets (groups) and their relationships. In this diagram, the grey rectangle is used to describe the sample space and the circle represents event A.

**Figure 4-1
Venn Diagram**

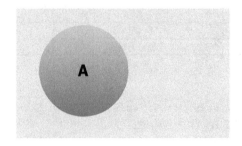

The part of the sample space that is outside of the circle is called the *complement* of A. The **complement of an event** A, denoted by A^c, is the event consisting of all points in the sample space that are not in A. Since the sum of probabilities of all events in a sample space is one (1), the probability of A^c can be computed as:

$$P(A^c) = 1 - P(A)$$

Now consider a second event, event B, which is 'Selecting an MBA graduate' out of the group of executives. Figure 4-2 presents both events A and B on the same diagram.

Figure 4-2
Venn Diagram of the Two Events

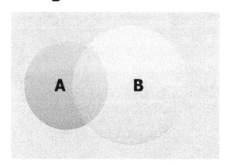

There are several areas of interest to us in this diagram. The grey rectangle is the sample space (the entire group of executives). Circle A is the group of female executives. Circle B is the group of executives with MBA degrees. The shaded area that is covered by both circles A *and* B is the *intersection* of the two events, i.e., female executives who have an MBA degree. The **intersection of events** A and B is the event containing all sample points that are *in both A and B*. This relationship is denoted as P(A and B), or as P(A∩B). Finally, the *total* area covered by circle A and circle B represents the *union* of the two events: here, including executives who are either female or MBA graduates. The **union of**

events A and B is the event containing all sample points that are *in A* or *in B* (in other words, *in either A or B*). This relationship is denoted as P(A or B) or as P(A∪B).

Now consider a third event, event C, defined as 'Selecting an executive without a Masters degree (MBA or other)'. Figure 4-3 shows event C in the sample space, alongside event B, which is defined as 'Selecting an executive with an MBA degree'.

Figure 4-3
Venn Diagram of Mutually Exclusive Events

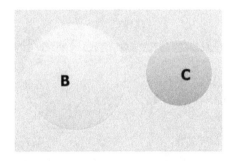

What do we learn from the above Venn diagram? First, we learn that there are more MBA graduates in the group than there are executives without a Masters degree (because circle B is larger than circle C). Second, we see that, unlike the circles in Figure 4-2, circles B and C do not share any points in common. They do not overlap. When two events have no common points we say that the events are *mutually exclusive*. Two events are thus said to be **mutually exclusive** if the occurrence of one event precludes the occurrence of the other. For example, cats and dogs are mutually exclusive. If you see an animal and it is a cat, then there is a zero percent probability that it is a dog. Returning to our example, since the MBA is a Masters degree, the two events defined as 'with' and 'without' a Masters degree are mutually exclusive.

A final relationship to mention is *independence* (not shown in a figure). Two events are said to be **independent** if the occurrence of one does not impact the probability of the other. As this concept may be a bit difficult to grasp at first, we will use several examples. Consider, first, the events 'Today is Meatless Monday' and 'It is raining outside'. It is quite clear that today being Meatless Monday will not affect the probability of rain (and vice versa) and, therefore, these events are independent of each other. There are many examples of independent events in classical probability. For example, obtaining a Heads in a one-coin toss will have no effect on the outcomes of subsequent coin tosses. Similarly, when rolling two dice, the outcome of one has no effect on the outcome of the other. And, there is no point in choosing past winning lottery numbers when selecting numbers for a lottery ticket, since the outcomes of draws are independent of each other. Of course, not all events are independent of each other. For example, the event 'It is raining' and the event 'Taking an umbrella' are not independent because the probability of taking an umbrella will change depending on whether or not it is raining outside.

We have so far defined the important relationships between events. We will now elaborate on how probability is computed for each of these relationships and how they are applied in common situations. Please make sure that the above concepts are clear before continuing to read this chapter, as they are important to understanding what follows.

Contingency Tables

Assume that we know that the overall size of the group of executives described in the previous section is 100, of which 40 are female executives, 75 hold an MBA degree, and 25 of the females hold an MBA degree. We can display this information in a table. Table 4-2 categorizes the executives by whether or not they have an MBA degree (top row) and by gender (Column 1). Such a table is called a **contingency table** (and also a *crosstab* table) and it displays *joint frequencies* of two variables.

Table 4-2
A Contingency Table – Executives Example

	MBA	No MBA	Total
Female	25	15	40
Male	50	10	60
Total	75	25	100

In creating the table, we begin with the information known to us and use this information to fill in the blanks. Specifically, we know that there are 40 female executives out of the 100 people in the group. Therefore, we know that the remaining 60 executives are male. Similarly, if there are 75 executives with an MBA degree, the remaining 25 do not have an MBA degree (they can have a different Masters degree or no Masters degree at all). Finally, if 25 of the 40 female executives have an MBA degree, the remaining 15 do not.

Note that the Total column in Table 4-2 sums up the total number of executives for each gender. Similarly, the Total row sums up the number of executives with and without an MBA degree. Finally, the total number of executives is shown in the bottom right cell and it is 100.

We can use Table 4-2 to compute various probabilities of events of interest. For example, if our experiment is 'Selecting any executive from the group', what is the probability that we have selected a male executive? From the table we know that there are 60 male executives out of 100 and, therefore:

$$P(\text{Male}) = \frac{60}{100} = 0.6 \text{ (or 60\%)}$$

What is the probability of selecting an executive who does not have an MBA degree?

$$P(\text{No-MBA}) = \frac{25}{100} = 0.25$$

The probabilities we have computed above are called **marginal probabilities** because they represent the probabilities at the *margins*, i.e., the 'Total' column and 'Total' row, of the table. The probabilities that are inside of the table are called **joint probabilities** and they represent the probability of two events occurring together. For example, the probability of selecting a male executive without an MBA degree is:

$$P(\text{Male } and \text{ No-MBA}) = \frac{10}{100} = 0.1$$

And, the probability of selecting a male executive with an MBA degree is:

$$P(\text{Male } and \text{ MBA}) = \frac{50}{100} = 0.5$$

Adding these two together will provide the probability of selecting a male executive (with or without an MBA degree):

$$P(\text{Male}) = P(\text{Male } and \text{ No-MBA}) + P(\text{Male } and \text{ MBA}) = 0.1 + 0.5 = 0.6$$

The Addition Rule

We previously defined the *union of events* as an event containing all sample points that are *in A or B or both*. To compute the probability of the union of two events we use the **addition rule**, which states that:

$$P(A \text{ or } B) = P(A) + P(B) - P(A \text{ and } B)$$

Graphically, we are adding circle A and circle B and then subtracting the overlapping area (A and B) since it was counted twice: once as part of circle A and once as part of circle B, as shown in Figure 4-4.

**Figure 4-4
The Addition Rule**

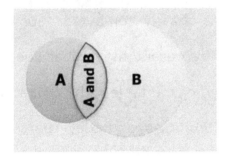

Using the information provided in Table 4-2, we can calculate the probability of the union of the two events 'Selecting a female executive' (event A) and 'Selecting an executive with an MBA' (event B):

$$P(Female \; or \; MBA) = P(Female) + P(MBA) - P(Female \; and \; MBA) =$$

$$= \frac{40}{100} + \frac{75}{100} - \frac{25}{100} = \frac{90}{100} = 0.9$$

We find that 90% of the group are either females or have an MBA degree (or both).

In order to make sure that the concepts being introduced are clear, let us consider another example that uses classical probability. Assume that there are 500 numbered and colored balls in a container. Our experiment is 'Selecting a single ball'. Table 4-3 provides information on ball colors and numbers.

Table 4-3
A Contingency Table – Ball Example

Ball Number	Ball Color			
	Red	Green	Blue	Total
1	15	50	35	100
2	10	60	30	100
3	5	40	55	100
4	40	50	10	100
5	30	50	20	100
Total	100	250	150	500

Let us review all of the probabilities we have discussed thus far. First, the right-most column and the bottom row provide the marginal probabilities for each color and number. For example:

$$P(\text{Green}) = \frac{250}{500} = 0.5$$

$$P(2) = \frac{100}{500} = 0.2$$

$$P(5) = \frac{100}{500} = 0.2$$

We also know the joint probabilities of different combinations of events. For example:

$$P(\text{Red and } 1) = \frac{15}{500} = 0.03$$

$$P(\text{Blue and } 3) = \frac{55}{500} = 0.11$$

Note that P(Red and Blue)=0 because these are *mutually exclusive* events. That is, a ball can be either red or blue, but not both.

We can also use the addition rule to compute the probabilities of unions of events. For example, the probability of selecting a ball that is *either* Red *or* 1 is:

$$P(\text{Red or } 1) = P(\text{Red}) + P(1) - P(\text{Red and } 1) =$$

$$= \frac{100}{500} + \frac{100}{500} - \frac{15}{500} = \frac{185}{500} = 0.37$$

In a similar fashion, we can compute the probability of selecting either a Red or a Blue ball:

$$P(Red \; or \; Blue) = P(Red) + P(Blue) - P(Red \; and \; Blue) =$$

$$= \frac{100}{500} + \frac{150}{500} - 0 = \frac{250}{500} = 0.5$$

The final part of this equation P(Red *and* Blue) is zero (0), because the two events are mutually exclusive, with the probability of both occurring at the same time being zero.

Conditional Probabilities

Returning to the Executives Example (Table 4-2), assume that we have selected one executive at random and she is a female. Given that we know the executive is a female, what is the probability that she holds an MBA degree? We know that 25 of the 40 women have an MBA degree. Therefore, we can compute the probability of a female MBA as: 25/40=0.625. What we have just computed is called a **conditional probability**, or the probability of event A *given that* event B had already occurred. Note that this is not the same as simply selecting a person that is both a female and holds an MBA degree. The difference lies in the order of the selection. If we first make one selection and then the other, we are dealing with a conditional probability because our second selection is conditioned on the outcome of the first selection.

By knowing that the person we have selected is a female, we reduce our sample space from the full group of 100 executives to the more limited group of 40 female executives. Therefore, the conditional probability is computed out of the more limited sample space of female executives only:

P(Selecting MBA given that we have already selected Female) =

$$= \frac{P(\text{Female and MBA})}{P(\text{Female})} = \frac{25}{40}$$

Or, more generally:

$$P(A|B) = \frac{P(A \text{ and } B)}{P(B)}$$

where P(A|B) reads 'the probability of A given B'.

To make sure this is clear, consider the following: in the example in Table 4-3, what is the probability of selecting a Red ball, given that our ball carries the number 1? Looking at the table, we know that there are 100 balls with the number 1, of which fifteen are Red. Therefore:

$$P(\text{Red}|1) = \frac{15}{100} = 0.15$$

Contrast this with the probability of selecting a single ball that is both Red and has the number 1:

$$P(\text{Red and } 1) = \frac{15}{500} = 0.03$$

Again, the sample space for the conditional probability case has been reduced from 500 (all balls) to 100 (balls with the number 1 only).

There is one final thing to consider before we move on to the next topic. Recall that we have defined *independent events* as the case where the occurrence of one event has no impact on the probability of the other event. Applying this notion to conditional probabilities, independence means that:

$$P(A|B) = P(A)$$

and

$$P(B|A) = P(B)$$

That is, if A and B are independent, then knowing that B has already occurred will not make any difference in determining A's probability. Using the one-coin toss example, the probability of obtaining Heads is: P(Heads)=0.5. The probability of obtaining Heads on a second toss *given that* we obtained Heads on the first toss is: P(Heads$_2$|Heads$_1$)=0.5, since P(Heads$_2$|Heads$_1$)=P(Heads$_2$) and P(Heads)=0.5 for any single toss. We can use this insight to determine if two events are independent of each other. If we find that, for any two events, P(A|B)=P(A), we can then conclude that the two events are independent of each other.

The Multiplication Rule

We have one more relationship to examine, namely the *intersection* of two events: P(A and B). In the previous sections we have used the contingency table to obtain P(A and B), but we can also compute it from already-known probabilities. Specifically, from the discussion on conditional probabilities we know that:

$$P(A|B) = \frac{P(A \text{ and } B)}{P(B)}$$

And, we can rearrange this formula to compute:

$$P(A \text{ and } B) = P(B)*P(A|B) = P(A)*P(B|A)$$

This formula is known as the **multiplication rule**. Note that in the above formula the order of the two components does not matter. We just have to make sure that the marginal probability we use (P(A) or P(B)) serves as the condition for the second component. That is, if we started with P(A), then our next component should be P(B|A); and, if we started with P(B), then our next component is P(A|B).

Let us see how this works in our two examples. First, assume that we know P(Male)=0.6 and P(MBA|Male)=0.833. We can compute:

$$P(\text{Male } and \text{ MBA}) = 0.6*0.833 = 0.5$$

Similarly, if we know that P(Red)=0.2 and P(2|Red)=0.1, we can compute

$$P(2 \text{ } and \text{ Red}) = 0.2*0.1 = 0.02$$

Finally, for independent events, the multiplication rule is simplified to P(A)*P(B). This is because P(B|A) is simply P(B).

Unit 2 Summary

- A **Venn diagram** uses circles to represent sets, with overlaps of the circles representing relationships between the sets. Venn diagrams are a commonly used tool to visualize events and their relationships.

- A **contingency table** (also known as a *crosstab* table) displays the joint frequencies of two variables. From the contingency table we can compute the **marginal probabilities** of single events, as well as the **joint probabilities** of two events occurring together.

- The **complement of an event** A, denoted by A^c, is the event consisting of all sample points that are not in A. The probability of the complement of A is computed as:

$$P(A^c) = 1 - P(A)$$

- The **union of events** A and B is the event containing all sample points that are *in A or B or both*. This relationship is denoted as P(A or B) or as P(A∪B). To compute the probability of the union of two events we use the **addition rule**, which states that:

$$P(A \text{ or } B) = P(A) + P(B) - P(A \text{ and } B)$$

- Two events are said to be **mutually exclusive** if the occurrence of one event precludes the occurrence of the other. For mutually exclusive events, the addition rule is simplified (because P(A and B)=0) to:

$$P(A \text{ or } B) = P(A) + P(B)$$

- A **conditional probability** is the probability of event A occurring *given that* event B has already occurred. This is denoted as P(A|B), read as 'the probability of A given B'. It is computed as:

$$P(A|B) = \frac{P(A \text{ and } B)}{P(B)}$$

- When two events are said to be **independent** of each other, the occurrence of one does not impact the probability of the other. Therefore, for independent events A and B:

$$P(A|B) = P(A)$$

and

$$P(B|A) = P(B)$$

- The **intersection of events** A and B is the event containing all sample points that are *in both A and B*. This relationship is denoted as P(A and B) or as P(A∩B). To compute the probability of the intersection of two events, we use the **multiplication rule**, which states that:

$$P(A \text{ and } B) = P(B)*P(A|B) = P(A)*P(B|A)$$

- For independent events, the multiplication rule is simplified to:

$$P(A \text{ and } B) = P(A)*P(B)$$

Unit 2 Exercises

1. There are four teams left in the NHL playoffs. An analyst assigns the following probabilities to each team to win:

   P(Toronto) = 0.3
   P(Buffalo) = 0.35
   P(Los Angeles) = 0.2
   P(Vancouver) = 0.15

   Determine the probabilities of the following events:

   a. Toronto losing

   b. An Eastern conference team winning (Toronto or Buffalo)

   c. Los Angeles, Buffalo or Toronto winning

2. Given the joint probabilities shown in Table 4-e2, calculate all marginal probabilities:

   Table 4-e2

	A_1	A_2
B_1	0.2	0.05
B_2	0.34	0.41

3. Given the probabilities shown in Table 4-e3, complete this contingency table:

Table 4-e3

	A_1	A_2	
B_1	0.11		
B_2		0.71	
	0.56		

Now, determine the probabilities of the following events:

a. Find $P(A_2 \text{ and } B_1)$

b. Find $P(A_1 | B_1)$

c. Find $P(B_2 | A_2)$

d. Find $P(B_2 \text{ or } A_1)$

4. Can two events be both mutually exclusive and independent of each other? Provide mathematical support for your argument using the probability rules we described in this unit.

5. You hire two programmers to write code for a new app you are developing. 60% of the code comes from programmer A and the remaining code is written by programmer B. On average, 5% of programmer A's code will have bugs (i.e., coding errors) in it compared to 7% of programmer B's code.

    a. Create a contingency table based on the data in the question.

    b. What is the probability of bugs in your app?

    c. What is the probability that the code has bugs and was created by programmer A?

    d. What is the probability that the code has bugs or that it came from programmer B?

    e. You found a bug in your code. What is the probability that it was generated by programmer B?

f. Are the bugs independent of programmers? Provide support using the probability rules we have learned in this unit.

6. Table 4-e4 is a contingency table for unsuccessful *Kickstarter* projects:[28]

Table 4-e4
A Contingency Table of Unsuccessful *Kickstarter* Projects

Category	Pledge Level Percentages						Total
	0%	1% to 20%	21% to 40%	41% to 60%	61% to 80%	81% to 99%	
Film & Video	8,440	21,614	3,281	1,091	351	155	**34,932**
Music	6,038	13,246	2,525	901	274	102	**23,086**
Publishing	5,876	13,536	1,891	718	232	91	**22,344**
Art	3,335	7,782	1,377	511	178	89	**13,272**
Games	1,981	11,002	1,829	886	407	276	**16,381**
Total	**25,670**	**67,180**	**10,903**	**4,107**	**1,442**	**713**	**110,015**

a. What is the probability that a selected project is a music project?

b. What is the probability that a selected project obtained 40% of funding or less?

c. What is the probability that a selected project is a music project and has obtained 0% funding?

d. You select one art project. What is the probability that it obtained 81% or more of funding?

e. A selected project obtained 30% funding. What is the probability that it is a games project?

f. What is the probability that a selected project is either an art project or a games project?

g. What is the probability that a selected project is both an art project and a games project?

h. Are a project's category and funding level independent of each other?

---

[28] Source: http://www.kickstarter.com/help/stats?ref=footer

UNIT 3

Bayes' Theorem

Thomas Bayes was an 18th century English reverend and mathematician who established a mathematical basis for probability. The resulting **Bayes' Theorem** is a formula for revising initial probability predictions based on relevant new evidence. Often, we begin probability analysis with initial, or prior, probabilities. For example, as you begin to study for an important assessment test (such as the SAT), you are likely to have some prediction about your probability of success on the assessment test. Then, you obtain new evidence. For example, you obtain information on the success probabilities of people who took a preparatory course for the test and then consider taking the course as a means of increasing your probability of success. Prior to registering for the preparatory course (and, prior to spending the money!), you might wish to calculate your likely expected improvement on the test. Bayes' theorem provides the method for this calculation. We will now review two examples in depth to understand Bayes' Theorem.

Suppose you would like to develop a new game app for the *iPhone* and plan to obtain the funding to develop and market the game via *Kickstarter*. Based on past data, you expect the probably of successfully funding a project on *Kickstarter* to be 36%. In other words, if we define event S to be 'Successful project funding', we know that

$$P(S) = 0.36$$

and, using the complement rule, we also know that the probability of failure is

$$P(S^c) = 0.64$$

where S^c denotes the complement of event S.

After studying the success of past projects, you see that there is a difference in the success rates of different types of projects: approximately 8% of the successfully-funded projects were games, whereas approximately 9% of the not-funded projects were games. Using probability language, we can write

$$P(G|S) = 0.08$$

and

$$P(G|S^c) = 0.09$$

where G stands for the event 'The project belongs to the games category'. We use conditional probability here to represent the knowledge we already have about the funding outcome (successful or not).

Armed with this new knowledge about the successful and unsuccessful funding of games projects, you can now figure out the success probability you can expect given that your project belongs to the games category. That is:

$$P(S|G) = ?$$

Let us start with summarizing what we know in a contingency table, as shown conceptually in Table 4-4a and, after using available information, in Table 4-4b. After plugging this information (highlighted in green) into Table 4-4b, we can compute the rest of the probabilities missing from the table. For example, P(G) can be found by adding P(G and S) and P(G and S^c), P(G^c and S) can be found by subtracting P(G and S) from P(S), and so on.

Table 4-4a
A Contingency Table – Bayes' Theorem Example

	Game (G)	Other Project (G^c)	Total
Successfully Funded (S)	P(G and S)	P(G^c and S)	**P(S)**
Not Funded (S^c)	P(G and S^c)	P(G^c and S^c)	**P(S^c)**
Total	**P(G)**	**P(G^c)**	**1**

Table 4-4b
A Contingency Table with Data – Bayes' Theorem Example

	Game (G)	Other Project (G^c)	Total
Successfully Funded (S)	P(G and S) = P(G|S)*P(S) = 0.08*0.36 = 0.0288	P(G^c and S) = 0.36 − 0.0288 = 0.3334	**0.36**
Not Funded (S^c)	P(G and S^c) = P(G|S^c)*P(S^c) = 0.09*0.64 = 0.0576	P(G^c and S^c) = 0.64 − 0.0576 = 0.5824	**0.64**
Total	**P(G) = 0.0288 + 0.0576 = 0.0864**	**P(G^c) = 0.3312 + 0.5824 = 0.9136**	**1**

Recall that we are looking for P(S|G). Let us use the probability rules covered in Unit 2 of this chapter to work out this conditional probability. We know that:

$$P(S|G) = \frac{P(S \text{ and } G)}{P(G)}$$

We further elaborate on the above equation by identifying that the *marginal probability*, P(G), can be computed as the sum of all the joint probabilities in its corresponding column in the contingency table:

$$P(S|G) = \frac{P(S \text{ and } G)}{P(S \text{ and } G) + P(S^C \text{ and } G)}$$

Finally, using Table 4-4b we can compute the probability P(S|G)

$$P(S|G) = \frac{0.0288}{0.0288 + 0.0576} = 0.3333$$

which is the success probability for games projects, just slightly lower than the overall success probability of 0.36.

What we just developed above is the Bayes' Theorem formula. The general format for Bayes' Theorem is:

$$P(B|A) = \frac{P(B \text{ and } A)}{P(A)} = \frac{P(B \text{ and } A)}{P(B \text{ and } A) + P(B^C \text{ and } A)}$$

Let us review another example to ensure that these concepts are clear. An event organizer is considering three locations for an upcoming concert: indoor, under the sky (or, outdoor) and under a tent. The locations will yield different payouts (determined as a function of both location capacity and the likely attendance for a concert at the location) depending on the weather that day, as shown in Table 4-5.

Table 4-5
Concert Planning Payoff (in Dollars) Example

	Good Weather (p=0.6)	Bad Weather (p=0.4)
Indoor	$5,000	$10,000
Outdoor	$10,000	-$3,000
Tent	$7,000	$2,000

Table 4-5 also shows that the event planner believes there is a 60% chance of good weather (event G), and 40% chance of bad weather (event G^c). Using this information, the planner can make an initial decision about which venue is preferable under expected weather conditions. Assume, however, that the cautious planner would like to obtain additional information on expected weather conditions, and consequently hires a weather specialist for that purpose. The specialist has a pretty good record of past predictions. Specifically, she has been able to correctly predict good weather 80% of the time when good weather actually occurred. Defining event PG as a prediction of good weather by the specialist, we know that:

$$P(PG|G) = 0.8$$

In other words, given that we know weather conditions were good, the specialist correctly predicted this weather outcome 80% of the time. We also know that the specialist correctly predicted bad weather (when bad weather actually occurred) 90% of the time:

$$P(PG^c|G^c) = 0.9$$

where PG^c is the event of a prediction of bad weather by the specialist.

Now for a key piece of information: assume that the specialist has just delivered her report, which predicts good weather conditions. Given this knowledge, what should be the planner's revised prediction of the weather conditions? In other words:

$$P(G|PG) = ?$$

To solve this question, we will use a different approach than that used in the previous example. Specifically, we will construct a **probability tree**, which is a diagram showing all possible events with their probabilities. Two important things to know about probability trees are: (1) the tree is shown sequentially left-to-right, meaning that events that occur first are presented to the left of later events, and (2) the sum of all branches coming out of a specific node should be one (this is the graphical equivalent of the complement rule). Figure 4-5 depicts our probability tree for this example.

Figure 4-5
The Probability Tree

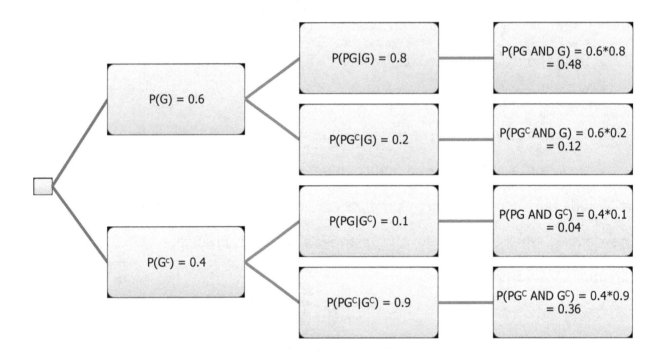

The probability tree begins with our initially believed probabilities about good and bad weather conditions, i.e., P(G) and P(G^C). We then place the conditional probabilities about the specialist's performance, e.g., P(PG|G)=0.8 and P(PG^C|G^C)=0.9, as the second level of the tree. These probabilities are placed at this level because they are conditioned on the first branches of the tree. We used the complement rule to compute the probabilities for the missing branches: P(PG^C|G)=0.2 and P(PG|G^C)=0.1. Finally, we use this information to compute all the joint probabilities at the right-hand side of the tree. For example, we can compute P(PG and G) by multiplying the branches leading up to this point: P(PG|G)*P(G)=0.8*0.6=0.48. Now we have all of the information needed to compute the new probability:

$$P(G|PG) = \frac{P(G \text{ and } PG)}{P(PG)} = \frac{P(G \text{ and } PG)}{P(G \text{ and } PG) + P(G^C \text{ and } PG)} =$$

$$= \frac{0.48}{0.48+0.04} = 0.923$$

Thus, using the report prepared by the specialist *and* knowledge about the specialist's past predictions, the concert planner can revise his estimate of weather conditions and then re-evaluate the selection of the best location for the concert.

Unit 3 Summary

- **Bayes' Theorem** is a mathematical formula that allows refining probabilities based on initial probability estimates and new evidence. It is computed as:

$$P(B|A) = \frac{P(B \text{ and } A)}{P(A)} = \frac{P(B \text{ and } A)}{P(B \text{ and } A) + P(B^C \text{ and } A)}$$

- A **probability tree** is a diagram consisting of nodes and branches. Nodes represent mutually exclusive events, and branches represent the probability associated with these events. The sum of probabilities on all branches coming out of the same node is one (1).

Unit 3 Exercises

1. Using the concert planning data described in the second example of Unit 3 (the concert location selection), find: $P(G^C|PG^C) = ?$

2. Assume that the *Kickstarter* project you are interested in is a publishing project. Based on past data, you know that 9,391 out of 107,610 successfully funded projects were publishing projects. You also know that 22,344 out of 191,853 unsuccessful projects were publishing projects. Earlier in the chapter, it was noted that the success probability of new *Kickstarter* projects was 36%. Using the same success probability (36%) and the information provided in this question, find the probability of a publishing project being successful.

3. The *yuppie flu* is now better known as Chronic Fatigue Syndrome (CFS), with symptoms including fatigue, loss of memory/concentration, unexplained muscle pain, and headache, among others.[29] After several weeks of feeling poorly, you are beginning to think that you might be suffering from CFS. An Internet search reveals that 3% of adults suffer from the disease at some point over their lifetime. Assume that a blood test exists that correctly identifies those with the disease 85% of the time and correctly diagnoses

---

[29] Source: http://www.mayoclinic.com/health/chronic-fatigue-syndrome/DS00395/DSECTION=symptoms

those without the disease 98% of the time.[30] You have just received a positive result, which indicates that you are suffering from the disease. What is the probability that you actually have the disease?

END-OF-CHAPTER PRACTICE

1. A study of accounting students asked whether the textbook they are using was bought new or used. The results are shown in Table 4-e5.

    Table 4-e5

	New	Used
ACTG100	0.28	0.07
ACTG200	0.15	0.12
ACTG300	0.10	0.13
ACTG400	0.06	0.09

    a. If a member of the study is selected, find the probability they bought a used book for ACTG100.

    b. If a member of the study is selected, find the probability they are enrolled in ACTG400.

    c. If a member of ACTG300 is selected, find the probability they bought a new book.

    d. If a student who bought a used book is selected, find the probability they are enrolled in ACTG200.

2. A game show records the cash winnings of each of their contestants, categorized by gender and amount won, shown in Table 4-e6. One contestant is randomly selected.

---

[30] The data for this question is fictional.

Table 4-e6

	Female	Male
$0 - $999	0.19	0.21
$1,000 - $9,999	0.26	0.23
$10,000 - $19,999	0.05	0.04
>$20,000	0.01	0.01

a. Find the probability that the contestant is a male.

b. Find the probability that the contestant won less than $20,000.

c. If the contestant is female, find the probability that she won between $1,000 and $9,999.

d. Find the probability that the contestant is male or won between $10,000 and $19,999.

3. The social media director for a company set up a *Facebook* page and *Twitter* profile. After a month, he finds that 65% of the Internet activity occurs on the company's *Facebook* page, 15% of the total activity happens before 6pm, and 80% of the company's *Twitter* activity occurs after 6pm. Find the percentage of activity that occurs on the company's *Facebook* page after 6pm.

4. A new drug to help prevent asthma is tested on 600 male and 400 female patients and recorded as effective or ineffective. The drug was effective 73% of the time overall and ineffective 15% of the time with females. Does gender play a role in the drug's effectiveness?

5. A parlay is a single bet that links together two or more individual wagers and is dependent on all of the linked-wagers winning together. There are much higher payoffs with a parlay than if the wagers placed each individual bet separately, since the difficulty of all the wagers winning is much higher. If any of the bets in the parlay lose, the entire parlay loses.[31] You engage in a parlay, betting that both *Manchester United* and *Chelsea* will win their matches (there are no ties). The published odds of *Chelsea* winning are 60% and the published odds of *Manchester* winning are 71%. Assuming independence between the matches, what is the probability of you winning the parley?

---

[31] Source: http://en.wikipedia.org/wiki/Parlay_%28gambling%29

6. Two equal-size groups of mutual fund managers (all the members of one group having graduated from *Harvard* and all the members of the second group having graduated from *Yale*) engage in a competition to try to outperform the market. Overall, 65% of the managers outperformed the market. If we select a manager at random, the probability of this manager being from *Harvard* and not outperforming the market is 19%. What proportion of the successes (i.e., outperforming the market) came from *Yale*?

7. Of those who have read e-books, 88% have also read printed books.[32] If 1 out of 6 people read e-books and 28% do not read printed books, what proportion reads either printed or e-books?

8. If Jim fails his calculus exam, he predicts there is a 70% chance he will fail linear programming, as well. There is a 30% chance he will fail calculus and a 60% chance he will fail linear programming. What is the probability that he will pass both exams?

9. Captain Anderson claims he could land a plane under a simulated difficulty level of 9, which is considered to be very high. He runs the simulation and lands successfully five out of seven times. Six other pilots each attempt this same landing seven times. If the plane was crashed 80% of the time for all the simulated landings, what proportion of the crashes are attributed to the other pilots?

10. Fred tells the truth 80% of the time. He rolls two dice and claims they summed to 6. What is the probability that his roll of the two dice did not actually produce a sum of 6?

11. There are two cookie jars. Jar #1 has ten peanut butter cookies, five chocolate chip cookies, and three oatmeal raisin cookies. Jar #2 has five peanut butter cookies, ten chocolate chip cookies, seven oatmeal raisin cookies, and one sugar cookie. If Sarah randomly reaches into one of the jars and takes out a cookie, what is the probability that she reached into Jar #1, if the type of cookie that she got was:

    a. A sugar cookie?

    b. A peanut butter cookie?

    c. A chocolate chip cookie?

    d. An oatmeal raisin cookie?

---

[32] Source: http://dailyinfographic.com/libraries-are-forever-e-books-print-books-can-coexist-infographic

12. A shady (yet completely honest) street performer offers you a chance to play his game for the low price of $10. His game involves you pushing two different buttons. One of the buttons, when pushed, has a 10% chance of winning you $40; and the other button, when pushed, has a 20% chance of winning you $25. You are allowed two button presses (either pushing the same button twice or pushing each button once) in a single game. Is it worth playing?

13. As a contestant on a game show, you are presented with the following game: you will spin a wheel that will indicate which of three boxes you will open. You will then receive the cash amount that is in the opened box. The host lets you choose between three alternatives, shown in Tables 4-e7, 4-e8 and 4-e9. Each table shows the probability of a box coming up on the wheel and the cash reward in that box. Given this information, which of the three alternatives should you choose to play?

Table 4-e7

Box	Prize	Probability
1	$100	0.33
2	$100	0.33
3	$100	0.34

Table 4-e8

Box	Prize	Probability
1	$10	0.80
2	$500	0.15
3	$1000	0.05

Table 4-e9

Box	Prize	Probability
1	$10	0.70
2	$200	0.20
3	$300	0.10

14. You have two unlabeled and sealed crates filled with red and yellow balls. You can unseal each crate and randomly pull out one ball from each crate. One box has 20% yellow balls and 80% red balls, and the other has 80% yellow balls and 20% red balls.

   a. If you pull out one red ball and one yellow ball, what is the probability that the red ball came from the crate with 20% yellow balls and 80% red balls?

b. What is the probability of pulling out two red balls?

c. What is the probability that you pull out a red ball from the predominantly yellow crate and then pull out a yellow ball from the predominantly red crate?

d. If you pull out a yellow ball, what is the probability that it is from the predominantly yellow crate?

15. The game *Plinko,* on *The Price is Right,* can be simplified and modeled by a triangle of pegs shown in Figure 4-e10.

Figure 4-e10

				X				
			X		X			
		X		X		X		
	X		X		X		X	
A		B		C		D		E

In this game, a disk is dropped from the top and has an equal probability of bouncing either left or right. In this scenario, the disk bounces four times and then ends up at position A, B, C, D or E.

a. What is the probability that the disk ends up at position C?

b. What is the probability that the disk: ends up at position B in a first game, and then at either position A or E in an immediately following second game?

16. Reconsider the game of *Plinko* as having a 0.6 probability of the disk bouncing right, and a 0.4 probability of bouncing left.

a. What is the probability that the disk ends up at position C?

b. What is the probability that the disk: ends up at position B in a first game, and then at either position A or E in an immediately following second game?

17. Finally, again considering the game of *Plinko,* what range of probabilities for the disk bouncing right must be used for position D to be the most common ending position?

CHAPTER 5: Discrete Probability Distributions

In Chapter 2 we discussed *frequency distributions,* which outline how data are spread (or distributed*)* across a range of data values. In Chapter 3, we used these distributions to provide context to our data by examining values such as the center of the distribution and the dispersion of values around this center. For example, we explained that a new graduate's consideration of an offered salary might involve learning how this salary compared to those of others with similar qualifications. In this chapter, we will build on the same notion of *distribution*. However, rather than looking at existing values of some variable of interest (that is, looking at the frequency distribution of the values), we will look at *expected* values of this variable and the probability of occurrence of these expected values. In doing this, we create a *probability distribution*.

Suppose, for example, that you have recently started a *Twitter* account and are interested in the number of followers you can expect. To determine the expected number of followers, we define a *random variable*[33], X, as the number of followers you may have on *Twitter*, based on your observations of the number of followers of your friends and colleagues. Table 5-1 displays possible values of X (we have grouped them into categories) alongside the probabilities of occurrence we expect for each of the values. Interpreting Table 5-1, there is a 29% chance that you will have no followers, a 52% chance that you will have between one and nine followers, and a 19% chance that you will have ten or more followers.

---

[33] Random variables can obtain different values at different probabilities. We will explain more about these variables in Unit 1 of this chapter.

Table 5-1
Probability Distribution Example

X	P(X)
0	0.29
1 to 9	0.52
10 or more	0.19

Using the information in Table 5.1, we can formulate our *expectation* about the number of *Twitter* followers you are likely to observe. We explain how this is done in this chapter (for discrete variables) and in Chapter 6 (for continuous variables).

Probability distributions, such as the one in Table 5-1, are important statistical tools that help us make *inferences* about a population of interest. For another example, consider the situation where a manufacturing manager believes that the daily percent of defects in the manufacture of a product follows the distribution shown in Table 5-2, where X is a random variable describing the percent of defects and P(X) is the probability attached to each value of X. Given such a defect rate distribution, the manager would likely be very concerned to find a defect rate of 20% for a sample of just-manufactured products. As a result, the manager is likely to bring in some engineers and technicians to examine the manufacturing process and machinery for problems.

Table 5-2
Probability Distribution for a Product's Defect Rate

X	P(X)
0%	0.02
1% to 5%	0.75
6% to 10%	0.229
More than 10%	0.001

Essentially, all of the data associated with a variable of interest follow some distribution. The distribution may be *discrete*, that is, describing the probabilities of values of a discrete (countable) random variable, such as the number of customers walking into a store during a specified period of time. Or, the distribution may be *continuous,* describing the probabilities of intervals of a continuous (measurable) random variable, such as income, time or distance. Further, many variables seem to follow distributions having similar shapes and properties. In this chapter and the one that follows, we will cover commonly-occurring distributions (including the binomial, the normal and the t distributions, each of which is used in later chapters to support inference testing). As previously mentioned, this chapter focuses on *discrete probability distributions*, which are used to understand the behavior of discrete variables. This chapter covers the following topics:

- Random Variables
- Probability Distribution
- Bivariate Probability Distributions
- The Binomial Distribution
- Binomial Probability Distribution in Excel
- Mean and Variance of the Binomial Distribution
- The Poisson Distribution

UNIT 1

Random Variables

A **random variable** is a variable that may obtain any value within our sample space, given the variable's probability distribution. For example, rolling a die is an experiment with six possible outcomes (the sample space). We can thus define a random variable, X, as describing the possible outcomes of rolling a die. Hence, X can obtain any number between 1 and 6, as shown on the first column of Table 5-3.

Table 5-3
Dice Example

X	P(X)
1	0.167
2	0.167
3	0.167
4	0.167
5	0.167
6	0.167

Of course, rolling a die is a probabilistic experiment, meaning that there are probabilities attached to obtaining each of the values of X. The numbers in the second column of Table 5-3 describe these probabilities (and, hence, represent the variable's probability distribution). In this dice example, 1/6 (or 0.167) is the probability of obtaining any of the values of X.

In the example above, X is a *discrete random variable*. Recall from Chapter 1 that *discrete* variables are the result of a counting process. Other examples of discrete random variables include a count of the number of customers that walk into a store in a given day, the number of errors in a computer program's code, or the number of product defects (per thousand products produced) associated with a product manufacturing process. In each of these examples, the variables' values are outcomes of counting processes.

Random variables can also be *continuous*, that is, they can describe the results of a measurement process. For example, we can define X as the time it takes to serve a bank customer, as the gas mileage of cars, as the height of children, as the income of residents of a city, or even as countable variables that have very large ranges, such as the number of Twitter followers or the number of monthly downloads for a specific mobile app. Again, we will focus on continuous random variables in Chapter 6.

Probability Distribution

A ***probability distribution*** describes the probabilities attached to all values of the random variable. They can be discrete or continuous, based on the random variable they describe. Probability distributions can be displayed as a table or a graph. In this chapter, we will be working mostly with tables, given the focus on discrete random variables. However, as Chapter 6 focuses on continuous variables and it is impractical to cover *all* values of a continuous variable in a table, Chapter 6 will be mostly using graphs to represent probability distributions.

Table 5-4 presents the probability distribution of the discrete random variable Vacation Days, based on a survey of 500 Americans conducted by the travel website Expedia in 2012.[34] In Table 5-4, we define X to be a random variable that indicates the number of yearly vacation days that survey respondents receive from their places of employment. While X might conceivably take on any value from zero to 365, X is grouped into six categories in Table 5-4. The second column in the table, (f_x), describes the frequency of responses for each category (for example, 120 of the survey respondents indicated that they receive between one and ten vacation days). You may recognize that the third column is, in fact, the *relative frequency* for each category, computed by dividing each cell's frequency by the

---

[34] Source: http://media.expedia.com/media/content/expus/graphics/other/pdf/Expedia-VacationDeprivation2012.pdf. More recent versions of the survey exist that compare vacation days in the US to other countries, if you'd like to learn more about this topic.

total number of responses. You should recall from Chapter 4 that relative frequency is one approach for obtaining probability values. Thus, we label this third column as P(X=x). Note that we use the uppercase X to represent the *name* of the variable and the lowercase x to represent a specific *value* of X (or, as is the case here, a range of values). Hence, P(X=0) is the probability that a person receives no vacation days, and P(1≤X≤10) is the probability that a person receives between one to ten vacation days.

Table 5-4
Vacation Days Probability Distribution

X	f_x	P(X=x)
0	91	18.2%
1 to 10	120	24%
11 to 20	145	29%
21 to 30	66	13.2%
Over 30	13	2.6%
Not sure	65	13%
Total	500	100%

You will further see, looking at the third column in Table 5-4, that the same requirements that were defined in Chapter 4 hold for our probability distribution. Specifically, that all probabilities are between zero and one, and that the sum of all probabilities in the table is one:

$$0 \leq P(X=x) \leq 1$$

$$\Sigma P(X=x) = 1$$

In a similar manner, other rules that we covered in Chapter 4 also hold for probability distributions. For example, the probabilities in Table 5-4 are mutually exclusive (recall: when constructing frequency tables, we ensure that our categories are mutually exclusive), and we can apply the addition rule to find the probability of more than one category

$$P(X \leq 30) = P(X=0) + P(1 \leq X \leq 10) + P(11 \leq X \leq 20) + P(21 \leq X \leq 30) =$$

$$= 18.2\% + 24\% + 29\% + 13.2\% = 84.4\%$$

or

$$P(1 \leq X \leq 20) = P(1 \leq X \leq 10) + P(11 \leq X \leq 20) = 24\% + 29\% = 53\%$$

and so on.

Let us now consider a different example to further demonstrate this unit's concepts. Assume that you have a great idea for a new financial app and decide to go ahead and develop it. You would like to get some idea of what you might expect in terms of downloads. Assume that the information in Table 5-5 is data you have obtained about the probability distribution of daily app downloads, where X represents the number of downloads and P(x) is the probability attached to each value of X.

Table 5-5
App Downloads Probability Distribution

X	P(x)
10	5%
50	35%
100	30%
500	25%
1500	5%

You can get a pretty good idea of what to expect just by looking at Table 5-5. However, for a more accurate estimate, you can compute the mean and variance of the probability distribution. Let us first recall two formulas for computing the mean and the variance from Chapter 3:

$$\text{Weighted mean} = \frac{\sum_{i=1}^{n} w_i x_i}{\sum_{i=1}^{n} w_i}$$

$$\text{Variance} = \sigma^2 = \frac{\sum_{i=1}^{N}(x_i - \mu)^2}{N}$$

The first equation above describes the computation of a weighted mean. We are using the weighted mean in order to account for what we already know about the probability of each value of the random variable X: values with higher probabilities of occurrence should *weigh in* more strongly in the computation of our expected value. Thus, to compute the mean of the probability distribution, we simply compute the sum of each value of X weighted by its corresponding probability:

$$E(X) = \mu = \sum_x xP(x)$$

In the above equation, E(X) represents the **expected value** of X, which is how the average is termed in a probability distribution. µ, as you may recall from Chapter 3, is the Greek notation used to represent the population mean. Finally, the formula itself is simply the average of each value, x, multiplied by its corresponding probability, P(x), accounting for all possible values of X. Therefore, using the numbers in Table 5-5, we compute

$$E(X) = \mu = \sum_x xP(x) = 10 * 0.05 + 50 * 0.35 + 100 * 0.3 + 500 * 0.25 + 1500 * 0.05$$

$$= 248$$

where 248 is the *expected* number of downloads based on Table 5-5.

The computation of the variance of the probability distribution is also similar to how we previously computed the variance. Specifically, recall that the variance is the average squared deviation from the mean. Hence

$$V(X) = \sigma^2 = \sum_x (x - \mu)^2 P(x)$$

where V(X) represents the variance of X, σ^2 is the corresponding Greek notation for the variance, and the expression inside of the summation is the weighted average of the squared deviation from the mean (μ). The standard deviation will be computed by taking the square root of the variance:

$$\sigma = \sqrt{\sigma^2}$$

Using the numbers in Table 5-5, we can therefore compute

$$V(X) = \sigma^2 = \sum_x (x - \mu)^2 P(x)$$

$$= (10 - 248)^2 * 0.05 + (50 - 248)^2 * 0.35 + (100 - 248)^2 * 0.3 + (500 - 248) * 0.25 + (1500 - 248)^2 * 0.05 = 117{,}376$$

and

$$\sigma = \sqrt{\sigma^2} = \sqrt{117376} = 342.6$$

Bivariate Probability Distributions

The distribution in Tables 5-4 and 5-5 are examples of a ***univariate distribution***, meaning the distribution of a single variable. In Chapter 4, we used contingency tables to describe the joint probabilities of two variables of interest. We apply this same approach to describe a ***bivariate distribution***, or the probability distribution of two variables.

We will use a new example to demonstrate this concept of a bivariate distribution. Conference calls are common in today's organizations. Conference calls are an effective means to involve many participants in a discussion but, unfortunately, can sometimes drag beyond their scheduled completion times. You have decided to track data on conference call size (number of participants) and length within your organization.

Table 5-6 displays the joint distribution of two variables: Number of Participants (X) and Call Length (Y). P(x,y) is the *joint probability* of specific values of X and Y; for example, P(4,60) is the joint probability of a call with four participants and lasting 60 minutes. From Table 5-6, we know that P(4,60)=0.28.

Table 5-6
Bivariate Probability Distribution

		Call Length			
		30 minutes	60 minutes	90 minutes	Total
Number of Participants	3	0.05	0.12	0.03	**0.2**
	4	0.03	0.28	0.09	**0.4**
	5	0.02	0.15	0.03	**0.2**
	6	0	0.08	0.02	**0.1**
	7	0	0.07	0.03	**0.1**
	Total	**0.1**	**0.7**	**0.2**	**1**

Similar to the univariate case, we can compute the average and standard deviation for each of these variables, using their *marginal probabilities*. Specifically, the average and variance of X are:

$$E(X) = 3*0.2 + 4*0.4 + 5*0.2 + 6*0.1 + 7*0.1 = 4.5$$

$$V(X) = (3-4.5)^2 * 0.2 + (4-4.5)^2 * 0.4 + (5-4.5)^2 * 0.2 + (6-4.5)^2 * 0.1$$
$$+ (7-4.5)^2 * 0.1 = 1.45$$

And, the average and variance of Y are:

$$E(Y) = 30*0.1 + 60*0.7 + 90*0.2 = 63$$

$$V(X) = (30-63)^2 * 0.1 + (60-63)^2 * 0.7 + (90-63)^2 * 0.2 = 261$$

We can also compute a very important and interesting measure called the **covariance** of X and Y, which is a measure of the *shared variation* in X and Y. Specifically, the covariance is computed as:

$$COV(X,Y) = \sigma_{xy} = \sum_x \sum_y (x - \mu_X)(y - \mu_Y) P(x,y)$$

Or, in our example:

$$COV(X,Y) = (3 - 4.5)(30 - 63)(0.05) + (4 - 4.5)(30 - 63)(0.03) + \cdots$$
$$+ (6 - 4.5)(90 - 63)(0.02) + (7 - 4.5)(90 - 63)(0.03) = 3.3$$

The covariance measures the extent to which values of X and Y move in the same direction (or move in opposite directions) as (x,y) data points are examined (that is, what happens to Y when X increases or decreases and vice versa). Consider, again, the formula given above for the covariance. If, for a specific data point (x,y), the value of X is above its mean *and* the value of Y is below its mean, then their multiplication will result in a negative number (which is then multiplied by the probability of data point (x,y) in creating a weighted average). On the other hand, if the values of X and Y are both above their respective means, or if both values are below their respective means, then their multiplication will result in a positive number. The weighted average thus shows whether or not, on average, X and Y tend to move in the same directions vis-à-vis their respective means. A positive covariance indicates that *either* higher than average values of one variable tend to be associated with higher than average values of the other variable *or* that lower than average values of one variable tend to be associated with lower than average values of the other variable. A negative covariance indicates that higher than average values of one variable tend to be associated with lower than average values of the other variable. Finally, if X and Y were *independent* of each other, their covariance would be zero.

The limitation of the covariance measure is that its magnitude (i.e., the strength of the relationship between X and Y) is not directly evident. For example, while a covariance of 3.3 indicates that the relationship between X and Y is positive, we still do not know the strength of this positive relationship. In order to determine the magnitude of the relationship between X and Y, we can divide the covariance by the two standard deviations, as shown below:

$$COR(X,Y) = \rho = \frac{COV(X,Y)}{\sqrt{V(X)}\sqrt{V(Y)}} = \frac{\sigma_{XY}}{\sigma_X \sigma_Y}$$

The measure we have just obtained is one you may have heard about and is termed the **coefficient of correlation**, which is a standardized measure of the strength of the relationship between X and Y. Values for the coefficient of correlation closer to +1 indicate a strong positive relationship, whereas values closer to -1 indicate a strong negative relationship. Values close to zero simply indicate a weak relationship between the two variables. Computing the coefficient of correlation for our example is shown below:

$$\rho = \frac{3.3}{\sqrt{1.45}\sqrt{261}} = 0.17$$

In other words, the value of 0.17 indicates that there is a weak but positive relationship between the number of participants at a conference call and the length of the call (or, the greater the number of participants, the longer the call).

These concepts of covariance and the coefficient of correlation are very important, and we will return to them in future chapters as we delve deeper into exploring relationships between variables. For now, let us summarize what we have introduced so far, practice these new concepts to make sure they are clear, and move on to discuss two commonly-used discrete probability distributions: the *binomial distribution* and the *Poisson distribution*.

Unit 1 Summary

- A **random variable** is a variable that may obtain any value within our sample space at a given probability. The variable is denoted using a capital 'X', with specific values represented using a lower case 'x'.

- A **probability distribution** describes the probabilities attached to all values of the random variable. Probability distributions can be discrete or continuous, based on the random variable they describe. Probability distributions can be displayed as a table or a graph.

- The following must hold for all probability distributions:
$$0 \leq P(X=x) \leq 1$$
$$\Sigma P(X=x) = 1$$

- The mean of X is called the **expected value** and it is computed as:
$$E(X) = \mu = \sum_x xP(x)$$

- The variance of X is computed as:
$$V(X) = \sigma^2 = \sum_x (x-\mu)^2 P(x)$$

- A **univariate distribution** describes the distribution of a single variable. A **bivariate distribution** describes the joint distribution of two variables.

- The **covariance** of X and Y is a measure of the shared variation in X and Y. Specifically, the covariance is computed as:
$$COV(X,Y) = \sigma_{xy} = \sum_x \sum_y (x-\mu_X)(y-\mu_Y)P(x,y)$$

- The **coefficient of correlation** is a standardized measure of the strength of the relationship between X and Y. It ranges between ±1, with values closer to 1 (or -1) representing a strong positive (or negative) relationship and values closer to 0 representing a weak relationship. It is computed as:
$$COR(X,Y) = \rho = \frac{COV(X,Y)}{\sqrt{V(X)}\sqrt{V(Y)}} = \frac{\sigma_{XY}}{\sigma_X \sigma_Y}$$

Unit 1 Exercises

1. Visit the *www.data.gov* website[35], select one or more of the provided data sets, and identify five variables for which data are available. For each of these variables, identify whether the variable is discrete or continuous.

2. Table 5-e1 presents data of the number of marriages in Sweden in 2009 by age of the bride.[36] Develop the probability distribution of the random variable defined as the number of marriages within the ranges and answer the following:

    a. What proportion of brides are 40 years of age or older?

    b. What is the probability of a bride being between the ages of 25 and 39?

    c. What proportion of brides are under twenty years of age?

Table 5-e1
Bride's Age – Sweden

Age Range	Number of Marriages
0 - 14	0
15 - 19	947
20 - 24	4,620
25 - 29	11,029
30 - 34	10,809
35 - 39	7,375
40 - 44	4,698
45 - 49	3,159
50 - 54	2,128
55 - 59	1,255
60 - 64	735
65 - 69	504

3. Repeat parts a, b, and c using the 2009 census data from Mexico (taken from the same source as the Swedish data), shown in Table 5-e2.

---

[35] Source: http://www.data.gov/
[36] Source: http://unstats.un.org/unsd/demographic/products/dyb/dybsets/2009-2010.pdf

Table 5-e2
Bride's Age – Mexico

Age Range	Number of Marriages
0 - 14	474
15 - 19	125,629
20 - 24	255,148
25 - 29	241,933
30 - 34	133,977
35 - 39	70,310
40 - 44	42,469
45 - 49	27,942
50 - 54	16,802
55 - 59	9,462
60 - 64	5,022
65 - 69	5,240

4. International recording artist Adele is nominated for six *Brit Awards*, the British equivalent of the *Grammy Awards*. Suppose there are four other nominees in each category and that the odds of a given nominee winning a category (for all categories) are equivalent. Prepare a chart to show the probabilities of Adele winning one *Brit*, two *Brits* and so on – all the way to sweeping all six categories.

5. The probability distribution of the number of students simultaneously using the ten computers in a computer lab is shown in Table 5-e3.

Table 5-e3
Number of Computer Users

X	P(X)
0	.00
1	.05
2	.10
3	.15
4	.20
5	.15
6	.15
7	.10
8	.05
9	.05
10	.00

a. What is the expected number of students in the lab at any given point-in-time?

b. What is the probability that exactly three students are using computers?

c. What is the probability that more than five computers are being used?

d. What is the probability that less than five computers are being used?

e. What is the probability that a group of three students would be able to work together in the computer lab, assuming that they each need a computer?

6. Table 5-e4 describes the joint probability distribution of two random variables: Party Size (number of customers per table) and Meal Value (expenditures in dollar per table).

Table 5-e4

		Meal Value			Total
		$50 or Less	$50 - $100	$100+	
Party Size	1	0.05	0.02	0	0.07
	2	0.2	0.15	0.05	0.4
	3	0	0.1	0.13	0.23
	4	0	0.1	0.02	0.12
	5	0	0.05	0.03	0.08
	6+	0	0	0.1	0.1
	Total	0.25	0.42	0.33	1

a. Are you able to compute the expected meal value in this restaurant?

b. Are you able to compute the expected party size?

c. What is the probability that a couple would spend more than $100 on a meal?

d. What is the probability that a single diner would spend less than $100 on a meal?

e. What is the probability that a table of eight diners would spend more than $100?

f. Given that the check was for $100, what is the probability that there were four diners in the party?

g. Is there a relationship between party size and meal value?

7. A guitar enthusiast counts how many chords are in each of the songs he knows how to play. The counts of chords are shown in Table 5-e5. Find:

   a. P(X=4)

   b. P(X<5)

   c. P(X≤3 or X≥5)

Table 5-e5

Chords per Song (X)	3	4	5	6	7
Count (f_x)	4	9	6	3	1

8. The Humanities courses at State University can be 3.00, 6.00 or 9.00 credits. The number of courses for each credit count is shown in Table 5-e6. Find the expected credit weight, the variance, and the probability of selecting a course that is less than 9.00 credits.

Table 5-e6

Credits	3.00	6.00	9.00
Number of Courses	80	101	41

9. The distribution of the number of computers in U.S. households (in millions) is as follows:[37]

   - Zero computers: 27.4
   - One computer: 46.9
   - Two computers: 24.3
   - Three computers: 9.5
   - Four computers: 3.6
   - Five computers: 2.0

   Create a probability distribution for the number of computers in U.S. households and find the expected value and the variance of this random variable.

10. The fuel consumption (gallons per 100 miles) of one company's car fleet is distributed as shown in Table 5-e7.

---

[37] Source: http://www.statista.com/statistics/187606/average-number-of-computers-per-household-in-the-united-states/

Table 5-e7

X	4	5	6	7
P(x)	.08	.35	.52	.05

a. Find P(X<6)

b. Find P(X≥7)

c. Find P(4≤X≤7)

11. The number of cylinders for each model of car in the company's car fleet was also recorded. A bivariate distribution for number of cylinders and fuel consumption is shown in Table 5-e8. Find the covariance and coefficient of correlation.

Table 5-e8

		Number of Cylinders				Total
		3	4	6	8	
Fuel Consumption	4	.06	.02	0	0	.08
	5	.05	.20	.10	0	.35
	6	.05	.15	.27	.05	.52
	7	0	0	0	.05	.05
	Total	.16	.37	.37	.10	1

12. A *Twitter* user creates a distribution of the number of hashtags she uses per tweet (Table 5-e9).

Table 5-e9

X	0	1	2	3	4
P(x)	.35	.30	.16	.10	.04

a. What is the expected number of hashtags in a given tweet?

b. This *Twitter* user also records the number of *favorites* each tweet receives (Table 5-e10 shows the bivariate distribution). Find the covariance and coefficient of correlation for these two variables.

Table 5-e10

		Hashtags					
		0	1	2	3	4	Total
Favorites	0	.25	.05	.06	.01	0	**.37**
	1	.05	.10	.08	.02	0	**.25**
	2	.04	.09	.02	.01	.01	**.17**
	3	.01	.06	.01	.04	.02	**.14**
	4	0	.02	0	.02	.03	**.07**
	Total	**.35**	**.32**	**.17**	**.10**	**.06**	**1**

UNIT 2

The Binomial Distribution

Imagine that you are about to take a quiz consisting of ten *True* or *False* questions. Assuming that you did not study for the quiz, your plan is simply to guess the answers (and, you have a 50% chance of being correct on any given question). The above scenario describes a basic **binomial experiment** (also known as a Bernoulli trial). A binomial experiment is characterized by a fixed number of trials n (the ten questions in our example) whose outcome can be either *success* or *failure*, with the probability of a *success* outcome given as p (50% in our example), where the outcome of individual trials are independent of each other. These characteristics are summarized in Figure 5-1.

Figure 5-1
Properties of Binomial Experiments

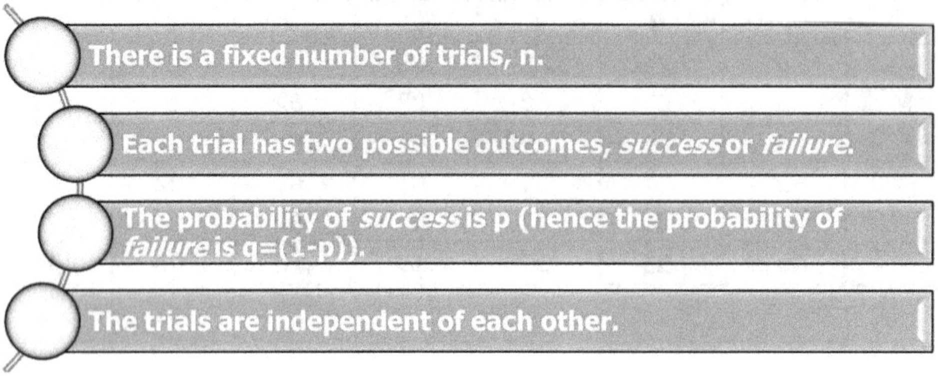

- There is a fixed number of trials, n.
- Each trial has two possible outcomes, *success* or *failure*.
- The probability of *success* is p (hence the probability of *failure* is q=(1-p)).
- The trials are independent of each other.

We can define a *random variable* X that counts the number of successes in a binomial experiment (for example, the number of correct questions after taking the quiz). Here, X is defined as a **binomial random variable**. Any experiment with the characteristics described in Figure 5-1 generates a binomial random variable, which is the count of successes in the n trials. There are many examples of such experiments in real life, for example: the number of satisfied customers of a company, the number of products that are defective in a manufacturing process (here, the *success* is in reality a *failure*), and the number of people who vote for a specific candidate.

The benefit of being able to work with binomial random variables is that we can use the parameters n and p to easily compute a range of probabilities for possible outcomes. Let us go through an in-depth example to illustrate this concept.

According to a report by *comScore*, a company focused on measuring the digital world, as of December 2015 nearly 43% of smartphone subscribers in the

United States have an *Apple* device.[38] Let us define a random variable X as the count of people with an *Apple* smartphone. Supposing that we take a random sample of twenty smartphone users, what is the probability that ten of them will have an *Apple* smartphone? In other words, what is the probability of P(X=10)?

The parameters of our binomial experiment are summarized in Figure 5-2:

Figure 5-2
The Smartphones Binomial Experiments

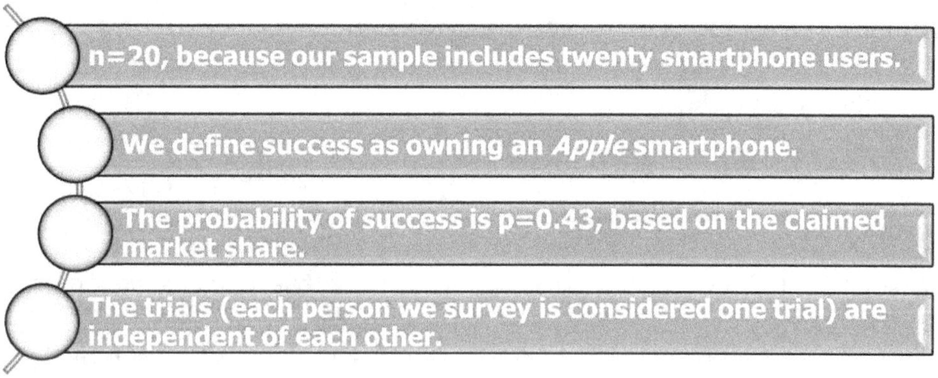

- n=20, because our sample includes twenty smartphone users.
- We define success as owning an *Apple* smartphone.
- The probability of success is p=0.43, based on the claimed market share.
- The trials (each person we survey is considered one trial) are independent of each other.

We can compute the probability of finding ten successes, P(X=10), in the twenty trials (n=20) using the binomial formula shown below:

$$P(x) = \frac{n!}{x!\,(n-x)!} p^x (1-p)^{n-x}$$

The binomial formula consists of two parts. The first part

$$\frac{n!}{x!\,(n-x)!}$$

is called the *counting rule for combinations* and it simply counts how many possible ways there are to obtain ten (x) successes in twenty (n) trials. The expression n! (n factorial) is computed as $n \times (n-1) \times (n-2) \times ... \times 1$. The second part

$$p^x (1-p)^{n-x}$$

---

[38] Source: https://www.comscore.com/Insights/Rankings/comScore-Reports-December-2015-US-Smartphone-Subscriber-Market-Share

refers to the probability of obtaining x successes (calculated as: p^x) *and* n-x failures, (calculated as: (1-p)^{n-x}). Together, these two parts allow us to compute the probability of obtaining exactly ten successes in twenty trials:

$$P(10) = \frac{20!}{10!\,(20-10)!} 0.43^{10}(1-0.43)^{20-10} = 0.1446$$

In other words, there is a nearly 14.5% chance that we will find ten *Apple* smartphones in our sample of twenty smartphone users, given the market share of 43%. Similarly, we can compute the probability of obtaining exactly fifteen successes in twenty trials:

$$P(15) = \frac{20!}{15!\,(20-15)!} 0.43^{15}(1-0.43)^{20-15} = 0.00296$$

That is, there is nearly no chance that we will find fifteen *Apple* smartphone owners in our sample of twenty smartphone users, given the market share of 43%.

Note that while we defined success above as *having* an *Apple* smartphone, we could have, just the same, defined success as *not* having an *Apple* smartphone. In such a case, we would have to remember that p is now 0.57 (the market share of all other devices) and continue with the same procedure as above. For example, we can reverse the above example of finding fifteen *Apple* smartphones (our previous definition of a success) and compute the probability of finding five people *without* an *Apple* smartphone (our new definition of a success) as:

$$P(5) = \frac{20!}{5!\,(20-5)!} 0.57^{5}(1-0.57)^{20-5} = 0.00296$$

While this problem restatement redefines the nature of a success, the probability being asked is exactly the same – resulting in an outcome that is the same as before. In other words, the definition of what comprises a success and what comprises a failure is in our hands. We just need to remember to properly assign the probability of success as per our definition of a success.

To ensure that you understand how to apply the binomial probability rules, we will now go over another example. Suppose that you run a specialty Web store that sells pieces of art. An important measure in online retailing is the cart conversion rate, which is the percentage of customers that complete an order after setting the order up in a shopping cart. Based on your research, you believe that a 5% conversion rate is reasonable. Framing this as a binomial experiment, we define X to be a random variable counting the number of website visitors that have converted their shopping cart into a purchase. What is the probability that, out of 100 site visitors who have placed items in their shopping carts, six of these visitors will complete the purchase? We can, again, use the binomial formula to compute this probability:

$$P(6) = \frac{100!}{6!\,(100-6)!} 0.05^6 (1-0.05)^{100-6} = 0.15$$

Thus, there is about a 15% chance that out of 100 site visitors who have placed items in their shopping carts, six of these visitors will complete the purchase.

Now consider a different question: what is the probability that, out of 100 site visitors who have placed items in their shopping cart, *between six to ten customers (inclusive)* will complete the purchase? Here, we are aiming to compute P(6≤X≤10), which can be computed as:

$$P(6 \leq X \leq 10) = P(6) + P(7) + P(8) + P(9) + P(10)$$

This can be computed by applying the binomial formula to compute each probability and then summing them all up. Clearly, while straightforward, this exercise would be quite time consuming. Consider, further, a question such as: what is the probability that, out of 100 site visitors who have placed items in their shopping cart, *at least* six will complete the purchase? Now we are aiming to compute P(X≥6), which would require adding up all probabilities from 6 to 100 successes!

Many statistics books include the *Binomial Probability Table*, which computes, for different values of n and p, the corresponding binomial probabilities so that we do not have to go through the lengthy computations. Because these tables are limited to specific n and p values, it is more convenient to use Excel - as virtually any values of n and p can be used. The next section explains how this is done.

Binomial Probability Distribution in Excel

=BINOM.DIST(number_s,trials,probability_s,cumulative) is an Excel function that takes as its arguments the number of successes (x), the number of trials (n) and the probability of success (p), and returns the corresponding binomial probability. The fourth argument, *cumulative*, takes a value of either 0 or 1. A value of 0 will compute the probability P(X=x), whereas a value of 1 will return the *cumulative* probability: P(X≤x).

Assume that n=10, p=0.4, and that we are looking for P(X=2). Using the binomial formula, we compute:

$$P(2) = \frac{10!}{2!\,8!} 0.4^2 0.6^8 = 0.1209$$

Using Excel we type: =BINOM.DIST(2,10,0.4,0) to obtain this same value of 0.1209.

Now assume that we are looking for P(X≤2). Using the binomial formula, we compute:

$$P(X \leq 2) = P(0) + P(1) + P(2) = \frac{10!}{0!\,10!} 0.4^0 0.6^{10} + \frac{10!}{1!\,9!} 0.4^1 0.6^9 + \frac{10!}{2!\,8!} 0.4^2 0.6^8$$

$$= 0.006 + 0.04 + 0.121 = 0.167$$

(Note that the value of 0! is simply 1). Alternatively, we can use Excel: typing =BINOM.DIST(2,10,0.4,1) will return a value of 0.167, the same as just computed. Note that the fourth argument of the function was 1 instead of 0. A value of 1 tells Excel to compute the cumulative probability.

We can also find $P(2 \leq X \leq 6)$. To do so, we add together all probabilities between 2 and 6: P(2)+P(3)+P(4)+P(5)+P(6). Alternatively, we can compute $P(X \leq 6)$ and then subtract from it $P(X \leq 1)$. Note that we are subtracting $P(X \leq 1)$ because we want the number 2 to still be included in the result. Hence:

$$P(2 \leq X \leq 6) = P(X \leq 6) - P(X \leq 1)$$

Let us return to our cart conversion example where we were looking to find the probability of *at least six* visitors to the site converting their shopping cart into a purchase: P(X≥6). We cannot compute this directly using Excel. The cumulative binomial function in Excel computes the probabilities *up to* a certain number of successes. Hence, when we type =BINOM.DIST(6,100,0.05,1), we obtain the sum of the probabilities from 0 to 6, or P(0)+P(1)+P(2)+P(3)+P(4)+P(5)+P(6), while, in fact, we are looking for the sum of the probabilities from 6 to 100 successes. To overcome this problem, we use the complement rule introduced in Chapter 4. Specifically, we know that P(0≤X≤100)=1. We also know that we can split P(0≤X≤100) into two parts: P(0≤X≤5)+P(6≤X≤100). Therefore, combining these two pieces of knowledge, we compute:

$$P(6 \leq X \leq 100) = 1 - P(0 \leq X \leq 5)$$

Using Excel, we find P(0≤X≤5): =BINOM.DIST(5,100,0.05,1) to obtain the probability of 0.616. Using this probability, we then compute:

$$P(6 \leq X \leq 100) = 1 - 0.616 = 0.384$$

This result means that there is about a 38% chance that at least six of 100 site visitors who place items in their shopping carts will complete the purchase.

Mean and Variance of the Binomial Distribution

We conclude this unit with the formulae for computing the mean and variance of the binomial distributions. As with all other distributions we have introduced thus far in this book, it is useful to be able to compute the mean and the variance of the binomial distribution. To do so, we use the following formulae:

$$\mu = np$$

$$\sigma^2 = np(1-p)$$

$$\sigma = \sqrt{np(1-p)}$$

In our cart conversion example:

$$\mu = 100 * 0.05 = 5$$

$$\sigma^2 = 100 * 0.05 * 0.95 = 4.75$$

$$\sigma = \sqrt{4.75} = 2.18$$

In other words, in a sample of 100 customers the expected number of cart conversions is five, with a standard deviation of 2.18.

Unit 2 Summary

- A **binomial experiment** is characterized by a fixed number of trials n whose outcome can be either *success* or *failure*, each with a given success probability p, independent of the outcomes of other trials.

- We can define a **binomial random variable**, X, that counts the number of successes in a binomial experiment.

- The binomial formula enables computing probabilities as:
$$P(x) = \frac{n!}{x!\,(n-x)!} p^x (1-p)^{n-x}$$

  The formula consists of two parts. The first part simply counts how many possible ways there are to obtain x successes in n trials. The second part incorporates the probability of success (p) and the probability of failure (1-p).

- The mean and variance of the binomial probability distribution are computed as:
$$\mu = np$$
$$\sigma^2 = np(1-p)$$
$$\sigma = \sqrt{np(1-p)}$$

- To compute binomial probabilities, use the approaches shown below:
    - P(X=x) ... Use the binomial formula or Excel with 0 for *cumulative*
    - P(X≤x) ... Use Excel with 1 for *cumulative*
    - P(X≥x) ... Use the complement rule to compute 1-P(X≤x-1)
    - P(x_1≤X≤x_2) ... Use P(X≤x_2)−P(X≤(x_1-1))

Unit 2 Exercises

1. *Angry Birds* used to be one of the most popular games on smartphones and tablets. According to one source, 76% of those who play *Angry Birds* downloaded the free version of the game. Suppose you surveyed a sample of 25 people about whether they have downloaded the free version of the game:

    a. What is the probability that fifteen of the 25 respondents have the free version?

    b. What is the probability that *at least* fifteen of the 25 respondents have the free version?

    c. What is the probability that *at most* fifteen of the 25 respondents have the free version?

    d. What is the probability that between five and ten people (inclusive) did not download the free version of the game?

2. A survey claims that 46% of parents are not concerned about their children's use of social media. Suppose that you surveyed 200 parents.

    a. What is the probability that 100 parents are concerned about their children's use of social media?

b. What is the probability that *less than* 100 parents are concerned about their children's use of social media?

3. According to the *New York Times*, a 2013 survey shows that "nearly eight in 10 say they know someone in their circle of family and friends who has lost a job" and "three in 10 said the economy would never fully recover from the Great Recession."[39] In a survey of 50 students at your school:

   a. What is the probability that seven students know someone who has lost a job?

   b. What is the probability that between three and six students (inclusive) *don't* know someone who has lost a job?

   c. What is the probability that more than two students believe the economy would recover?

4. This chapter's Excel file includes a worksheet titled 'Cumulative Binomial Table'. The worksheet includes the cumulative binomial table for n=5. The first column (Column A) provides the different values that X (the number of successes) can obtain. Thus, X ranges from zero to five when n=5. The first row of the table (Row 5) provides different possible values of p, the probability of success. Inside the table are the cumulative binomial probabilities for the different number of successes, computed using the =BINOM.DIST() function. Using the provided worksheet as a guide, create similar tables for n=10 and for n=15.

5. The statistics provided below were found in various surveys reported on *CNN.com*. Assuming these statistics are true, you then poll 100 people about the same questions. For each part of this question, first state what a *success* is and then compute the expected number of successes, as well as its variance.

   a. "In 2015, 30.4% of Americans 20 and older said they were obese" (May 25, 2016)

   b. "Over the past year, the survey indicates, 28% of employees have called in to work sick when they were feeling well" (October 24, 2014)

   c. "Four out of 10 of Web users have been harassed online, according to a survey released Wednesday by the Pew Research Center." (October 22, 2014)

---

[39] Source: http://www.nytimes.com/2013/02/07/business/profound-weight-of-layoffs-seen-in-survey.html

d. "Seven in ten Americans think the Founding Fathers would be disappointed by the way the United States has turned out." (July 4, 2013)

e. "Fifty percent of adults in the United States say they would support government funding for all federal campaigns while banning private contributions." (June 24, 2013)

f. "44% of Americans say their financial situation is worse today than it was a year ago." (June 19, 2013)

g. "Just two out of every five Americans, 40%, said that global climate change was a major threat to the United States." (June 25, 2013)

h. "The top three sins seducing most Americans are procrastination, overeating and spending too much time on electronic media. 60% of Americans admitted that they're tempted to worry too much or procrastinate; 55% said they're tempted to overeat; and, 41% said they're tempted by sloth, or laziness." (February 8, 2013)

6. There are ten computers in a PC lab. The probability that any of the computers is used at any given moment is 15%. Answer the following:

    a. What is the probability that exactly three computers are being used?

    b. What is the probability that more than five computers are being used?

    c. What is the probability that less than five computers are being used?

    d. What is the probability that a group of three students would be able to work together in the PC lab assuming they each need a computer?

    e. What is the expected number of computers being used at any given moment? What does this number mean?

7. *Pokémon*, the world's second most successful media franchise based on a video game, was originally a role-playing game for the *Nintendo Game Boy*. There are 649 *Pokémon* grouped into seventeen types, including Water and Grass, which represent 16.8% and 11.4% of all *Pokémon*, respectively.[40] If you have twenty *Pokémon* cards:

    a. What is the probability of having three Water types?

---

[40] Source: http://visual.ly/17-pokemon-types

b. What is the probability of having more than one Grass type?

c. What is the probability of having exactly four cards that are either Water or Grass types?

8. Space exploration has literally skyrocketed over the last 50 years, with nearly 200 missions sent throughout the solar system. Of the 183 missions, an astonishing 40% were to the moon.[41] In a random sample of ten space missions, what is the probability of exactly five being to the moon? What is the probability of six or more being to the moon?

9. A farmer keeps 40% of his cattle while selling the remainder. Of ten randomly selected cows, find:

   a. The probability of eight being kept.

   b. The probability of eight being sold.

   c. The probability of more than four being sold.

   d. The probability of all ten of the cows being sold.

UNIT 3

The Poisson Distribution

The Poisson distribution is similar in nature to the binomial distribution in the sense that it also focuses on the number of successes in a given experiment. The difference is that while a binomial random variable is a count of successes in a given number of trials, the **Poisson random variable** is a count of successes in a given time or space interval. Examples of Poisson random variables include the number of customers that walk into a store in an hour, the number of accidents on a stretch of highway in a week, and various sports statistics (e.g., the number of goals per game). Figure 5-3 summarizes the properties of the **Poisson experiment**.

---

[41] Source: http://visual.ly/fifty-years-exploration

Figure 5-3
Properties of Poisson Experiments

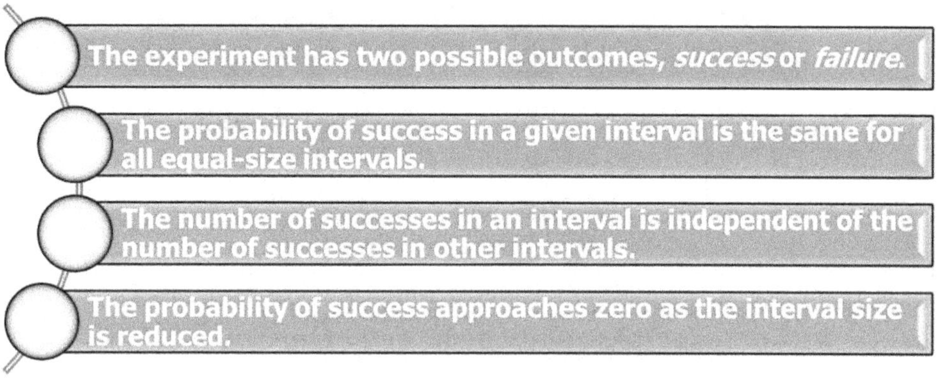

- The experiment has two possible outcomes, *success* or *failure*.
- The probability of success in a given interval is the same for all equal-size intervals.
- The number of successes in an interval is independent of the number of successes in other intervals.
- The probability of success approaches zero as the interval size is reduced.

To compute the probability of a specific number of successes in a given interval, we can use the following equation

$$P(x) = e^{-\lambda} \frac{\lambda^x}{x!}$$

where x is the number of successes, e is the base of the natural logarithm and λ (lambda) is a parameter of the distribution, used to represent the mean number of successes in a given interval (hence, $E(X)=\mu=\lambda$). Note that in the Poisson distribution, the following also holds: $V(X)=\sigma^2=\lambda$. That is, both the mean and variance of the Poisson distribution are equal to the Poisson parameter λ.

Let us use an example to demonstrate how the Poisson distribution is applied. In one city, an average of three 911 calls are accepted every minute. We can say that the rate of 911 calls follows a Poisson distribution with a parameter λ of three. We define X as the average number of calls that come into the call center during any given minute. What is the probability that within a given minute there will be exactly one call coming into the call center?

$$P(1) = e^{-3} \frac{3^1}{1!} = 0.149$$

What is the probability that there will be at most five calls?

$$P(X \leq 5) = P(0) + P(1) + P(2) + P(3) + P(4) + P(5)$$
$$= e^{-3}\frac{3^0}{0!} + e^{-3}\frac{3^{10.}}{1!} + e^{-3}\frac{3^2}{2!} + e^{-3}\frac{3^3}{3!} + e^{-3}\frac{3^4}{4!} + e^{-3}\frac{3^5}{5!} = 0.916$$

As with the binomial distribution, the latter computation would be more easily accomplished using Excel. We can use the Excel function =POISSON.DIST(x,mean,cumulative) to find this probability, where x is the number of successes (5 in our example), mean is the mean number of successes per interval (the λ parameter, which is 3 in our example), and cumulative is assigned either 0 or 1 in a manner similar to that of the fourth argument of the binomial distribution. In our example we are, indeed, looking for the cumulative probability (at most five) and, thus, we type =POISSON.DIST(5,3,1) to obtain the answer of 0.9161.

Now consider a different example to demonstrate the difference between the Poisson distribution (the above example) and the binomial distribution. According to the local newspaper, 40% of all 911 calls are accidental, i.e., inadvertent or *pocket-dialed*. In a sample of ten calls, what is the probability that five will be accidental? In order to answer this question, we would need to go back to using the binomial distribution, as we are looking for x successes (where a success is defined as an accidental 911 call) in n trials:

$$P(5) = \frac{10!}{5!\,(10-5)!} 0.4^5 (1-0.4)^{10-5} = 0.2$$

Thus, there is a 20% chance that out of the ten 911 calls, five are accidental.

Often, when working with hourly, daily or yearly data, we end up having to work with very large numbers. For example, a website might easily receive an average of 1,200 visitors per day. It would be difficult to work with this scale of data in the Poisson formula and the mathematics involved can be eased by reducing the given scale of data to a smaller scale. The best way to achieve this is to reduce the size of the interval studied. Thus, rather than analyzing daily visits to the website, we can work with the average hourly visits (1200/24=50 visits per hour) or even average shorter intervals (e.g., 25 visits per 30 minutes, etc.) to still answer important questions about the website's traffic. For example, the calculation for determining the probability that the website would receive 40 or more visits in a given hour is:

$$P = 1 - P(X \leq 39) = 0.936$$

We obtain this probability using Excel by typing: =1-POISSON.DIST(39,50,1).

Unit 3 Summary

- A **Poisson random variable** measures the number of successes in a given interval of time or space. Its probability follows the Poisson distribution.

- The **Poisson experiment** is characterized by the following: (1) the experiment has two possible outcomes, 'success' or 'failure'; (2) the probability of success is the same for all equal-size intervals; (3) the number of successes in any given interval is independent of the number of successes in other intervals; and, (4) the probability of success approaches zero as the size of the interval approaches zero.

- The λ (lambda) parameter is the average rate of the distribution. λ is proportionate to the interval and can be divided by the number of units to obtain the average rate of success in various interval sizes.

- Both mean and variance of the Poisson variable are equal to λ :
$$E(X) = \mu = \lambda$$
$$V(X) = \sigma^2 = \lambda$$

Unit 3 Exercises

1. Compute the following Poisson probabilities using the Poisson formula:

    a. P(X=3) if λ=3

b. P(X=4) if $\lambda=5$

c. P(X<2) if $\lambda=3$

d. P(X≥3) if $\lambda=4$

2. Within a computer program, the number of bugs (i.e., coding errors) per lines of code has a Poisson distribution with an average of fifteen bugs per 1,000 lines.

    a. Find the probability that there will be exactly eight bugs in 1,000 lines of code.

    b. Find the probability that there will be at least eight bugs in 1,000 lines of code.

    c. Find the probability that there will be at least one bug in 1,000 lines of code.

    d. Find the probability that there will be no more than one bug in 1,000 lines of code.

    e. Find the variance of the number of bugs in 1,000 lines of code.

    f. Find the standard deviation of the number of bugs in 1,000 lines of code.

3. A store manager is considering cutting costs by keeping the store closed during the morning, as customer traffic in the morning is much lower than the rest of the day. Based on past data, the store manager knows that the number of customers entering the store between 9am and 12pm follows a Poisson distribution with an average of ten customers during this time period. The store manager believes that she needs at least two sales to justify keeping the store open and she further knows her conversion rate (visitors to sales) is 20%.

    a. What is the probability that she will break even if she keeps the store open in the morning?

    b. Another option is to open the store just a little later, say at 10am. What is the probability that at least eight customers will enter the store between 10am and 12 noon?

4. This chapter's Excel spreadsheet has a worksheet titled 'Cumulative Poisson Table'. In this worksheet, we have created a table of Poisson probabilities by varying values of x (the number of successes) for a given value of λ.

Change the value of λ in cell B4 to a variety of different values in order to understand how the Poisson probabilities change.

5. Use the Poisson function in Excel to compute the following probabilities for X when λ=7:

   a. P(X≤6)

   b. P(X=5)

   c. P(X≥3)

   d. P(7≤X≤10)

6. Again, use the Poisson function in Excel, but change the value of λ to 12 and compute the following probabilities:

   a. P(X≤13)

   b. P(X=15)

   c. P(X≥23)

   d. P(X≤28)

   e. P(20≤X≤27)

7. A factory manager needs to decide which machine (either machine A, B or C) to buy. The manager will not purchase a machine unless the probability of having more than three errors per shift is less than 10%. Table 5-e11 shows the number of errors that each of the machines produced in two shifts (in trials). Which machine (or machines) could be bought?

   Table 5-e11

A	B	C
2	3	4

8. The number of customers arriving at a downtown hotel who desire valet parking is six per hour. One employee working alone claims to have parked twenty cars in the last two hours. What is the probability of his claim?

9. In the United States, 7,000 kids drop out of high school every day (or roughly five kids every minute).[42] Find the probability that in a stretch of two minutes, less than eight kids drop out.

10. The *Leonardo da Vinci-Fiumicino Airport* in Rome is the sixth busiest airport in Europe with 3 million passengers monthly.[43] Find the probability that passenger traffic will exceed 11 million passengers over a span of three months.

END-OF-CHAPTER PRACTICE

1. Irene is taking a four-question multiple-choice quiz, with each question containing four choices. Assume there is a 75% chance of her getting each individual question right. Find the probability of the following:

    a. Irene scores over 50%.

    b. Irene scores at least 75%.

    c. Irene answers none of the questions correctly.

    d. Irene scores 100%.

2. According to 2011 data, Asia is the most populous continent with its 4.2 billion inhabitants accounting for approximately 60% of the world population, with Africa being in second place with 15%, and North America at 5%.[44] Suppose a random sample of 25 people from all regions of the world is taken. Find the following probabilities:

    a. Fifteen or less of the sampled people are from Asia.

    b. None of the sampled people are from North America.

    c. More than ten of the sampled people are from Africa.

    d. Less than twenty of the sampled people are from either Africa or North America.

3. The probability distribution of the number of customers waiting to speak to a customer representative of a cable company is shown in Table 5-e12.

---

[42] Source: http://www.unitedwaytriangle.org/blog
[43] Source: http://en.wikipedia.org/wiki/Leonardo_da_Vinci-Fiumicino_Airport
[44] Source: http://esa.un.org/unup/Analytical-Figures/Fig_overview.htm

Table 5-e12

X	P(X)
0	.00
1	.00
2	.00
3	.05
4	.20
5	.15
6	.15
7	.15
8	.15
9	.05
10	.10

a. What is the expected number of customers waiting on the line?

b. What is the probability that exactly six customers are waiting on the line?

c. What is the probability that more than two customers are waiting to speak to a customer representative?

d. What is the probability that less than three customers are waiting on the line?

e. What is the probability that, if you call the cable company, you'd be the seventh customer to wait on the line to speak with a customer representative?

4. There are 25 tables in a restaurant. The probability that any particular table is occupied at any given moment is 20%. Answer the following:

   a. What is the probability that 24 tables are occupied?

   b. What is the probability that more than one table is not occupied?

   c. What is the probability that more than five tables, but less than twenty tables, are occupied?

   d. What is the probability that, if you showed up without a reservation, you'd be able to immediately get a table at the restaurant?

5. An up-and-coming fashion designer must pitch her collection for a spot at *New York Fashion Week*. The decision-making board must select at least eight pieces for her to earn a spot on the runway. She believes there is a

70% chance of a piece being selected. If there are ten pieces in her collection, determine the probability that she will win a spot.

6. Table 5-e13 presents the joint probability distribution of refurbished phones' age (in years) at the time of a repurchase and the phones' remaining lifespans (in months).

Table 5-e13

		Phone Age (In Years)			
		0.5	1	1.5	Total
Remaining Lifespan (In Months)	18	0.02	0.05	0.2	0.27
	24	0.05	0.23	0.05	0.33
	30	0.2	0.1	0.1	0.4
	Total	0.27	0.38	0.35	1

a. What is the probability that a year-old refurbished phone would last 30 months?

b. What is the probability that a 1.5-years old refurbished phone would last 18 months?

c. What is the probability that a randomly selected refurbished phone is six months old and would last 30 months?

d. What is the probability that a randomly selected refurbished phone is twelve months old?

e. What is the probability that a randomly selected refurbished phone has a lifespan of two years?

f. Given that a randomly selected refurbished phone has lasted for 24 months, what is the probability that it was twelve months old when it was re-sold?

7. A typical secretary types an average 70 words per minute (wpm) with an average of two typing errors per page. A company wishing to encourage employee improvement announces a bonus to each secretary who can type faster than the average pace or have fewer errors per page. What is the probability that a secretary will receive the bonus?

8. Two realtors are going head-to-head in a friendly competition to see who can sell more houses within a month of taking a listing. Realtor 1 takes on ten clients and has a 50% success rate in selling a house in less than a month. Realtor 2 takes on fifteen clients and has a 40% success rate in

selling a house in less than a month. Who was more likely to sell at least five houses?

9. A disputed aspect of *Major League Baseball* is that there is no video replay. The umpire's decision, based on the naked eye, is undisputable. Suppose that when calling stolen bases, the umpire makes the wrong call 5% of the time. In a sample of 25 runners, what is the probability that more than five were wrongfully called out? What is the probability that exactly one was wrongfully called out?

10. You designed a new financial app and are distributing it on a trial version through your personal website. Suppose that your app becomes so popular that it receives, on average, 7,200 downloads per day. Assuming that your server cannot handle more than ten downloads per minute, what is the probability that it will crash?

11. A softball team is playing their next ten games at home. The probability of winning a home game is 0.6.

    a. What is the probability that they will win at least eight of these games?

    b. What is the most realistic number of games they can expect to win?

12. The number of callers waiting on the line to speak to a customer representative follows a Poisson distribution with Lambda=4.

    a. Find the probability that when you call there are exactly five people ahead of you.

    b. Find the probability that you will be the third caller waiting to speak.

    c. What is the probability that you do not have to wait at all?

    d. What is the expected number of callers waiting on the line at any given moment?

13. If a hospital admits new pneumonia patients at a rate of about five per day, what is the probability that during the next week:

    a. Exactly 35 pneumonia patients will be admitted?

    b. Exactly 40?

    c. 45 or more?

14. Given that a small business hires a new employee about every three months, what is the probability that:

    a. A new employee is hired within two months?

    b. Two new employees are hired within three months?

    c. No new employees are hired within five months?

15. A student can write an average of one essay page per hour. Assuming she only has seven hours a week to write essays, what probability does she have of completing all her weekly work if she is assigned essay pages according to the information in Table 5-e14?

    Table 5-e14

Number of Total Essay Pages	Probability
0	0.2
3	0.3
7	0.3
9	0.2

16. John is actively looking for a job. If each job application has a .02 chance of getting a job, what is the probability that:

    a. John will get a job on his very next application?

    b. John will not get a job within his next 50 applications?

    c. John can improve his application success rate from .02 to .03 if he skips applying for jobs and instead spends the time polishing up his resumé. How many of the next twenty jobs can John skip before the probability of getting a job within the next twenty job opportunities becomes less than if John did not spend the time improving his resumé?

17. A *Twitter* post has a probability of 0.4 of getting re-tweeted. If I wanted to get at least 30 posts re-tweeted with an 80% or more chance of success, how many tweets should I make?

18. A book publisher claims he only makes money on one out of every eight books he publishes. If he publishes:

    a. 30 books, what is the probability that he makes money on five of them?

b. 40 books, what is the probability that he makes money on only one of them?

c. 80 books, what is the probability that he makes money on exactly 40 of them?

d. 80 books, what is the probability that he makes money on 40 or more?

19. In a recent poll, 60% of people voiced concerns about their privacy when using the Internet. If this group is representative of the whole population, what is the probability that two randomly chosen people will both respond that they are not concerned?

20. A factory produces faulty watches according to a Poisson distribution with an average of two faulty watches per 1,000 total watches. What is the probability that given the scenario above, the number of faulty watches in 4,000 total watches will be:

a. Exactly eight watches?

b. Greater than or equal to five watches, but less than eight watches?

c. Ten watches or less?

CHAPTER 6: Continuous Probability Distributions

In the previous chapter, we introduced the concept of *random variables* with their accompanying *probability distribution*. The probability distribution provides information on the values of the random variable that we might expect to observe. Chapter 5 focused on *discrete* random variables, which are the results of counting processes. Many important variables, however, are *continuous* and in this chapter we cover some of the common probability distributions of continuous random variables.

Recall our example of the Internet art store from Chapter 5, in which we used the *binomial* distribution to predict the expected number of shopping cart conversions. Specifically, we defined X as a discrete random variable, counting the number of customers who made the conversion from putting pieces of art into a shopping cart to actually purchasing these art pieces, and we studied the probability of different numbers of customers making purchases from their shopping carts. There are many other variables we may be interested in learning about, such as the amount of time a single website visit lasts or the amount of money spent by each shopper. Both of these are examples of continuous variables.[45]

Suppose you sample one customer at random and measure the amount of money this person spent while shopping at the website. What do you expect this amount to be? If you know that an average customer spends $300 per purchase, then you already have some idea of what to expect in terms of the amount of

---

[45] We conceptually treat monetary variables as the result of a continuous measuring process even though, in a physical sense, money is not strictly continuous due to the fact that it is limited by the bills and coins available to us at a point-in-time.

money a random customer will spend. If you also know the standard deviation of the amount of money spent is $70, then you may have an even better idea of what to expect. But, if only the mean and standard deviation of a random variable are known, many interesting questions cannot be answered, such as: What is the probability that this random customer will spend $400 or more? What is the probability that she will spend between $250 and $450? What is the probability that she will spend exactly $326.76? Questions such as these require knowledge of the *probability distribution* of the variable and this is what Chapter 6 is about. This chapter covers the following topics:

- Continuous Probability Distributions
- The Uniform Distribution
- The Normal Distribution
- The t-Distribution
- The χ^2 Distribution
- The F-Distribution
- Working with the Normal Distribution
- Determining Normal Distribution Probabilities with Excel and with the Standard Normal Distribution Table
- P-Value
- Inverse Probability Calculations

UNIT 1

Continuous Probability Distributions

A **continuous random variable**, X, can take on any value at a given probability. As in previous chapters, all probabilities must be between zero and one, and the sum of all probabilities is one. In fact, because we are dealing with a continuous variable, the probability of any *specific* value of the variable is practically zero. For example, what is the probability that you will sleep for *exactly* seven hours, five minutes, nine seconds and three milliseconds tonight? Essentially zero! In practice, however, when working with continuous random variables, we usually look at cumulative probabilities across intervals of values (e.g., what is the probability that you will sleep less than seven hours, or more than eight hours, or between seven and nine hours?).

To describe the probability distribution of continuous random variables, we use a line graph that represents the distribution's **probability density function**. The probability density function, or *density function* for short, describes the likelihood that the values of X, the continuous random variable, fall within a given interval. An example of a density function is shown in Figure 6-1.

Figure 6-1
Example of a Probability Density Function

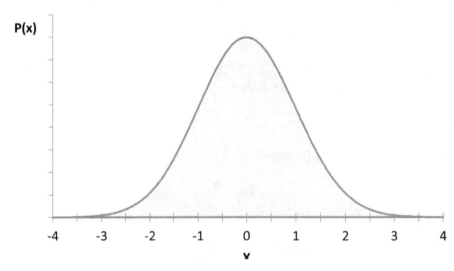

The total area captured between the line and the x-axis is equal to one, because the sum of all probabilities in a sample space is one. We can compute the probability of X falling in a specific interval along the x-axis by computing the partial area underneath the density function for this interval. Mathematically, this requires the use of calculus. In practice, and in this book, we use statistical software (specifically, Excel) to derive the values needed in our analyses.

The calculation of probabilities depends on the specific shape of the distribution. We begin this chapter with a review of the most commonly-used probability distributions – the ones that you are likely to use in future analyses. We provide only a general overview of these distributions here and will discuss them again in future chapters that require their use. The remainder of the chapter is then spent on the *normal distribution*, which plays a critical role in many statistical analyses.

The Uniform Distribution

Continuing with our Internet art store example, suppose that the company runs a flash sale once each day, at different times. You logged on to the website at 8am and wonder what the probability is that the flash sale will take place before 8pm. Let us define X, a random variable, as measuring the time of day when the flash sale takes place. Let us further assume that X can take on any value in a 24-hour interval (from 12am on one day to 12am on the next day), with equal probabilities for any of the possible values to occur. Here, X follows a ***uniform***

distribution[46], which assigns an equal likelihood of occurrence to all possible values of X. This uniform distribution is shown graphically in Figure 6-2.

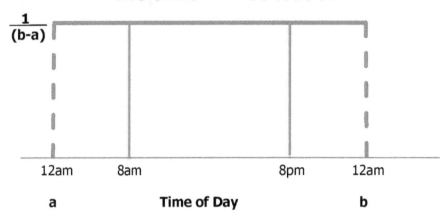

Because the uniform distribution has a rectangular shape, the probability of X taking a specific range of values, i.e., between x_1 and x_2, can be computed as

$$P(x_1 \leq X \leq x_2) = (x_2 - x_1) * \frac{1}{(b-a)}$$

where x_1 and x_2 define the interval of interest and a and b, respectively, are the lower (0) and upper (24) values of the random variable. Thus, we can compute the desired probability in our example as:

$$P(8am \leq X \leq 8pm) = (\text{number of hours between 8am and 8pm}) * \frac{1}{24} = \frac{12}{24} = 0.5$$

Given this result, there is a 50% chance that the flash sale will take place during the twelve-hour time period.

The Normal Distribution

The ***normal distribution*** is a family of distributions that share the same shape and attributes, but vary in their parameters; namely, the mean and the

---

[46] The uniform distribution can also be *discrete,* as in the case of rolling a die (each outcome has the same probability). Here we focus on the *continuous* uniform distribution.

standard deviation. You may have heard about the normal distribution in your past studies or at work, and you may have heard or seen depictions of the 'bell curve' shown in Figure 6-3. The bell curve represents the shape of a normal distribution.

**Figure 6-3
The Normal Distribution Curve**

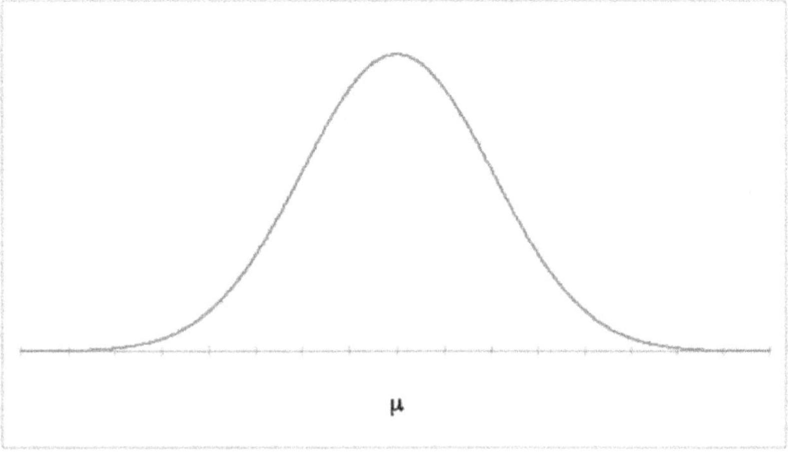

μ

The normal distribution is **symmetric**; that is, each half is a mirror image of the other. The location of its peak is determined by the mean, whereas the standard deviation determines the width of the distribution. Two examples of normally distributed data include: the height of people (the height of women in the US is normally distributed with a mean of 63.8 inches and a standard deviation of 4.25 inches[47]) and the IQ of people (IQ is normally distributed with a mean of 100 and a standard deviation of fifteen[48]).

The normal distribution is very important in statistics because its properties enable, or support, a wide range of analytics (that is, there are many different calculations we are able to conduct if we know that a variable is, or is close to being, normally distributed). For this reason, we sometimes use various variable

---
[47] Source: http://www.cdc.gov
[48] Source: https://en.wikipedia.org/wiki/Intelligence_quotient?dur=3432

transformations in attempts to *normalize* a data set. For example, we may take the square root of each data point or the log of each data point, both of which are commonly-used transformations for normalizing data. We will learn how to work with the normal distribution and to compute probabilities in Unit 2 of this chapter, and then build on this knowledge in subsequent chapters.

The t-Distribution

The normal distribution requires either that large samples are available or that the population standard deviation (σ) is known. In most cases, one or both of these conditions are not met. In these situations, we work with a similar distribution called the t-distribution. The **t-distribution** is similar in shape to the normal distribution (specifically, to a normal distribution with a mean of zero and a standard deviation of one), but its shape also depends on the sample size, n.

More precisely, the shape of the t-distribution depends on a sample's number of **degrees of freedom**. Degrees of freedom refer to the number of data values that are free to vary in a data set. For each parameter (e.g., μ or σ) that we are estimating, we lose one degree of freedom. In order to better understand what this means, consider the sample X_1, X_2 and X_3, where the X values can obtain any possible value. In this example we have three degrees of freedom. Suppose that $X_1=2$, $X_2=4$, and that *the average of the three values has to be 3*. Given this requirement, the only possible value that X_3 can obtain is 3. Thus, we have lost one degree of freedom. X_1 and X_2 can still obtain any possible value, but X_3 will then be forced to be a value that brings the sample average to 3. When working with the t-distribution, because we are working with an estimated value of the standard deviation (s rather than σ), we have n-1 degrees of freedom.

As the sample (and thus the number of degrees of freedom) becomes large, the t-distribution approaches the normal distribution. This is illustrated in Figure 6-4, below. We will explore, in depth, the t-distribution and its uses in Chapter 10.

Figure 6-4
The t- Distribution Curve

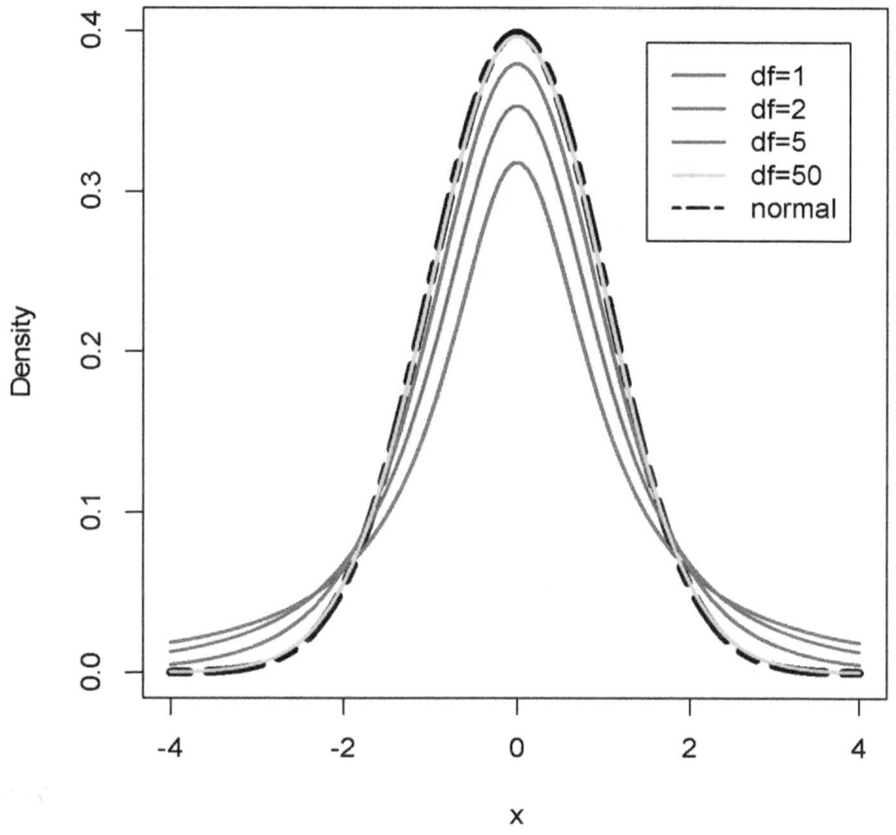

The χ^2 Distribution

The χ^2 *distribution* (*chi-squared distribution*) is the distribution of the sum of squares of n-1 variables (n being the sample size). You may recognize the term *sum of squares* from previous chapters, where we have used it to describe the concept of *variance*. Indeed, the chi-squared distribution is used to study the probability distribution of the variance, among its other uses. The distribution is not symmetrical; it is positively skewed, meaning that its peak (i.e., the majority of values) is closer to the left side of the scale and it has an elongated tail on the right side. It also has no negative values. This makes sense if we consider that a sum of squares cannot take on negative values (all values are squared). As with the t-distribution, the precise shape of the χ^2 distribution is a function of the number of its degrees of freedom. The χ^2 distribution is shown in Figure 6-5 and will be covered in more depth in Chapter 9.

Figure 6-5
Chi-Square Distribution

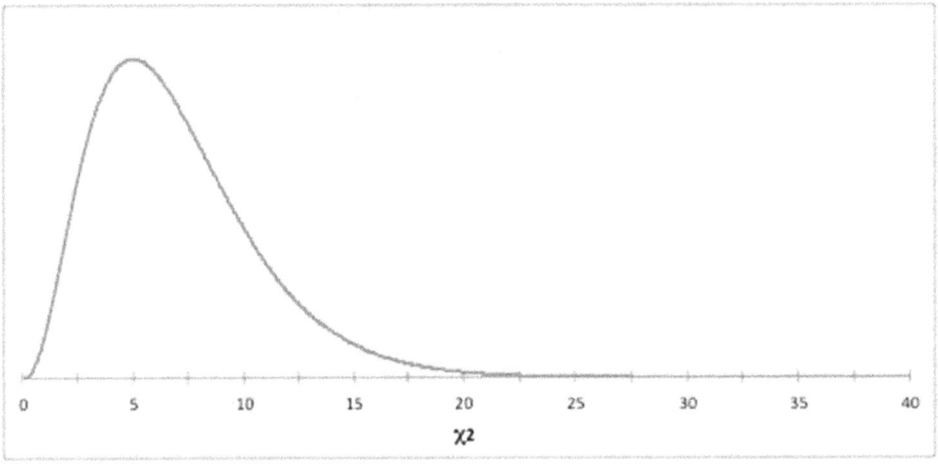

The F-Distribution

The final distribution we will introduce in this chapter is the **F-distribution**, which is the distribution of the ratio of two variances: σ^2_1/σ^2_2. Like the χ^2 distribution, the F-distribution is not symmetrical, can obtain only positive values, and its precise shape depends on the number of degrees of freedom (in this case, the degrees of freedom of both the numerator and denominator; remember, we are looking at a ratio). The F-distribution is shown in Figure 6-6 and will be covered in more depth in Chapter 10.

Figure 6-6
The F-Distribution

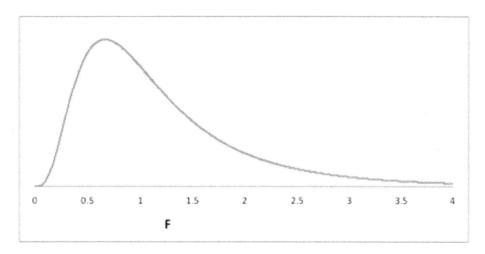

Unit 1 Summary

- A **continuous random variable**, X, can take on any value at a given probability. All probabilities must be between zero and one and the sum of all probabilities is one. Because we are dealing with a continuous variable, the probability of any *specific* value of the variable is essentially zero.

- The ***probability density function*** describes the likelihood that the value of X falls within a given interval. Probabilities are expressed as the area between the curve and the x-axis. We use statistical tables and software to compute these probabilities.

- The continuous **uniform distribution** assigns an equal likelihood of occurrence to all possible values of X.

- The **normal distribution**, represented as a bell curve, comprises a family of distributions that share the same shape and attributes, but that vary in their parameters; namely, the mean and the standard deviation. The normal distribution is **symmetric**; that is, each half is a mirror image of the other.

- The **t-distribution** is similar in shape to the normal distribution and used when working with small samples. It can also be used when the population standard deviation (σ) is unknown and we are using the sample standard deviation (s) instead. Its precise shape depends on the number of degrees of freedom. **Degrees of freedom** refer to the number of data values that are free to vary in a data set.

- The ***χ^2 distribution*** (chi-squared distribution) is the distribution of the sum of squares of n-1 variables (n being the sample size). The distribution is positively skewed rather than symmetrical. It does not include negative values and its precise shape also depends on the number of degrees of freedom associated with a sample.

- The **F-distribution** is the distribution of the ratio of two variances: σ^2_1/σ^2_2. Like the χ^2 distribution, the F-distribution is not symmetrical, can only obtain values that are positive and its precise shape depends on the number of degrees of freedom associated with both the numerator and denominator.

Unit 1 Exercises

1. List five continuous random variables from any domain of interest to you.

2. What can you say about the probability distribution each of these variables is most likely to follow?

3. In your own words, explain what is meant by the term *probability distribution*.

4. What are the key differences between *discrete* and *continuous* probability distributions?

5. In the *Olympic Games'* 100m freestyle swimming competition, what is the probability of an exact tie between two swimmers?

UNIT 2

Working with the Normal Distribution

We mentioned in the previous section that the normal distribution is a family of distributions, each with its own mean and standard deviation. When a variable, say X, is normally distributed, we denote it as:

$$X \sim N(\text{mean, standard deviation}) \text{ or } X \sim N(\mu, \sigma)$$

Returning to the Internet art store example, an important variable for website store owners is the amount of time that visitors spend on their websites. If we believe that the time a visitor spends on the art store website follows a normal distribution with a mean of five minutes and a standard deviation of one minute, then we would write this as:

$$X \sim N(5,1)$$

Now, suppose we wish to compute the probability that a randomly chosen visitor will spend less than four minutes on the site? Recall that earlier in this chapter we mentioned that probabilities are computed as the area under the probability density curve, which is shown for this example in Figure 6-7. Here, the x-axis is measured in minutes and you can see that the value 5, which we know is the mean value of the distribution, corresponds to the distribution's peak. Also note the area under the curve that is shaded red and that is bounded on the right by the value 4. This shaded area represents the probability of obtaining a value less than 4 with this distribution. How much of the area under the curve do you think this shaded area represents?

Figure 6-7
Working with the Normal Distribution

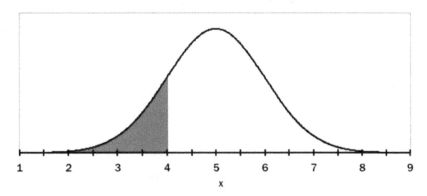

We mentioned earlier in the chapter that calculus is needed to calculate this area, but we can also use statistical software, such as Excel, to compute probabilities. Specifically, Excel's function that allows us to calculate this area (and, hence, the desired probability) is:

=NORM.DIST(x,mean,standard_dev,cumulative)

Using the values and parameters from our example, this function would be written

=NORM.DIST(4,5,1,1) = 0.1586

where 4 is the value of x we are interested in learning about, 5 is the mean time people spend on the website, 1 is the standard deviation of the time people spend on the website, and the final 1 indicates that we are interested in the cumulative probability (that is, the probability of four minutes or less, depicted as the area shaded red in Figure 6-7).[49] Hence, the probability that a randomly chosen visitor will spend less than four minutes on the site is 0.1586, or 15.86%.

While Excel enables us to work with any normal distribution using the =NORM.DIST() function, it is still more common, and often easier (especially for someone just learning statistics) to understand how to use the normal distribution if the normal distribution is converted into the *standard normal distribution*, also known as the *Z distribution*.

The **standard normal distribution** is the distribution of a special random variable Z, which is *normally* distributed with a mean of zero and a standard deviation of one:

$$Z \sim N(0,1)$$

Any normal distribution can be converted into the standard normal distribution by converting a normally-distributed variable X into its corresponding **Z score** as follows:

$$Z = \frac{X - \mu}{\sigma}$$

Note that we do not convert the distribution as a whole, but rather individual values of X. This is illustrated by the examples in Table 6-1, which demonstrate how values from different normal distributions are converted into Z scores. The first

---

[49] If a value of 0 is entered for this fourth parameter, i.e., the *cumulative* argument, Excel will treat the variable as a discrete variable and will return the probability of X=4, rather than the probability of X≤4. In this book, we only work with cumulative probabilities for our continuous distributions and thus will always select a value of 1 for this final argument.

column provides information on the distribution of X (the mean and standard deviation), the second column is the probability we are interested in finding, and the third column contains the corresponding Z scores.

Table 6-1
Converting X Values to Z Scores

Distribution	P(x)	Corresponding Z Scores
X~N(10,2)	P(X≤7)	$P\left(Z \leq \dfrac{7-10}{2}\right) = P(Z \leq -1.5)$
X~N(3,0.4)	P(X≥3.5)	$P\left(Z \geq \dfrac{3.5-3}{0.4}\right) = P(Z \geq 1.25)$
X~N(35,5)	P(30≤X≤40)	$P\left(\dfrac{30-35}{5} \leq Z \leq \dfrac{40-35}{5}\right) = P(-1 \leq Z \leq 1)$
X~N(4,0.2)	P(X≥6)	$P\left(Z \geq \dfrac{6-4}{0.2}\right) = P(Z \geq 10)$
X~N(50,25)	P(30≤X≤35)	$P\left(\dfrac{30-50}{25} \leq Z \leq \dfrac{35-50}{25}\right) = P(-0.8 \leq Z \leq -0.6)$

Figure 6-8 (parts a, b and c) shows the information in the first three rows of Table 6-1 in a graphical form. With each part of the figure, the bottom graphs show the distribution of X with the desired probability shaded in the graph; and, the top graphs show the corresponding standard normal (Z) distributions. As you can see, the only difference between the bottom and top graphs is the labeling of the x-axis, with the bottom graphs centered on the mean value of each respective distribution and *all* top graphs centered on a value of 0 (remember, the standard normal distribution always has a mean of zero).

Figure 6-8
Converting X Values to Z Scores

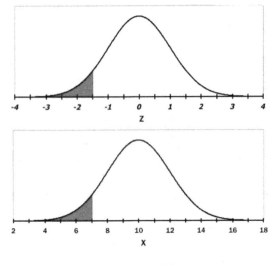

Part a: X~N(10,2)
P(X≤7) same as P(Z≤-1.5)ig

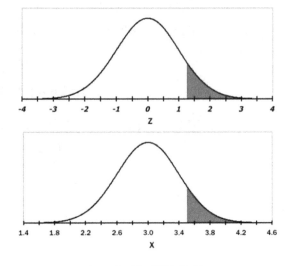

Part b: X~N(3,0.4)
P(X≥3.5) same as P(Z≥1.25)

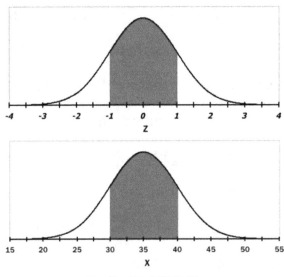

Part c: X~N(35,5)
P(30≤X≤40) same as P(-1≤Z≤1)

Z scores have a special meaning that is important to understand. For each value of X (x_1, x_2, etc.), the corresponding Z score is a measure of *how far* – in terms of the number of (the distribution's) standard deviations – the value of X is from the distribution's mean value. Look at any of the Z distributions in Figure 6-8.

Negative values of Z correspond to values of X below (i.e., to the left of) the distribution's mean, while positive values of Z correspond to values of X above (i.e., to the right of) the distribution's mean. Further, a Z score of -1.5 means that the corresponding X value lies exactly 1.5 standard deviations below the mean. Similarly, a Z score of 1.25 means that the corresponding X value lays exactly 1.25 standard deviations above the mean. Intuitively, you can see that the further a value is from the mean, the larger the Z score (in absolute value), and the smaller the probability of sampling this value out of the distribution.

Now, let's return to our Internet art store example. Recall that we were looking for the probability that a customer will spend less than four minutes on the site. The value of a customer visit of four minutes is exactly one standard deviation below the mean (the value of X is 4, the distribution's mean is 5 and the distribution's standard deviation is 1). Indeed, you can calculate that the corresponding Z score for a four-minute visit is -1 (that is, one standard deviation *below* the mean):

$$Z = \frac{4-5}{1} = -1$$

The same conversion formula is applied for the examples in Table 6-1. For example, a value of 7 in a distribution with a mean of 10 and standard deviation of 2 is located one and a half standard deviations below the mean and, indeed, the corresponding Z score (first row of Table 6-1) for this value is -1.5.

The reason it is important to understand this meaning of a Z score is because it is essential in understanding how probabilities are derived with a normal distribution. Recalling the *empirical rule* introduced in Chapter 3, there is a smaller probability of finding values that are farther away from the mean; and, in statistics the most common way to represent how far a value is from its distribution's mean is in terms of the number of standard deviations. Specifically, the empirical rule states that for normal distributions, approximately 68% of all the values in the distribution lie within one standard deviation from the mean. Approximately 95% of all values lie within two standard deviations from the mean, and nearly all values (99.7%) lie within three standard deviations from the mean. This is illustrated graphically in Figure 6-9:

Figure 6-9
The Empirical Rule

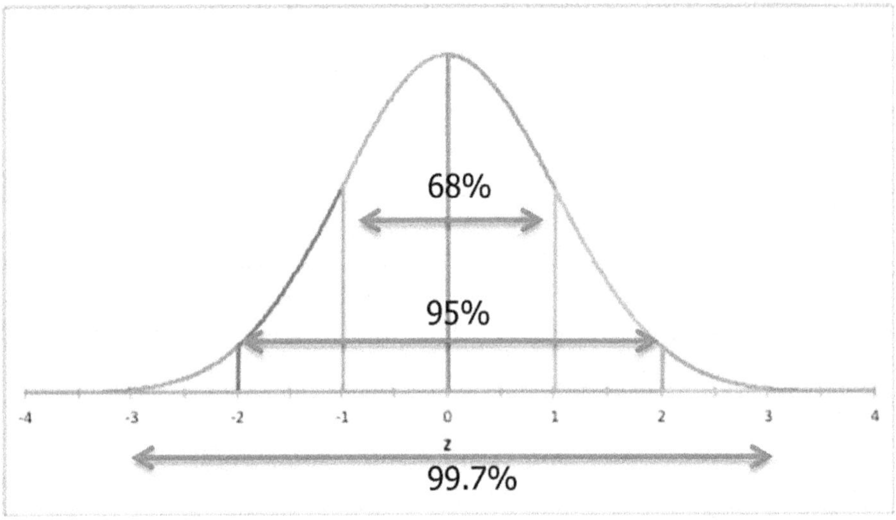

Determining Normal Distribution Probabilities with Excel and with the Standard Normal Distribution Table

The Excel function =NORM.S.DIST(z,cumulative) is the formula for computing any standard normal probability. Choosing a value of '1' for cumulative will return the *cumulative probability*, that is: $P(Z \leq z)$. For example, the Excel function =NORM.S.DIST(0.94,1) returns a value of 0.8264. In other words, approximately 83% of all possible Z score values are less than or equal to a Z score value of 0.94.

A second way to determine probabilities associated with the normal distribution is to use what is referred to as the standard normal distribution table. This table is provided in this chapter's Excel file in the worksheet titled 'Standard Normal Table'. To match this table with similar tables you might find in other statistics books, it is constructed in two parts: one (the upper part) for the negative side of the standard normal distribution and one (the lower part) for the positive side of the standard normal distribution. This table is similar to the binomial distribution table created in Chapter 5. The standard normal table provides the probability of commonly-used Z scores and is especially useful when you do not have access to computer software to obtain these probabilities (for example, if you are writing a paper-based exam and are asked to determine probabilities). Both parts of the standard normal distribution table provide the cumulative probability

$$P(Z \leq z)$$

where z corresponds to a particular Z score value.

In the table, the Z score values are shown in the first column and the first row *combined*. For example, in cell A4 you will find the value -2.8. In cell E1 you will find the value 0.03. By combining -2.8 and 0.03, a Z score of -2.83 results, and the probability of a Z score of -2.83 is located in the meeting point of Row 4 and Column E, or in cell E4. As another example, the probability shown in cell C5 (0.0034) is for $P(Z \leq -2.71)$. If you track the row and column that meet at cell C5, you will see that Row 5 refers to the Z value of -2.7 and Column C refers to the value of 0.01. Together these two values comprise a Z score of -2.71, the probability of which is located in cell C5.

Let us return to the examples in Table 6-1, replicated in Table 6-2, to practice finding different Z distribution probabilities in Excel.

Table 6-2
Finding Standard Normal Probabilities

P(Z)	Convert to	Compute
$P(Z \leq -1.5)$	Leave as is	0.0668
$P(Z \geq 1.25)$	$1 - P(Z \leq 1.25)$	0.1056
$P(-1 \leq Z \leq 1)$	$P(Z \leq 1) - P(Z \leq -1)$	0.6827
$P(Z \geq 10)$	$1 - P(Z \leq 10)$	0
$P(-0.8 \leq Z \leq -0.6)$	$P(Z \leq -0.6) - P(Z \leq -0.8)$	0.0239

Consider the first row in the table:

$$P(Z \leq -1.5)$$

To find this probability, simply type:

=NORM.S.DIST(-1.5,1) = 0.0668

This probability can also be obtained from cell B17 within the Excel worksheet labeled 'Standard Normal Table'. Cell B17 is where the Z values of -1.5 (Row 17) and 0.0 (Column B) meet.

Now consider the example in the second row of Table 6-2, also shown in Figure 6-10: $P(Z \geq 1.25)$.

Figure 6-10
The Example in the Second Row of Table 6-2

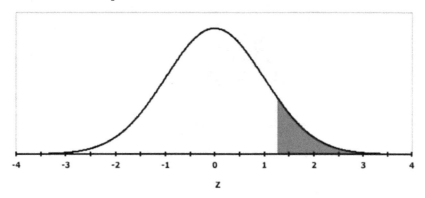

The area we are looking for is to the *right* of 1.25. Remember, however, that the Excel function provides the *cumulative probability*, or the area to the *left* of 1.25. There are two approaches to determining this probability. First, you can use the complement rule and compute:

$$P(Z \geq 1.25) = 1 - P(Z \leq 1.25)$$

Alternatively, you can use the *symmetry* of the Z distribution to compute

$$P(Z \geq 1.25) = P(Z \leq -1.25)$$

because both sides of the distribution are mirror images of each other. Then, by either applying the Excel function (=NORM.S.DIST(-1.25,1)) or by using the standard normal distribution table (cell G20), you will find this probability to be: 0.1056.

Next, consider the example in the third row of Table 6-2, illustrated in Figure 6-11: $P(-1 \leq Z \leq 1)$.

Figure 6-11
The Example in the Third Row of Table 6-2

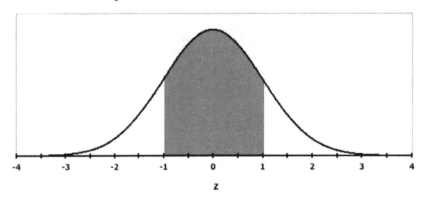

To find the area captured *between* two Z score values (here, between -1 and 1), we subtract one area under the curve from another area under the curve. Specifically, we first find $P(Z \leq 1)$, which covers all the area to the left of the value 1, and we then find $P(Z \leq -1)$, which covers all the area to the left of the value -1. Then, subtracting the second probability from the first probability gives us the sought probability (i.e., just the shaded area of Figure 6-11):

$$P(-1 \leq Z \leq 1) = P(Z \leq 1) - P(Z \leq -1) = 0.8413 - 0.1587 = 0.6826$$

Again, the cumulative probabilities needed above can be obtained by using either the Excel function or the standard normal distribution table.

We can tie this third example back to the empirical rule, since we calculated that the probability of a value falling within ±1 standard deviations from the mean is about 68%. Can you repeat this exercise to find probabilities that lie within ±2 standard deviations from the mean? Now, recall that the empirical rule also states that practically all values (99.7%) lay within ±3 standard deviations from the mean. You can check the standard normal distribution table to see that, indeed, probabilities beyond Z=±3 are nearly zero. For this reason, the probability being sought in the example given in the fourth row of Table 6-2, $P(Z \geq 10)$, is practically zero.

We leave it to you to compute on your own the sought probability in the example provided in the fifth row of Table 6-2 to make sure you understand these types of calculations.

P-Value

The **p-value** is a specific term used to describe a probability that is computed as

$$P(Z \geq |z|)$$

which is the probability of obtaining a Z score value *farther away from the mean* than a *specified* Z score value, shown in the above probability expression as the absolute value of z.[50] Since we are using the absolute value of z, *farther away* can mean being either to the right of a positive Z score value or to the left of a negative Z score value. This probability is given a special name, i.e., *p-value*, because it plays an important role in the statistical inference tests that are covered in later chapters. We will be discussing the p-value in more depth later, but are introducing it here to describe its computational aspects.

Consider first the left panel of Figure 6-12, which shows a shaded area to the left of the Z score value of -2. Recognizing that what is being sought is a cumulative probability, we can compute (using either Excel or the standard normal distribution table) the p-value for a Z score of -2 as:

$$P(Z \leq -2) = 0.0228$$

---

[50] The notation |z| refers to the *absolute value* of Z, which measures the distance the number is from zero without regard to direction (positive or negative).

Now consider the right panel of Figure 6-12, which shows a shaded area to the right of the Z value of 1.7. Remembering that the associated Excel function and the standard normal distribution tables provide cumulative probabilities, we use the complement rule to compute the sought p-value as:

$$P(Z \geq 1.7) = 1 - P(Z \leq 1.7) = 0.0446$$

**Figure 6-12
P-Value**

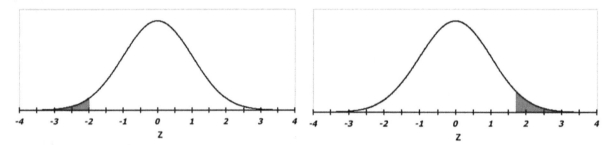

Inverse Probability Calculations

A final technique we will cover before concluding this chapter is finding the Z score value that corresponds to a given probability. This is referred to as an *inverse probability calculation* because we are given a probability value and asked to determine the associated Z score value (i.e., the inverse of being given a Z score value and being asked to determine the associated probability). Specifically, given a probability value

$$P(Z \leq z^*) = 0.05$$

what is the value of z^*, the sought Z score value?

To find this value, we use the Excel function =NORM.S.INV(probability). This function uses a single argument, *probability*, and returns the Z score value for which the following holds:

$$P(Z \leq z^*) = \text{probability}$$

It is important to recognize that this probability refers to a *cumulative* probability, i.e., the probability of obtaining a value for Z that is *less than or equal* to z*. For example:

=NORM.S.INV(0.05) = -1.645.

In other words, 5% of the area under the probability distribution curve lies to the left (i.e., the cumulative probability) of a Z score value of -1.645. This is illustrated in the left panel of Figure 6-13.

Figure 6-13
Inverse Probability Calculation

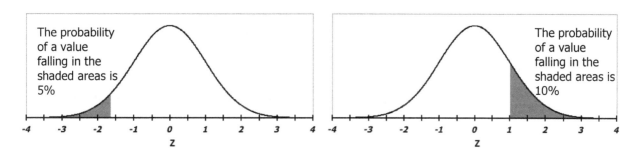

Now, consider the right panel of Figure 6-13. We are now looking for the Z score value for which 10% of the area under the probability distribution curve lays to the right of the Z score value. In other words, we are looking for the value of z^* for which:

$$P(Z \geq z^*) = 0.1$$

We cannot solve this directly, since both Excel and the standard distribution table provide cumulative probability values (i.e., the area under the distribution curve to the left of any given Z score). We can, however, use the complement rule, which tells us that the overall area under the curve should equal one:[51]

---

[51] Note that it is okay to use ≤ as well as ≥ in this equation, because in continuous distributions the value of P(Z=z*) is zero.

$$P(Z \geq z^*) + P(Z \leq z^*) = 1$$

We can rearrange the above to find:

$$P(Z \leq z^*) = 1 - P(Z \geq z^*) = 1 - 0.1 = 0.9$$

Hence, for the same z* value for which $P(Z \geq z^*) = 0.1$ holds, $P(Z \leq z^*) = 0.9$ also holds. This latter probability expression can be easily found using Excel, by typing:

$$=\text{NORM.S.INV}(0.9) = 1.282$$

This result indicates that 90% of the area under the probability distribution curve lays to the left of a Z score value of 1.282 and, hence, that 10% of the area under the curve lays to the right of this Z score value.

As a final example, let us return to the Internet art store example. Recall that the time a visitor spends on the art store website follows a normal distribution with a mean of five minutes and a standard deviation of one minute (X~N(5,1)). Suppose that we wish to find the amount of time spent on the website of the bottom 10% of our visitors. In other words, we are looking for a number of minutes (x*) such that the probability a visitor spent less time that it on our website is 10%, or P(X≤x*)=0.1. In order to find this value, we use the inverse probability function in Excel, i.e., =NORM.S.INV(0.1), to obtain a Z score of -1.282. We then compute x* as:

$$\frac{x^* - 5}{1} = -1.282 \rightarrow x^* = 3.72$$

That is, 10% of our visitors spend less than 3.72 minutes on the site.

A more direct way to obtain this result is to use the =NORM.INV() function, which does not require the conversion to a Z score: =NORM.INV(0.1,5,1) returns the value of 3.72.

Unit 2 Summary

- When a variable X is normally distributed, we denote it as:
$$X \sim N(\mu, \sigma)$$

- The **standard normal distribution** is the distribution of Z, which is normally distributed with a mean of zero and a standard deviation of one:
$$Z \sim N(0,1)$$

- Any normal distribution can be converted into the standard normal distribution using the following conversion:
$$Z = \frac{X - \mu}{\sigma}$$

- Z has a special meaning that is important to understand: For each value of X, the corresponding **Z score** is a measure of how far *in terms of number of standard deviations* this value is from the mean.

- The **p-value** is a term used to refer to the probability of a value falling *farther away from the mean* than a specific z value:
$$P(Z \geq |z|)$$

- We can use **inverse probability calculations** to find Z score values that correspond to specific probabilities.

Unit 2 Exercises

1. The eight variables in Table 6-e1 are normally distributed with the means and standard deviations shown in the table.

    a. Use the =NORM.DIST function to find the probability P(X≤x).

    b. Compute the Z scores corresponding to the specific x values in Column 3.

Table 6-e1

μ	σ	x	P(X≤x)	Z Score
6	0.5	5		
15	5	18		
15	5	21		
40	10	30		
5	0.4	6.2		
100	25	150		
1000	138	743.5		
10	3	10		

2. Find the following probabilities using Excel:

   a. P(Z≤-2.06)

   b. P(Z≥-0.69)

   c. P(Z≤3.8)

   d. P(-2.61≤Z≤-2.48)

   e. P(Z=1.44)

   f. P(Z≤-2.57)

   g. P(Z≥-2.21)

   h. P(Z≥3.09)

   i. P(-1.32≤Z≤2.05)

   j. P(0.81≤Z≤1.68)

   k. P(Z≥0.85)

   l. P(Z≤-2.13 or Z≥3.06)

   m. P(Z≤0.01)

   n. P(-0.64≤Z≤0.96)

   o. P(0.89≤Z≤1.34)

   p. P(Z≤-1.96 or Z≥1.96)

3. Find the p-value for the following Z scores:

    a. Z=2.3

    b. Z=-2

    c. Z=1.645

    d. Z=-1.96

    e. Z=0

    f. Z=4

    g. Z=-1.5

    h. Z=-3

4. Find the Z score values that correspond to the following probabilities:

    a. $P(Z \leq z^*) = 0.1$

    b. $P(Z \leq z^*) = 0.5$

    c. $P(Z \leq z^*) = 0.02$

    d. $P(Z \leq z^*) = 0.9$

    e. $P(Z \geq z^*) = 0.3$

    f. $P(Z \geq z^*) = 0.6$

    g. $P(Z \geq z^*) = 0.95$

    h. $P(Z \geq z^*) = 0.05$

    i. $P(-z^* \leq Z \leq z^*) = 0.68$

    j. $P(-z^* \leq Z \leq z^*) = 0.5$

    k. $P(-z^* \leq Z \leq z^*) = 0.99$

5. In the example of the Internet art store, consider the variable representing time spent on the website. We mentioned that this variable follows a normal distribution with a mean of five minutes and standard deviation of one

minute. If a customer is chosen at random, what is the probability that they spend more than three minutes on the site?

6. Weekly demand for a popular art piece offered by the Internet art store is normally distributed with a mean of 50 and standard deviation of five. The store usually keeps 60 pieces in stock and re-orders them once a week.

    a. What is the probability that average demand will not be met on a given week?

    b. How many of this popular art piece should you keep in inventory to ensure that you do not run short of demand more than 10% of the time?

7. IQ is normally distributed with a mean of 100 and a standard deviation of fifteen. Answer the following questions related to IQ:

    a. *MENSA* is an organization that accepts people who obtained a score on an IQ test that is within the upper two percent of the population. What is the cut-off IQ score for *MENSA*?

    b. A gifted child has an IQ of 130 or higher. What is the probability that a randomly selected child is gifted?

    c. If you have a sample of 40 children, what is the probability that four of these children are gifted? (Hint: What is your variable? Is it still normally distributed?)

END-OF-CHAPTER PRACTICE

1. Prior to the 1930s, airports used a two-letter abbreviation. Around 1947, the aviation industry was growing so fast that three letters were needed and *Los Angeles Airport* (LA) became LAX. The letter X has no specific meaning in this identifier, but LAX became one of the busiest airports in the world. The daily number of passengers at LAX is normally distributed with a mean of 163,000 and a standard deviation of 10,000.[52]

    a. What is the probability of LAX having over 175,000 passengers on a given day?

    b. What percentage of the time does LAX get between 148,000 and 178,000 daily passengers?

---

[52] Source: http://en.wikipedia.org/wiki/Los_Angeles_International_Airport#The_.22X.22_in_LAX

2. Every year, over 9 million Americans are victims to identity theft. The theft amount per person is normally distributed with a mean of $6,383 and a standard deviation of $2,000.[53] Suppose you are a victim of identity theft:

   a. What is the probability that you were robbed of less than $5,000?

   b. You are told your theft incident is in the top 1.5% of thefts in terms of amount stolen. What does this mean?

3. 3% of adults aged 25 to 29 send over 200 text messages per day.[54] If we know that the number of daily text messages sent by adults 25 to 29 is normally distributed with a standard deviation of 50, what is the mean number of daily text messages?

4. Suppose you are working on a pitch for a new project involving multiple hotel properties, and you need to convince your company's investment board that the investment will pay off. Based on company records, the board typically approves projects with at least an 85% chance of profit. If the profit associated with projects is normally distributed with a mean of $1,000,000 and standard deviation of $50,000, what must the minimum projected profit be for approval?

5. Let us take two of the top ten most expensive cities in the world, Tokyo and Moscow, and compare their rent prices. The average cost of a monthly two-bedroom living space (in $US) is $4,436 in Tokyo and $3,600 in Moscow.[55] Suppose rent prices are normally distributed with standard deviations of $200 for Tokyo and $320 for Moscow. Is there a larger probability of rent in Tokyo costing over $4,538 or rent in Moscow costing over $3,740?

6. With climate change becoming a bigger concern, there is an emphasis on carbon emissions from automobiles. The average car burning one gallon of gas produces twenty cubic pounds of CO_2.[56] If carbon dioxide emissions are normally distributed and it is known that 99% of cars produce more than one cubic pound of CO_2, then:

   a. Find the standard deviation of carbon emissions.

   b. Find the proportion of cars emitting between 20 and 35 cubic pounds per gallon of gas burned.

---

[53] Source: http://visual.ly/identity-theft-facts-and-figures
[54] Source: http://visual.ly/adults-and-mobile-phones
[55] Source: http://visual.ly/most-expensive-cities
[56] Source: http://visual.ly/automobiles-and-environment

7. The number of pitches thrown per game by the starting pitcher is normally distributed with a mean of 85 and a standard deviation of ten. The manager decides to take out the starter once he has thrown 100 pitches. What is the probability of this occurring for a randomly selected game for which this pitcher is the starting pitcher?

8. The *Deepwater Horizon* oil spill in April 2010 was one of the largest petroleum accidents in history. It lasted 87 days, claimed eleven lives and has spawned over 130 lawsuits. The average total discharge was estimated at 4.9 million barrels of oil. Assuming that the estimate of total discharge is normally distributed with a standard deviation of 0.16 million barrels,[57] what is the probability of the total discharge being under 4.5 million barrels?

9. In Ontario, the per capita household debt in 2012 was normally distributed with a mean of $17,621 and a standard deviation of $3,000.[58] What is the probability of a household having a debt of between $12,000 and $15,000?

10. The time to fully charge an *Android*-based phone is normally distributed with a mean of 45 minutes and a standard deviation of five minutes.

    a. What is the probability that an *Android*-based phone would be fully charged in less than 30 minutes? Less than 45 minutes? Less than 60 minutes?

    b. What is the probability that it would take more than 50 minutes to fully charge the phone?

    c. What percentage of phones would be fully charged in less than 55 minutes?

11. The lifespan of a refurbished *iPhone 4s* is normally distributed with a mean of 24 months and a standard deviation of one month.

    a. What is the probability that a refurbished *iPhone 4s* you purchased today will last more than two years?

    b. What is the probability that it would last six weeks less than the mean?

    c. What is the probability that it would *die* exactly 730 days from now?

    d. What percentage of refurbished phones would make it past 26 months?

---

[57] Source: http://www.uscg.mil/foia/docs/dwh/fosc_dwh_report.pdf
[58] Source: http://visual.ly/canadian-debt-infographic-economy-debt-and-you

CHAPTER 7: Introducing Hypothesis Testing

We often use statistics to test theories. A theory is a prediction, or a group of predictions, about how people (e.g., individuals, groups, communities, etc.), physical entities (e.g., weather elements, geological elements, chemical elements, etc.), and built devices (e.g., machinery, buildings, computers, etc.) behave. For instance, I can theorize that the satisfaction a person derives from using a purchased product depends on whether or not her pre-purchase expectations were met (referred to as *Expectations Confirmation Theory*) or I can theorize that people are more reluctant to making risky decisions when they face potential losses than when they face potential gains (referred to as *Prospect Theory*).

Theories begin as predictions, which are then repeatedly tested in various settings to either strengthen or refute them. Such testing often involves *statistical inference*, defined as the drawing of conclusions about a population of interest based on findings from samples obtained from that population. For example, I can test the link between expectations confirmation and customer satisfaction by using a sample of 50 shoppers who have purchased some product. If I find sufficient statistical evidence supporting the existence of this link, I can *infer* that the link exists for the population of shoppers of this product. Through repeated testing in different settings, on different products and with different samples, a theory is strengthened when statistical results support the theory and weakened when statistical results refute the theory.

As with customer satisfaction, statistics is very much about expectations. Recall from Chapter 1 that we distinguished between a *population* and a *sample*, with the population being the full group we are interested in learning about (e.g., all human beings, all residents of a city, all lines of code in a software program, etc.), and a sample being a thoughtfully-selected subset of this population. With statistical inference, we aim to test specific expectations that we have about the population's parameters using sample statistics.[59] We call these expectations *hypotheses*.

A hypothesis is a specific claim about the world that we wish to support (or refute). Here is an example, taken from an article in the *Wall Street Journal*.[60] Based on the article, companies now need an average of 23 days to screen and hire new employees. Assume that you work at the human resources department of a large company and your manager had just read this article. She believes this number is too low and that companies actually spend more than 23 days on the hiring process. It is important for her to determine if the article's claim is true, because she is sure to hear about it from her clients. Thus, she *hypothesizes* that the average company spends more than 23 days screening and hiring employees for a position. Next, she asks you to collect data and to test the *Wall Street Journal*'s claim. This process is called *hypothesis testing* and it is at the heart of inferential statistics. In explaining hypothesis testing in this book, we will differentiate between the logic of the test, which is applicable to many variables and cases, and the actual process of testing, which may vary based on the specific parameters and relationships of interest to us.

---

[59] Recall from Chapter 1, that a *parameter* was defined as an attribute of the population (e.g., the population's mean), while a *statistic* is a measurement obtained from sample data (e.g., the sample's mean). *Sampling error* reflects the difference between these values.
[60] Jan. 19, 2016, "How to Deal with a Long Hiring Process".

This introductory chapter provides the foundations needed to understand the logic of hypothesis testing, while future chapters will focus on describing different types of hypothesis tests and the application of these different tests. This chapter covers the following topics:

- Stating Hypotheses
- Properly Formulating Hypotheses
- Sampling Distribution of the Mean
- The Central Limit Theorem
- Working with the Sampling Distribution of the Mean
- Solving Problems by Integrating Concepts from Multiple Chapters
- The Logic of Hypothesis Testing
- Type I and Type II Errors

UNIT 1

Stating Hypotheses

A ***hypothesis*** is some specific claim that we wish to test. We differentiate between the *research hypothesis* and the *null hypothesis*. Let us start with a very simple example. A bank manager argues that, on average, people carry $50 or more in their wallet. This claim is the ***null hypothesis***. The ***research hypothesis*** contains the other side of this claim, that is – that people carry less than $50. We can also write it as

H_0: Average amount of money ≥ $50

H_1: Average amount of money < $50

where H_0 is used to notate the null hypothesis and H_1 (sometimes denoted H_A) is used to notate the research hypothesis, commonly referred to as the **alternative hypothesis**.

When conducting research, there are two claims that you are testing side-by-side: one is the existing status quo, which is the null hypothesis. The other is what you (the researcher) believe is true, which is your research, or alternative, hypothesis.

Here is another example. A survey finds that, on average, people have 200 or fewer friends on *Facebook*. However, you believe that *Facebook* users, in fact, have more than 200 friends. You can set your hypotheses as

H_0: Average number of friends ≤ 200

H_1: Average number of friends > 200

In this case, you are setting the alternative hypothesis to reflect what you believe to be true.

Let us look at one more example. A statistics professor wants to know if her section's grade average is different than that of other sections of the course. The average for all other sections is 75. To help the professor learn if her section's grade average is different than that of other sections, we need to set up the following hypotheses:

H_0: Section's grade average = 75

H_1: Section's grade average ≠ 75

Note that in all of the above examples, our hypotheses focused on some *average value*. We hypothesized about the average amount of money in people's wallet, about the average number of friends on *Facebook*, and about the average section grade performance in statistics courses. Hypothesis testing is not limited to testing means (averages). We can also hypothesize about other population parameters, such as standard deviations and proportions. For now, however, we will stay with the basic example of hypothesis testing of a population mean, introducing other parameters in later chapters.

Finally, note that there are three different sets of hypotheses in the above examples, as summarized in Figure 7-1. Comparing these three sets, you will see that the difference lies in the direction of the alternative hypothesis. The first set (the left part of Figure 7-1) tests whether the average is *lower than* a specific value. Hence, it is called a **lower-tail test**. The second set tests whether the average is *higher than* a specific value and is called an **upper-tail test**. Finally, the last set of hypotheses tests whether the average is simply *different than* a specific value (either higher or lower) and is called a **two-tail test**. We will explain the meaning and execution of these three tests later in this chapter.

Figure 7-1
Three Types of Hypothesis Tests

Lower-Tail Test
H_0: Average amount of money $\geq$ $50
H_1: Average amount of money < $50

Upper-Tail Test
H_0: Average number of friends $\leq$ 200
H_1: Average number of friends > 200

Two-Tail Test
H_0: Section average = 75
H_1: Section average $\neq$ 75

Properly Formulating Hypotheses

We have so far defined the null and alternative hypotheses. When we test hypotheses, our logic is as follows: we assume the null hypothesis to be true and then we try to refute it. Therefore, whatever claim you are looking to disprove should become the null hypothesis and the claim you would like to support will be your alternative hypothesis. Mathematically, this means that the equal sign (=, ≤ or ≥) should always be in the null hypothesis, with the respective alternative hypotheses being: ≠, > or <.

We now discuss two important conditions: hypotheses should be (1) *exhaustive* and (2) *mutually exclusive*. **Exhaustive** means that the hypotheses should cover every possible option. Consider again the opening example for this chapter about a company spending, on average, 23 days on its hiring process. The *Wall Street Journal*'s claim is the null hypothesis, whereas the human resource manager's belief (i.e., that the actual time companies spend on hiring (the population parameter) is higher) is the alternative hypothesis. Therefore, we write the hypotheses as:

H_0: Average days on hiring process = 23

H_1: Average days on hiring process > 23

However, the above set of hypotheses does not cover every possible option. Specifically, we have left out the *average days on hiring process < 23* option. To write the hypotheses properly, they need to be exhaustive[61], meaning:

H_0: Average days on hiring process ≤ 23

H_1: Average days on hiring process > 23

The other required condition for hypotheses is that they *are **mutually exclusive***, meaning there is no overlap between them. The following hypotheses

---

[61] You may find books and articles that use non-exhaustive hypotheses (H_0: Average days on hiring process = 23). There is an implicit assumption in using this partial form that the *< 23* option is irrelevant, given that your aim is to test only the *> 23* claim. We put this as a note for now, as it will become more clear later in the chapter.

are, therefore, *incorrect* because the = *23* option is in both the null and alternative hypotheses:

$$H_0: \text{Average days on hiring process} \geq 23$$

$$H_1: \text{Average days on hiring process} \leq 23$$

Unit 1 Summary

- **Hypotheses** are claims we wish to test and are formulated as statements about the parameter of interest (so far we have only discussed the mean). A hypothesis represents your belief about the value of that parameter.

- The **null hypothesis** is the one you wish to disprove and the **alternative hypothesis** (or **research hypothesis**) is the one *you believe* to be true. When setting up the hypotheses, remember this: you will never be able to *accept* (or *prove*) the null hypothesis. The only result from the hypothesis test will be to either *reject* or *do not reject* the null hypothesis. Mathematically, this means that the *equal sign should always be in the null hypothesis*.

- Hypotheses can take one of three forms that correspond to three types of tests:
    - **Lower-tail test**: the alternative hypothesis includes the *less than* relationship
    - **Upper-tail test**: the alternative hypothesis includes the *greater than* relationship
    - **Two-tail test**: the alternative hypothesis includes the *not equal* relationship

- Hypotheses should be **exhaustive**, covering every possibility in the test domain, and **mutually exclusive**, allowing no overlaps in the hypotheses.

Unit 1 Exercises

For each of the scenarios below, write the null and alternative hypotheses.

1. *Angry Birds* used to be one of the most popular games on various gaming platforms. According to one source, people collectively spent 300 million minutes per day playing *Angry Birds*;[62] and, a different source claims it cost the US economy billions of dollars in lost work time.[63] Suppose you believe this number (300 million minutes) was too high of an estimate. You conduct

---

[62] Source: http://www.ibtimes.com/articles/233613/20111018/angry-birds-stats-ipo-rovio-entertainment-andrew-stalbow-peter-vesterbacka-david-maisel.htm
[63] Source: http://www.geek.com/articles/games/businesses-are-losing-1-5-billion-due-to-employees-playing-angry-birds-on-the-job-20110914/

some more research to find out that, at its peak, there were roughly 30 million daily active users of *Angry Birds*, which means that, according to the claim, on average each player spent ten minutes per day (300 million minutes divided by 30 million users) playing *Angry Birds*. You plan to use a sample of current *Angry Birds* players at your school to test this claim. Set up the hypotheses you would use.

2. *C25K* is a running program designed to get people from the couch to running five kilometers in nine weeks.[64] Before signing up, you decide to test the effectiveness of the program. You assemble a sample of 50 adults (non-runners) who have signed up for the program. You measure how far they can run at the end of the nine weeks (Hint: first identify your parameter of interest and then think of what level of accomplishment with this parameter would indicate that the running program had accomplished its goal).

3. A dissatisfied customer can be harmful to any business. A consumer group claims that an average dissatisfied customer will tell between nine and fifteen people about their negative experience. The manager at the *We Don't Care Comic Book Store* thinks that even nine people is too high of an estimate. He believes that dissatisfied customers only tell a few friends about their experience. State the hypotheses to test the manager's belief.

4. The assistant manager at the *We Don't Care Comic Book Store* is actually quite concerned about dissatisfied customers. He knows that many customers will simply post the experience on their *Facebook* page, reaching a number much higher than fifteen people. Set up the hypotheses to test his concern that the average dissatisfied customer will let more than fifteen people know about the experience.

5. The data in Table 7-e1 is taken from the website of the *American Bureau of Labor Statistics*[65] and it describes time use on an average weekday for full-time university and college students:

---

[64] Source: http://www.c25k.com/
[65] Source: http://www.bls.gov/tus/charts/

Table 7-e1
Students' Time Use

Activity	Hours
Sleeping	8.7
Leisure and Sports	4.1
Educational Activities	3.3
Working	2.4
Other	2.3
Traveling	1.4
Eating and Drinking	1.0
Grooming	0.8
Total	24.0

Using Table 7-e1 as your guide for the three types of tests (lower-tail, upper-tail, and two-tail), write the hypotheses for the following tests:

a. A two-tail test for the average number of sleep hours

b. An upper-tail test for the average number of hours spent grooming

c. A lower-tail test for the average number of hours spent on educational activities

d. A two-tail test for the average number of hours spent eating and drinking

e. A two-tail test for the average number of hours spent on leisure and sports

f. An upper-tail test for the average number of hours spent working

UNIT 2

Sampling Distribution of the Mean

As described in the introduction to this chapter, in hypothesis testing we have a certain expectation, or hypothesis, about the value of a parameter of a population of interest and we aim to test this expectation using a sample of collected data. In this chapter, specifically, the parameter we are looking at is the population mean (μ) and the corresponding statistic is the sample mean ($\bar{X}$). In order to use the sample mean to infer conclusions about the population mean, we need to know more about the possible values that our sample mean may obtain and about their probabilities. Let us first use an example to demonstrate *why* we need to know more about the sample mean, and then move on to explain *how* we find this information.

Suppose that we are considering whether or not to make an investment in a used textbook store and are interested in determining the amount of money that customers are likely to spend when shopping at the store. Based on historical data provided by the current owner, the average customer making purchases at the store spent $100 with a standard deviation of $35. We are concerned that sales may have decreased over the last year (because of people's increasing preference for digital books rather than paper-based books) and would like to test this concern using hypothesis testing. The variable of interest in this test is average spending and we would like to find out if spending has decreased. Consequently, we set up our hypotheses as follows:

H_0: Average spending $\geq$ 100

H_1: Average spending $<$ 100

The above hypotheses can also be written as:

H_0: $\mu \geq 100$

H_1: $\mu < 100$

with μ being the hypothesized average spending (the parameter of interest) for the population (all customers who have made purchases at the store).

After setting up the hypotheses, we collect data on twenty random customers and find that the mean spending in our sample is $93.27. At the outset, it is easy to see that our sample mean is indeed lower than the hypothesized mean. But, should we reject the null hypothesis based on this one sample? After all, it is common for sample means taken from the same population to differ slightly from each other and from the population mean. The question we need to ask ourselves is: how possible is it for us to take a random sample of size n=20 out of a population with a mean of $100 and a standard deviation of $35 and obtain a sample mean of $93.27? If such an outcome is reasonably possible, then we have no reason to reject the null hypothesis and we can attribute the difference to sampling error. However, if the chances of finding this sample mean are extremely low, then we should conclude that the null hypothesis is likely false and we should reject it - as we have confirmed our belief that the *true* average spending is likely lower than $100. So, just how possible is it that we could obtain a sample mean of $93.27 given a population mean of $100 and population standard deviation of $35?

In order to answer the question, we need to know more about the possible values that our sample mean may obtain and their probabilities. Specifically, we need to know the *sampling distribution of the mean*. Recall from Chapters 5 and 6 that a probability distribution describes the range of values that a variable may obtain and their corresponding probabilities. We can construct a similar distribution for the sample mean. A **sampling distribution** refers to the probability distribution of any sample statistic. The **sampling distribution of the mean** describes the probabilities attached to all values of the mean of samples that are repeatedly taken from the same population. Let us continue with our used textbook store example to better understand this concept.

Based on the owner's past data, we believe that the average store customer spends $100 with a standard deviation of $35. Recall that we first took a random sample of twenty customers (n=20), then measured how much each of the customers spent, and finally computed the mean spending for that sample to be $93.27.

Now, let us repeat this sampling procedure 200 times, each time taking a different sample of twenty customers (n=20), measuring how much each of the customers spent and computing the mean spending for each sample. The results of this procedure are partially shown in Table 7-1[66] and are then graphed in Figure 7-2. This chapter's Excel file provides the data used to create this example, in the worksheet titled 'Spending Example'. At the bottom of Table 7-1, we compute the *mean of means*; that is, we averaged the means of the 200 samples, as well as computed the standard deviation of these 200 sample means. Figure 7-2 provides a graphical presentation of the data using a histogram (remember that we learned how to create histograms in Chapter 2). The x-axis displays classes (groups) of spending amounts and the y-axis provides the frequency of each class.

---

[66] To create Table 7-1, we averaged the twenty responses obtained in the first sample of customers and recorded the number ($93.27) under *Sample 1*. We then averaged the twenty responses obtained in the second sample and recorded the number ($104.34) under *Sample 2*, and so on.

Table 7-1
Sample Means of 200 Spending Samples

Sample (n=20)	Sample Mean ($\bar{X}$)
1	$93.27
2	$104.34
3	$100.49
4	$104.77
5	$96.82
6	$84.00
7	$92.53
8	$96.51
9	$94.14
10	$108.35
...	...
...	...
...	...
190	$97.80
191	$111.12
192	$100.98
193	$96.69
194	$81.69
195	$112.92
196	$108.38
197	$81.31
198	$76.97
199	$96.65
200	$111.65
Average over 200 samples:	*$99.04*
Standard Deviation:	*$11.14*

**Figure 7-2
Histogram of Means of 200 Spending Samples**

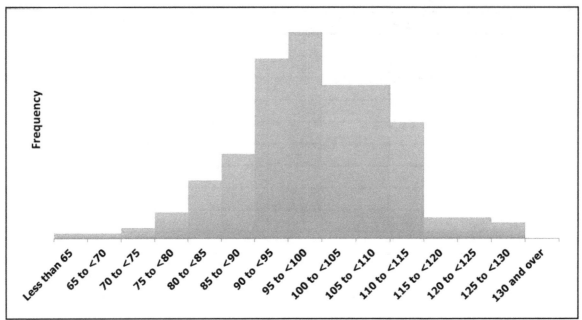

Figure 7-2 depicts the distribution of values of the means of samples of size n=20 that are repeatedly taken from the same population. As can be seen from Table 7-1, the sample means differ (to varying extents) from one another. The *sampling distribution of the mean* is created in the same manner followed in creating Figure 7-2, but accounting for *all possible values* of the mean (our example above was limited to only those values obtained in these 200 samples). The general theorem that guides us in understanding the sampling distribution of the mean is called the *central limit theorem*.

The Central Limit Theorem

The **central limit theorem** states that the sampling distribution of the mean of a random sample of any size (n) drawn from a normally distributed population[67] also follows a normal distribution with a mean of μ and a standard deviation of $\frac{\sigma}{\sqrt{n}}$. This standard deviation ($\frac{\sigma}{\sqrt{n}}$) is called the **standard error of the mean**.

Returning to the used bookstore spending example, we know that X, the random variable representing the amount spent by a single customer, is normally distributed with a mean of $100 and a standard deviation of $35, or

$$X \sim N(100, 35)$$

Next, the central limit theorem tells us that the *mean* ($\bar{X}$) of a *sample* of size n taken from this population of customers is also normally distributed with a mean of 100 and a standard deviation of $\frac{35}{\sqrt{n}}$. In our example with n=20:

$$\bar{X} \sim N(100, \frac{35}{\sqrt{20}}) \quad \text{or} \quad \bar{X} \sim N(100, 7.83)$$

You can refer back to our example data for an illustration of the above. Figure 7-3 was constructed in a manner similar to Figure 7-2, but this time we repeated the sampling procedure 1,000 times. The distribution illustrated in Figure 7-3 has a mean (the mean of all the sample means) of $99.83 and a standard deviation of $7.73. These numbers are close to the mean of 100 and standard deviation of 7.83 obtained by using the central limit theorem.

---

[67] This is just one part of the theorem. Later in this section, we extend the discussion to samples taken from non-normal populations.

Figure 7-3
Histogram of Means of 1,000 Spending Samples

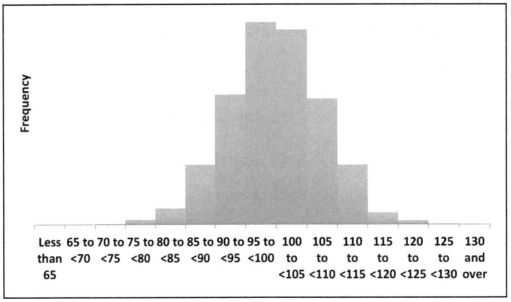

As we *both* increase further the number of samples that we take and reduce the class size of the frequency distribution, the shape of this sampling distribution will move closer and closer to being a continuous line and will eventually converge to that defined by the central limit theorem. Knowing about and applying the central limit theorem saves us from having to otherwise go through very complex calculations.

Following the logic of what has just been explained, if we know that a population's data values follow the normal distribution and if we know the population's mean and standard deviation, we can apply the central limit theorem to determine the sampling distribution of the mean. For example, if a different random variable Y follows a normal distribution with a mean of 50 and a standard deviation of 4, and we took a sample of size n=16, then we can use the central limit theorem to know that the sample mean ($\bar{Y}$) also follows a normal distribution with a mean of 50 but with a standard deviation of $\frac{4}{\sqrt{16}}$, or 1.

But, what if we know that this population *does not follow the normal distribution*? Let us consider a dice-rolling example to see what happens. When we roll a single six-sided die, we can obtain any one of the values from 1 to 6, each with a probability of 1/6. Defining a random variable as the outcome of a single die roll, this variable follows the uniform and discrete distribution (as introduced in Chapter 5) shown in Figure 7-4.

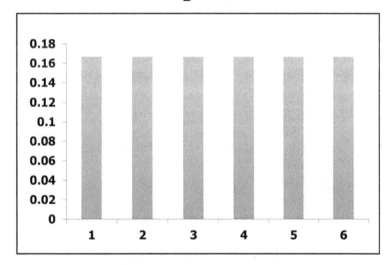

**Figure 7-4
Distribution of Single Die Roll Outcomes**

Now let us tweak our example a little. Instead of rolling a single die, we will roll *two* dice. And, instead of looking at the outcome on the individual die, we will define our random variable as the *average* of the values obtained from two rolled dice. For example, if one die rolled on 3 and the other on 5, then we will record the outcome of this roll as 4. In order to create the probability distribution for this new random variable, we need to define the sample space of the above experiment.

Remember, a sample space outlines all possible outcomes of an experiment. The sample space for the average of two rolled dice is detailed in Table 7-2.

Table 7-2
Sample Space for a Two Dice Roll

Die 1	Die 2	Average
1	1	1
1	2	1.5
1	3	2
1	4	2.5
1	5	3
1	6	3.5
2	1	1.5
2	2	2
2	3	2.5
2	4	3
2	5	3.5
2	6	4

Die 1	Die 2	Average
3	1	2
3	2	2.5
3	3	3
3	4	3.5
3	5	4
3	6	4.5
4	1	2.5
4	2	3
4	3	3.5
4	4	4
4	5	4.5
4	6	5

Die 1	Die 2	Average
5	1	3
5	2	3.5
5	3	4
5	4	4.5
5	5	5
5	6	5.5
6	1	3.5
6	2	4
6	3	4.5
6	4	5
6	5	5.5
6	6	6

Now we can create the probability distribution for the two dice roll average by counting the frequency of each value of the average. This is shown in Table 7-3 and in Figure 7-5. Now, compare Figures 7-4 and 7-5. The first thing to notice is that the distributions are not the same. This should not be surprising, given that each figure is associated with a distinct random variable: Figure 7-4 with the random variable representing the outcome of a single die roll and Figure 7-5 with the random variable representing the average of the values obtained from two rolled dice. Importantly, note that what we have actually done with Figure 7-5 is determined the *sampling distribution of the mean* of a sample size of n=2 (because we rolled two dice each time).

Table 7-3
Probability Distribution of the Average of Two Rolled Dice

Average	Frequency	Probability
1	1	1/36
1.5	2	2/36
2	3	3/36
2.5	4	4/36
3	5	5/36
3.5	6	6/36
4	5	5/36
4.5	4	4/36
5	3	3/36
5.5	2	2/36
6	1	1/36
	Σ=36	Σ=1

Figure 7-5
Probability Distribution of the Average of Two Rolled Dice

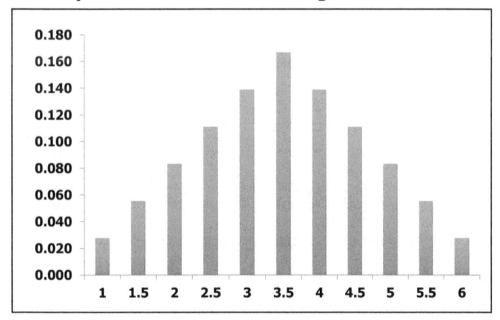

Now consider what happens as we increase the sample size to n=10 (rolling ten dice). As there are 6^{10} different combinations of rolling the ten dice, we will not even try to determine by hand the resulting probability distribution. Instead, we can take what we know about this calculation – the average of these rolls still ranges between 1 (if all dice fall on 1) and 6 (if all dice fall on 6) – and work out the distribution of the average of ten rolled dice with the help of a computer. Using increments of 0.1, Figure 7-6 shows a graph of the calculated probability distribution.

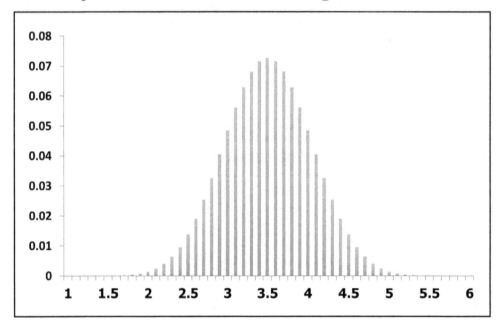

Figure 7-6
Probability Distribution of the Average of Ten Rolled Dice

Although Figure 7-6 still presents a discrete probability distribution (note that the columns do not touch each other), you should be able to detect a change in shape between Figure 7-5 and Figure 7-6, with the latter having a greater resemblance to the normal distribution. You can also imagine that by increasing the number of dice being rolled, the resulting probability distributions will become closer and closer to the normal distribution.

Indeed, the *central limit theorem* also tells us that *even when the population is not normally distributed*, the mean will follow an approximately normal distribution for *large enough* samples. What is meant by *large enough* depends on the distribution of the underlying population, with larger samples needed for populations that deviate to a greater extent from the normal distribution. However, a commonly-used rule of thumb is that a sample size of thirty (n≥30) is generally large enough.

Hence, regardless of the distribution of the underlying population, the sampling distribution of the mean of sufficiently large samples (n≥30) will follow an approximately normal distribution, such that:

$$\bar{X} \sim N(\mu, \frac{\sigma}{\sqrt{n}})$$

Figure 7-7 summarizes the above discussion of the central limit theorem.

Figure 7-7
The Central Limit Theorem

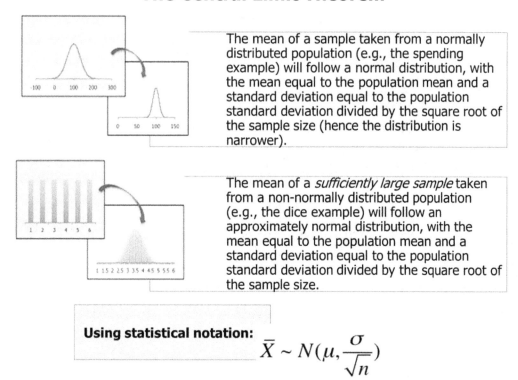

Working with the Sampling Distribution of the Mean

Before returning to the process of conducting hypothesis tests, it is important to ensure that the use of sampling distributions is clear. Consider the following example: commuting to work is a problem many Americans face on a daily basis. Indeed, *Forbes* magazine has published interesting data on the ten cities with the worst commute and the highest percentage of mega-commuters in 2013 (a mega-commuter refers to someone traveling 90 minutes or more and at least 50 miles to get to work each day).[68] Topping the list were San Francisco, New York, and Washington D.C. More recently (in 2015), *USA Today* published similar data, identifying Washington D.C. as the most congested city in the country.[69] Suppose that X is a random variable measuring commute time, in minutes, in Washington, D.C. Assume that X is normally distributed with a mean of 37 minutes and standard deviation of ten minutes. What is the probability that a randomly sampled commuter travels more than 45 minutes to work?

In order to answer this question, we need to go back to our discussion of continuous probability distributions in Chapter 6 and compute:

$$P(X > 45) = ?$$

Since we know that X is normally distributed, we can compute a Z score and find this probability:[70]

$$P(X > 45) = P\left(Z > \frac{45 - 37}{10}\right) = P(Z > 0.8) = 0.21$$

---

[68] Source: http://www.forbes.com/sites/jennagoudreau/2013/03/05/cities-with-the-most-extreme-commutes/
[69] Source: http://www.usatoday.com/story/money/2015/09/04/24-7-wallst-10-cities-worst-traffic/71701622/
[70] We can use Excel in one of two ways to compute this probability: '=1-NORM.S.DIST(0.8)' once we compute our Z score, or using the original distribution '=1-NORM.DIST(45,37,10,1)'.

That is, there is a 21% chance that a person who works in Washington, D.C. will commute more than 45 minutes to work.

Now consider a different question: what is the probability that in a sample of ten commuters the *sample mean* of commute time will be greater than 45 minutes? Although the wording of this question is similar to the one above, a key difference is that we are not asking about the commute time of a single person, but about the mean commute time of a sample of ten commuters. To answer this question, we need to apply the central limit theorem, which tells us that:

$$\bar{X} \sim N(37, \frac{10}{\sqrt{10}})$$

We are looking for:

$$P(\bar{X} > 45) = ?$$

which we can compute as:

$$P(\bar{X} > 45) = P\left(Z > \frac{45 - 37}{\frac{10}{\sqrt{10}}}\right) = P(Z > 2.53) = 0.0057$$

That is, there is less than a 1% chance of the *sample mean* of commute time being greater than 45 minutes.

Let us review these two commute time examples side-by-side to discuss some important differences.

Table 7-4
Comparison of Commute Time Examples

	First Commute Time Example	*Second Commute Time Example*
Interested In	A single value	Sample mean
Looking for	$P(X > 45) = ?$	$P(\bar{X} > 45) = ?$
Distribution	$X \sim N(37, 10)$	$\bar{X} \sim N(37, \frac{10}{\sqrt{10}})$
Number of Standard Deviations Above the Mean (Z)	0.8	2.53
Probability	0.21	0.0057

As shown in Table 7-4, in the first example we looked for the probability of a specific value falling above the mean of the distribution. In the second example, we looked for the probability of an observed mean of a specific sample falling above the distribution's mean. The most notable difference is how much smaller the probability was in the second example (less than 1% compared with 21%). The reason is simple: the standard deviation of the sampling distribution of the mean is much smaller than that of the original random variable. This is because sample means cluster more closely around the distribution's mean than do individual values.[71]

Table 7-5 explains when to use the distribution of a single variable and when to use the sampling distribution of the mean.

---

[71] Take some time to reflect on this and try to verbalize to yourself why the variance of sample means is smaller than that of individual values.

Table 7-5
When to Use the Distribution of a Single Variable and
When to Use the Sampling Distribution of the Mean

Use the Distribution of X to Answer:	Use the Sampling Distribution of $\bar{X}$ to Answer:
What is the probability that a specific value of X is greater than ...?	What is the probability that the mean of a sample of size n is greater than ...?
What is the probability that a specific value of X is less than ...?	What is the probability that the mean of a sample of size n is less than ...?
What is the probability that a specific value of X is between ... and ...?	What is the probability that the mean of a sample of size n is between ... and ...?

Solving Problems by Integrating Concepts from Multiple Chapters

In the commute time example just used, we applied knowledge gained from a prior chapter (i.e., the discussion of continuous probability distributions from Chapter 6). As the book progresses, we will increasingly be relying on what has been covered in previous chapters. Thus, it is very important that you understand the concepts covered in earlier chapters when tackling the content of later chapters. To illustrate this, consider the next example.

The average ice time per shift[72] for NHL players is 44.07 seconds, with a standard deviation of 3.86 seconds.[73] Assuming that ice time follows a normal distribution, let us define X as a random variable representing ice time per shift, such that X~N(44.07, 3.86). Can you compute the proportion of players who averaged over 48 seconds of ice time per shift? The key first step in answering any statistics question is to understand exactly what you are being asked to determine. Here, we are asked to find a proportion, or a *percentage* of players. If X is our random variable representing ice time per shift (measured in seconds), then we want to know what percent of X is greater than 48. In other words:

---

[72] Players generally come on and off the ice during a game. The average ice-time per shift refers to the average amount of time they spend on the ice between substitutions.
[73] Source: http://www.sportingcharts.com/nhl/stats/average-ice-time-per-shift/2015/

$$P(X > 48) = ?$$

We can solve this as:

$$P(X > 48) = P\left(Z > \frac{48 - 44.07}{3.86}\right) = P(Z > 1.0182) = 0.1543$$

Thus, we have determined that around 15% of *NHL* players averaged more than 48 seconds per shift on the ice. Again, we are using what we learned previously in Chapter 6 to answer this question.

Now, let us consider a related, but different, question: What is the probability that a sample of 50 players has a mean ice time over 48 seconds? Here we are not asked about the ice time of a single player, but rather about the mean ice time of a sample of 50 players. First, we will use the central limit theorem to find the sampling distribution of the mean:

$$\bar{X} \sim N\left(44.07, \frac{3.86}{\sqrt{50}}\right)$$

Then, we compute the sought probability as before:

$$P(\bar{X} > 48) = P\left(Z < \frac{48 - 44.07}{\frac{3.86}{\sqrt{50}}}\right) = P(Z > 7.1993) \cong 0$$

We have determined that it is highly unlikely (nearly zero probability) to find a sample mean (after sampling the ice time per shift of 50 randomly-selected *NHL* players) of more than 48 seconds.

Finally, consider a very different question. Suppose that we sampled five players. What is the probability that all of these players averaged over 48 seconds of ice time per shift? This is a new question that we have not encountered before, so let us make sure we understand exactly what we are being asked to determine.

We are asked about the probability that in our sample of five players *each* averages over 48 seconds of ice time per shift. There is an element of success in this question and this indicates that we are working with a *binomial distribution*. We are looking for five successes (x) in five trials (n), and the only piece of information absent is the probability of a single success (p). We can find the probability of a single player averaging over 48 seconds of ice time per shift. In fact, we computed it above:

$$P(X > 48) = P\left(Z > \frac{48 - 44.07}{3.86}\right) = P(Z > 1.0182) = 0.1543$$

Thus, 15.43% is the probability of finding a single player averaging over 48 seconds of ice time per shift. Next, we can use what we learned from Chapter 5 (i.e., the binomial probability distribution) to continue solving this question: we are looking for five successes in five trials, with the probability of success being 15.43%:

P(each of five players with over 48 seconds of ice time per shift) = 0.1543^5 = 0.0000875.

Thus, there is less than a 1% probability that every player in a sample of five players gets over 48 seconds of ice time per shift.

A final and very important comment to be made about this example is that we have made the assumption that players' ice times per shift are independent of each other, which is one of the conditions for working with the binomial distribution. That is, we have assumed that the probability of one player averaging over 48 seconds of ice time per shift will have no effect on the probability of other players averaging over 48 seconds of ice time.

The above examples provide a range of probability-related questions that can be answered by combining the knowledge gained in Chapters 5 through 7. The important things to remember when solving such questions are to: (1) carefully read a question to identify the parameter of interest, the distribution that is involved, and the specific probability that you are being asked to find; and, (2) select an appropriate technique (or set of techniques) for analyzing problems involving this parameter, this distribution and this probability.

Unit 2 Summary

- A **sampling distribution** refers to the probability distribution of a sample statistic. It provides information about possible values of the sample statistic and their corresponding probabilities.

- The **sampling distribution of the mean** describes the probabilities attached to all values of the mean of samples that are repeatedly taken from the same population.

- The **central limit theorem** states that the sampling distribution of the mean of a sufficiently large sample size (generally, n≥30) drawn from any population is approximately normally distributed with a mean of μ and a standard deviation of $\frac{\sigma}{\sqrt{n}}$. As a special case, if the underlying population follows a normal distribution, then the mean of a sample of any size will also be normally distributed.

- The standard deviation ($\frac{\sigma}{\sqrt{n}}$) of the sampling distribution of the mean is called the **standard error of the mean**.

Unit 2 Exercises

1. Based on the central limit theorem, what can you say about the sampling distribution of the means of the following samples?

    a. A sample of size 50 taken from a normally distributed population with a mean of 40 and a standard deviation of ten?

    b. A sample of size 50 taken from a non-normally distributed population with a mean of 40 and a standard deviation of ten?

c. A sample of size five taken from a normally distributed population with a mean of 25 and a standard deviation of one?

d. A sample of size ten taken from a normally distributed population with a mean of 1,000 and a standard deviation of 500?

e. A sample of size seven taken from a non-normally distributed population with a mean of three and a standard deviation of 0.2?

f. A sample of size 100 taken from a non-normally distributed population with a mean of three and a standard deviation of 0.2?

2. Use Excel to simulate and graph the sampling distribution of the average of rolling five dice 500 times. Use the following information to help you, as needed:

   a. You can treat each column in Excel as a single die and each row as a single roll. Hence, if you wish to roll five dice 500 times, you will need a 5 by 500 matrix.

   b. Use the =RANDBETWEEN(1,6) function to generate random integers between one and six, simulating dice. (Note that the values generated by the function change with each change to the worksheet. If you wish to keep them constant, you will need to copy them, then choose 'Paste Special', and paste them as 'Values Only'.)

   c. Compute the average for each set of five rolls (row averages).

   d. Create a frequency distribution of these row averages.

   e. Use the column chart to graph your distribution.

   f. What can you say about the distribution of the averages of rolling five dice 500 times? How will the graph change if you used 10 dice or 50 dice instead of just 5? Why?

   g. Continue working with the five dice data that you have created, but instead of 500 repetitions use 50 or 100 repetitions. How does the graph change? Why?

3. A random variable X is normally distributed with a mean of 500 and a standard deviation of 25. A sample of size n=100 was taken from this population. Find the following probabilities:

   a. $P(\bar{X} < 495) =$

b. $P(\bar{X} > 505) =$

c. $P(493 < \bar{X} < 497) =$

d. $P(X < 490) =$

e. Repeat the above four questions (a, b, c and d) using a sample size of n=25.

f. Repeat the above four questions (a, b, c and d) using a standard deviation of ten, i.e., X~N(500,10).

4. A random variable X is normally distributed with a mean of ten and a standard deviation of five. A sample of size n=4 was taken from this population.

   a. What is the probability that the sample mean is greater than twelve?

   b. Would your ability to answer the question change if you were told that X is not normally distributed? Why?

5. Table 7-e2 provides the mean and standard deviation of salaries of players in the *NHL, NFL*, and *MLB*:[74]

Table 7-e2
Player Salaries in the *NHL, NFL* and *MLB*

League	Mean Salary	Standard Deviation
NHL	$2.4 million	$1.1 million
NFL	$1.9 million	$0.38 million
MLB	$3.4 million	$1.3 million

For this question, assume that player salaries are normally distributed and independent of each other. Answer the following questions for each of these three leagues (that is, answer each question three times, once for each league):

a. What is the proportion of players paid over $9 million?

b. What is the probability that a sample of 50 players has a mean salary under $4 million?

---

[74] Source: http://visual.ly/does-money-buy-championships

c. What is the probability that in a sample of five players, all of them are paid under $4 million? (Assume player salaries are independent of one another in all three professional leagues.)

UNIT 3

The Logic of Hypothesis Testing

Now that we have an understanding of the sampling distribution of the mean, we return to hypothesis testing to better understand the logic of the testing procedure. We begin with an example.

A consumer group claims that wireless companies are worsening their customer service, with the time that customers are placed on hold when contacting the company reaching an average of 4.4 minutes, with a standard deviation of 54 seconds (or 0.9 minutes). You are a lobbyist for wireless providers and your clients believe that their customer service response times are much better than the average response time claim made by the consumer group. You are tasked with supporting your clients' position using hypothesis testing.

The variable of interest in this hypothesis test is average hold time (measured in minutes) and you would be comparing your data against the claim made by the consumer group. As we do not know much at all regarding the nature of the distribution of hold times, we will need to take a large-enough sample ($n \geq 30$) in testing the claim. For now, let us focus on the hypotheses, which are defined as:

H_0: Average hold time $\geq$ 4.4 (consumer group's claim)

H_1: Average hold time < 4.4 (what you wish to show)

The consumer group's claim is the null hypothesis, which you are trying to refute.

Hence, your claim is the alternative hypothesis. The above hypotheses can also be written as:

$$H_0: \mu \geq 4.4$$

$$H_1: \mu < 4.4$$

with μ being the hypothesized average hold time (the parameter of interest) for the population. This is a *lower-tail* test, since we are trying to show that the true mean is, in fact, lower than some hypothesized value.

After setting up the hypotheses, you collect data on 50 random customer service calls (a *large-enough* sample according to the central limit theorem). The data are shown in Table 7-6, with a sample mean of 4.0 minutes.

Table 7-6
Data Set: Wireless Call Hold Times Example

Customer Service Calls: Hold Times				
2.0	4.3	7.4	2.5	4.4
0.5	1.9	7.4	5.2	2.0
3.3	6.0	4.4	6.0	1.9
3.2	1.3	7.0	2.0	0.1
3.7	6.1	6.5	0.8	5.0
6.8	4.2	6.8	4.8	4.9
6.6	4.9	3.7	2.4	5.4
0.1	4.8	5.4	2.9	3.1
0.5	6.9	5.0	3.9	2.4
2.6	3.0	4.6	4.1	6.3

Let X be a variable representing hold time (in minutes) on calls. If we assume the null hypothesis to be true, then X has a mean of 4.4 minutes and a standard deviation of 0.9 minutes. Thus, the *sampling distribution of $\bar{X}$* (the sample mean) is:

$$\bar{X} \sim N(4.4, \frac{0.9}{\sqrt{50}})$$

(Recall that the standard error of the sample mean is the standard deviation of X divided by the square root of the sample size.)

The graph in Figure 7-8 shows the distribution of $\bar{X}$ (the blue line), as well as the hypothesized mean of 4.4 minutes and your sample mean of 4.0 minutes. From this graph, it is easy to see that this sample mean is indeed lower than the hypothesized mean. But, is it sufficiently low (that is, a sufficient distance from the hypothesized mean) for you to reject the null hypothesis?

**Figure 7-8
Sampling Distribution: Wireless Call Hold Times Example**

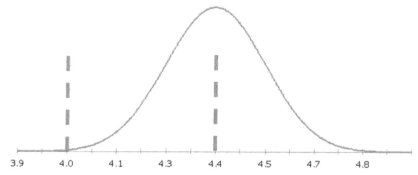

Recall that the question being asked in hypothesis testing is: "How possible is it to take a random sample of size n=50 out of a population with a mean of 4.4 minutes and standard deviation of 0.9 minutes and obtain a sample mean of 4.0 minutes?" If it is reasonably possible, then there is no reason to reject the null hypothesis. However, if the probability of finding this sample mean is extremely low (in our example, if the sample mean is positioned in the extreme *lower-tail* of the distribution – hence, the name of the test), then it could be deduced that the null hypothesis is false and we should reject it, concluding instead that the *true*

average hold time is lower than 4.4 minutes.

It is important that you firmly grasp the logic that underlies hypothesis testing as we will be applying these concepts throughout the remaining chapters of this book. To self-assess your understanding, consider an extreme example: assume that you meet a group of 1,000 people and that you are told that 990 of them are between the ages of twenty and 30, and ten are children younger than ten years old. You take a random sample of twenty people from this population, finding that the average age of the people sampled is fifteen. Do you still believe that what you were initially told about the group's distribution of ages is correct? Why (or why not)?

Returning to our hold-time examples, to calculate the probability of obtaining a sample mean with such a low value, we bring back a concept that was introduced in Chapter 6, namely the *p-value*. As defined in Chapter 6, the p-value refers to a specific probability

$$P(Z \geq |z|)$$

which is the probability of obtaining a Z score *farther away from the mean* than a specific Z score value (either to the right of positive Z score values or to the left of negative Z score values). Thus, the p-value provides a direct answer to our question: "How reasonable is it for us to take a random sample of size n=50 from a population with a mean of 4.4 minutes and a standard deviation of 0.9 minutes and obtain a sample mean of 4.0 minutes?"

To compute our p-value, we start with our sample mean and convert it into a Z score, based on its sampling distribution:

$$Z = \frac{4.0 - 4.4}{0.9/\sqrt{50}} = -3.14$$

This converted Z score value is called our **test statistic**. This test statistic is a value that computes the distance of our sample mean from the hypothesized population mean given the distribution of the test, in this case the Z distribution.[75] Then, we use Excel to calculate the probability of finding a Z score value at least as low as the value of our *test statistic* by typing =NORM.S.DIST(-3.14,1) and obtain:

$$P(Z \leq -3.14) = 0.0008$$

We now have our answer. The probability of obtaining a sample of size n=50 with a mean of 4.0 minutes, given the hypothesized distribution with a mean of 4.4 minutes and a standard deviation of 0.9 minutes, is practically zero (0.0008). Because this probability is extremely low, we can confidently conclude that the null hypothesis (H₀: μ ≥ 4.4) is false and *reject* it. In rejecting a null hypothesis, we are concluding that the alternative hypothesis (H₁: μ < 4.4) is true. Applying this logic to our example, we conclude from the findings of our sample of 50 random customer calls that the average customer hold time is less than 4.4 minutes.

As a probability of 0.0008 is very low, we would probably never question the conclusion to reject the null hypothesis. However, there may be borderline cases where, for example, we obtain a p-value of 0.02, 0.05, 0.1, or even 0.5. The rather obvious question that arises, then, is: "Where should I draw the line?" That is, at what probability level should you begin to say: "This is a very low probability and therefore I should reject the null hypothesis"? The answer to this question is

---

[75] In future chapters, we will learn how to compute test statistics when working with other distributions.

determined by the person conducting the test, based on a desired *significance level* referred to as α *(alpha)*.

In statistics, when we say that a finding is *statistically significant*, it means that the finding is unlikely to have occurred by chance. Our **level of significance** is the *maximum chance probability* we are willing to tolerate. What do we mean by maximum chance probability? When we take a sample of, say, customer calls and measure customer hold times, our sample mean can obtain any value with a given probability, determined by the sampling distribution of the mean. For example, in the above example, the probability of finding a sample mean of 4.0 minutes is 0.0008 – if we believe the null hypothesis and assume the true mean to be 4.4 minutes with a standard deviation of 0.9 minutes. Given this probability, we can draw one of two conclusions: (1) that the hypothesized mean is true and we were just very lucky to have found such an unlikely sample, or (2) that the hypothesized mean is not the true mean, which is likely lower than 4.4 minutes and closer to the value of 4.0 minutes that we observed with our sample. The maximum chance probability (or our significance level) is the probability at which we would switch from conclusion (1) to conclusion (2). In the social sciences, the most commonly used threshold for level of significance, or α, is 5% (a probability of 0.05), with marginal significance often considered to be 10% (a probability of 0.1) and any finding having a probability of occurrence greater than 10% generally considered non-significant. Hence, we form the following *decision rule* for our hypothesis testing context: for a derived p-value smaller than α (a selected level of significance), we reject the null hypothesis and conclude that it is false. As the derived p-value for our customer hold time example is 0.0008 (a value far smaller than any α value likely to be selected), we reject the null hypothesis.

The customer hold time example involved a lower-tail test: the alternative hypothesis was about the mean being *lower* than some hypothesized value. Let us go through a different example, this time for an upper-tail test.

A baseball player claims that the average speed of his fastball pitch is greater than that of his rival, who averages 90 mph. He collects data on the speed of 100 fastballs and finds his average speed is 90.8 mph. Assuming that we know the standard deviation of a fastball pitch to be 3.85 mph and that fastball pitches are normally distributed, what can we conclude about the pitcher's claim?

We begin by formulating the hypotheses. The pitcher would like to prove that his fastball has a higher speed than that of his rival. Therefore, the null hypothesis is that it is *not* higher (i.e., the same speed or lower) and the alternative is that it *is* higher:

$$H_0: \mu \leq 90$$

$$H_1: \mu > 90$$

with μ being the hypothesized average pitch speed.

Let X be a variable representing pitching speed (in mph). If we assume the null hypothesis to be true, then we know that X is normally distributed with a mean of 90 mph and a standard deviation of 3.85 mph. We also know that the *sampling distribution of* $\bar{X}$ (the average pitching speed) is:

$$\bar{X} \sim N(90, \frac{3.85}{\sqrt{100}})$$

To obtain the test statistic, we convert our sample mean of 90.8 to a Z score using the hypothesized distribution:

$$Z = \frac{90.8 - 90}{3.85/\sqrt{100}} = 2.078$$

Next, we need to determine the p-value for this test statistic. This is an *upper-tail test*. Here, we are looking for the probability of finding a value *at least as high* as our test statistic:

$$P(Z \geq 2.078) = 0.0188$$

(In Excel, remember to type =1-NORM.S.DIST(2.078,1) because you are looking for the upper-tail.) This p-value is illustrated using the red shaded area of Figure 7-9. Next, we need to apply our decision rule based on the desired level of significance (α). Assume for this question that α has been set at 1%. Our decision rule is to reject the null hypothesis if the derived p-value is smaller than 1%. We compare the derived p-value of 1.88% to the selected α of 1%. Since the derived p-value (the probability of finding a sample mean of 90.8) is greater than α, we *do not reject* the null hypothesis. Thus, we conclude that the player's fastball pitches are not faster than his rival's fastball pitches.

Figure 7-9
Test Statistic and P-Value: Baseball Player Example

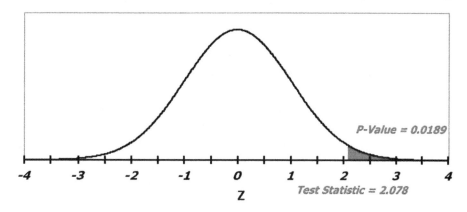

But what if we used a different α value in this example, say 5%? Well, for α=0.05, our conclusion actually changes and we end up *rejecting the null hypothesis*. This is because the probability of finding a sample mean of 90.8 is less than a 5% level of significance. It is not uncommon to find situations such as this; that is, where you would reject the null hypothesis for some α values, but not for other α values. Because of this, we often use the p-value to infer *the maximum significance of the test*. With this baseball player example, our test is significant at the 5% level (we were able to reject the null hypothesis), but it is *not* significant at the 1% level (we cannot reject the null hypothesis). The highest level of significance for which we would still reject the null hypothesis is 1.88%. Going back to our previous example of the wireless consumer group, the p-value of 0.008 means that our test is significant at both the 5% and 1% levels of significance. The lower the p-value, the greater the significance of our finding.

Let us review one final example, one introduced earlier in Unit 1 of this chapter, where a statistics professor wishes to determine if her section's grade average is different (that is, either greater than or less than) from the grades in the other sections of the course. The average grade for all of the other course sections is 75 with a standard deviation of 4. The professor collected test scores from 25 of her students and found a sample average of 72. She wishes to conduct a hypothesis test at the 5% level of significance (that is, α=0.05). To test the professor's claim, we need to set up the following hypotheses:

$$H_0: \mu = 75$$

$$H_1: \mu \neq 75$$

with μ being the hypothesized average grade.

Let X be a variable representing grades and assume we know that X is normally distributed. If we assume the null hypothesis to be true, then we believe X to be normally distributed with a mean of 75 and a standard deviation of 4. We also know that the sampling distribution of $\bar{X}$ (the section grade averages) is:

$$\bar{X} \sim N(75, \frac{4}{\sqrt{25}})$$

We convert our sample mean of 72 to a Z score using the hypothesized distribution:

$$Z = \frac{72 - 75}{4/\sqrt{25}} = -3.75$$

and, this Z score becomes our *test statistic*. Using the Excel function =NORM.S.DIST(-3.75,1), we find that the probability of obtaining a Z score of -3.75 is practically zero:

$$P(Z \leq -3.75) = 0.000088$$

But, we are not yet ready to reach a conclusion. Recall that this is a *two-tail test*. This means that we would reject the null hypothesis if our sample mean is either much higher than *or* much lower than the hypothesized value of 75. In the case of a two-tail test, the p-value used to determine the significance of the found sample mean must account for *both tails* of the probability distribution. Our p-value, thus, is calculated as:

$$\text{p-value} = 2 * P(Z \leq |test\ statistic|)$$

with the p-value for our two-tail test becoming:

$$\text{p-value} = 2*0.000088 = 0.000176$$

suggesting that our hypothesis test is significant at just about any level of α. Certainly, then, for an α of 5% we reject the null hypothesis (as the p-value<α), concluding that the grade average of the professor's section is indeed different (specifically, it is lower) from the grades in the other sections of the course.

In this unit, we have covered the logic of hypothesis testing and the use of probabilities (p-value and α) to reach a conclusion. In each of this unit's examples, we followed the same five steps: (1) we formulated the hypotheses, (2) we computed the test statistic based on the sample data and then used this test statistic to derive the p-value, (3) we identified a specific decision rule (reject the null hypothesis if the derived p-value is less than α) based on a selected significance level α, (4) we compared the derived p-value to the selected level of significance to draw a conclusion, and (5) we interpreted the conclusion with respect to the original example. These five steps are summarized graphically in Figure 7-10.

Figure 7-10
Steps in Hypothesis Testing

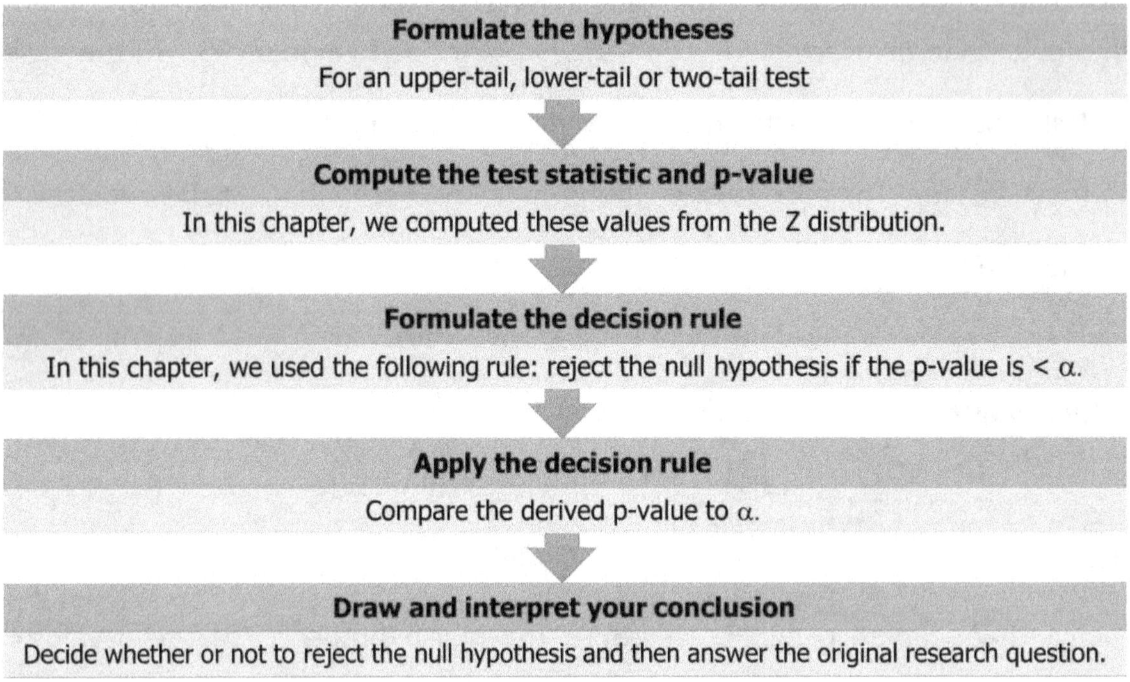

Formulate the hypotheses
For an upper-tail, lower-tail or two-tail test

Compute the test statistic and p-value
In this chapter, we computed these values from the Z distribution.

Formulate the decision rule
In this chapter, we used the following rule: reject the null hypothesis if the p-value is < α.

Apply the decision rule
Compare the derived p-value to α.

Draw and interpret your conclusion
Decide whether or not to reject the null hypothesis and then answer the original research question.

In later chapters, we will learn how to conduct tests for statistical parameters other than the mean and when using different types of distributions. We will also cover hypothesis testing techniques (with different decision rules) other than the p-value approach used in this chapter. However, prior to concluding this introduction to statistical inference, we need to cover one more important topic: the two types of errors that can arise with hypothesis testing.

Type I and Type II Errors

Two possible errors may result from hypothesis testing. The first one, called the *type I error*, occurs when you decide to reject the null hypothesis, but it is, in reality, true. In other words, you were wrong to reject it. The second one, called the *type II error*, occurs when you do not reject the null hypothesis when it is, in fact, false. These two types of errors are illustrated in Figure 7-11.

Figure 7-11
Type I and Type II Errors

Hypothesis Testing Indicates that You Should:	In Actuality	
	H_0 is True (Do not reject H_0)	H_0 is False (Reject H_0)
Not reject H_0		 Type II Error
Reject H_0	 Type I Error	

We'll use an example to illustrate these two types of errors. Assume you heard a claim that the average wireless phone customer receives no more than eight spam text messages each day, with a standard deviation of 0.8. Given your own experiences, you tend to doubt this claim. To test the claim, you collect data from 60 wireless customers and record the number of spam messages each customer received during a single day. Your sample average is 8.2. The stated hypotheses are:

$H_0: \mu \leq 8$

$H_1: \mu > 8$

You compute the test statistic:

$$Z = \frac{8.2-8}{0.8/\sqrt{60}} = 1.94$$

Testing at the 5% level of significance, you compute the p-value associated with the test statistic:

$$P(Z \geq 1.94) = 0.026$$

and conclude that the null hypothesis should be rejected, because 0.026<0.05. In other words, you conclude that the average customer receives more than eight spam text messages per day.

Let us review the logic of the test to understand the two types of errors. You assumed the null hypothesis to be true and computed the probability of finding the sample mean of 8.2. Assuming the null hypothesis is true, the probability of finding a sample of n=60 with a mean of 8.2 is quite low:

$$P(Z \geq 1.94) = 0.026$$

As previously mentioned, given this probability, you can draw one of two conclusions: either (1) the distribution is as was hypothesized and you were extremely lucky to have found such a sample, or (2) the true distribution is not the one described by the null hypothesis.

Assume that the null hypothesis is, in fact, true. If you drew the second conclusion above, then you would commit an error, specifically *a type I error*, by rejecting a true null hypothesis. Because our rejection rule is based on α (the level of significance), α is also the probability of committing a type I error. This means that, if we wish to reduce the chances of a type I error, we need to use smaller values of α. In the above example, if you use $\alpha=0.05$, you end up rejecting the null hypothesis and committing a type I error. However, using $\alpha=0.01$ would not have led you to reject the null hypothesis; and, by not rejecting the null hypothesis, you avoid committing a type I error.

At this point, it may seem strange to choose high values of α and risk committing a type I error. Why not choose very low α values each time we test a hypothesis? As with most things in life, we face a trade-off. Specifically, we need to consider another probability called β, which is the probability of committing a type II error. A type II error would occur if you decided not to reject the null hypothesis, but, in reality, the hypothesis was false (and you should have rejected it). In other words, in the above spam example, the null hypothesis is based on an assumption that people receive at most eight spam messages daily, but the truth is that they receive more than eight spam messages daily. Remember, all you know in carrying out the hypothesis test is what you observe from the sample you collect – you do not know what the true population parameter is.

The probabilities α and β are inversely related. This means that higher levels of α imply lower levels of β and vice versa. Therefore, in determining α, we need to consider which type of error is costlier: a type I error or a type II error. If a type I error is costlier, then we should choose a low value for α to avoid making the error. If a type II error is costlier, then we'll choose a higher α value to ensure β is low.

To understand this better, consider the following example. People who suffer from diabetes are often concerned with the Glycemic Index (GI) of foods, which measures the effect of food on blood glucose. A new energy bar manufacturer claims that its bar has a GI of 101 (which is fairly high). The accuracy of the manufacturer's claim is critical to diabetes patients, as they need to carefully monitor their carbohydrate intake and the GI provides a means for doing this. When conducting a hypothesis test about the accuracy of the manufacturer's claim (a two-tail test with $H_0: \mu=101$), we would therefore try to minimize the probability of making a type I error. Thus, we would select a lower value for α, such as 0.01.

Type II errors occur when we fail to reject what is, in reality, a false null hypothesis. For example, airlines are very concerned with the average weight of carry-on bags for the purpose of fuel consumption calculations. Suppose an airline currently uses an assumption that the average weight of a carry-on bag is 22 lbs. Given the importance of this assumption to flight safety, airlines periodically assess the continued validity of the assumption by testing to see if the average weight of a carry-on bag is still no more than 22 lbs ($H_0: \mu \leq 22$). In this example, a type II error would occur if the airline does not conclude that, on average, carry-on bags weigh more than 22 lbs, when in fact they do (thereby failing to reject a false null hypothesis). Such error would result in underestimating fuel requirements; depending on the extent of underestimation, this can be a grave error. Here, the airline would try to minimize the probability of a type II error by selecting a higher value for α, such as 0.1.

Unit 3 Summary

- α is the **level of significance** that is used in hypothesis testing and is determined by the researcher. Commonly used α values are 1%, 5% and 10%.

- A **test statistic** is a value calculated from the sample statistic (here, the sample mean) by converting it to a Z score (or a value from another distribution as will be seen in later chapters). The test statistic is used along with a p-value in assessing whether a hypothesis should be accepted or rejected.

- **P-value** is the probability of finding a test statistic (here, for the sample mean) as extreme as the one found, given that the null hypothesis is true. The smaller the p-value, the more evidence we have against the null hypothesis and the more likely we are to reject it. P-value calculations for the sample mean are shown in Figure 7-12 for upper-tail, lower-tail and two-tail hypothesis tests.

Figure 7-12
P-Value Calculations for Sample Mean Hypothesis Testing

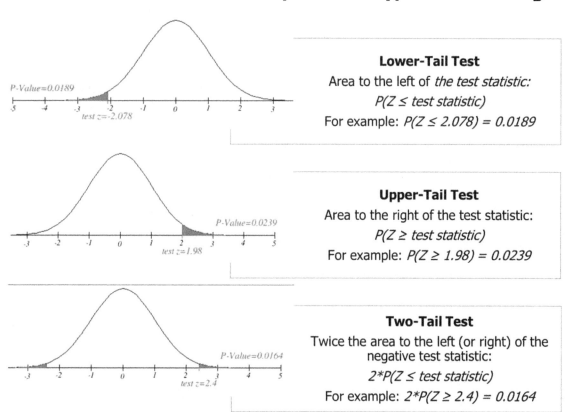

Lower-Tail Test
Area to the left of *the test statistic*:
$P(Z \leq \text{test statistic})$
For example: $P(Z \leq 2.078) = 0.0189$

Upper-Tail Test
Area to the right of the test statistic:
$P(Z \geq \text{test statistic})$
For example: $P(Z \geq 1.98) = 0.0239$

Two-Tail Test
Twice the area to the left (or right) of the negative test statistic:
$2*P(Z \leq \text{test statistic})$
For example: $2*P(Z \geq 2.4) = 0.0164$

- P-Value Decision Rule: Reject the null hypothesis when the derived p-value is less than α.

- A **type I error** is committed when a true null hypothesis is rejected. The probability of a type I error is α. A **type II error** is committed when a false null hypothesis is not rejected. The probability of a type II error is β. α and β are inversely related (i.e., as α increases, β decreases).

Unit 3 Exercises

1. Find the *p-value* for each situation.

    a. z = -1.82; lower-tail test

    b. z = 0.74; upper-tail test

    c. z = 2.07; two-tail test

    d. z = 3.01; upper-tail test

    e. z = -0.6; lower-tail test

    f. z = 0.6; two-tail test

2. The data in Table 7-e3 describes time use for an average weekday for full-time college students. The first column provides various types of activities, the second column provides the hypothesized number of hours a typical student spends on each activity (taken from the website of *the American Bureau of Labor Statistics*[76]), and the third column provides the population standard deviation (σ) for each activity. To test whether these data are reflective of student time use at your school, you selected a sample of 50 of your classmates and asked them to provide data on their time use for an average weekday. The sample means you calculated for each activity are provided in the fourth column. Using the information provided in Table 7-e3, find the *p-value* for each of the tests posed in questions 2.a.-2.h., using the provided value of α to draw your conclusion of whether or not to reject the null hypothesis.

---

[76] Source: http://www.bls.gov/tus/charts/

Table 7-e3
How Typical Students Spend Their Time

Daily Activities	Hypothesized Hours	Known standard deviation	Sample mean (n=50)
Sleeping	8.7	2.0	7.8
Leisure and Sports	4.1	0.5	4.2
Educational Activities	3.3	3.0	2.4
Working	2.4	2.0	2.9
Other	2.3	0.7	3.1
Traveling	1.4	0.2	1.4
Eating and Drinking	1.0	0.7	1.2
Grooming	0.8	0.3	1.0
Total	24.0		24.0

a. A two-tail test for the average number of sleep hours (use $\alpha=0.1$)

b. An upper-tail test for the average hours spent on leisure and sports (use $\alpha=0.1$)

c. A lower-tail test for the average hours spent on educational activities (use $\alpha=0.05$)

d. A two-tail test for the average hours spent working (use $\alpha=0.01$)

e. An upper-tail test for the average hours spent on other activities (use $\alpha=0.01$)

f. A two-tail test for the average hours spent traveling (use $\alpha=0.05$)

g. An upper-tail test for the average hours spent eating and drinking (use $\alpha=0.05$)

h. An upper-tail test for the average hours spent grooming (use $\alpha=0.05$)

3. By comparing a test statistic with the appropriate p-value, test the following claims (use $\alpha=0.05$). Describe what the type I and type II errors would be in each case:

   a. The makers of a new *Wii* workout game claim that, if you play the game for 30 days, you will average a weight loss of eight pounds. If the next 50 players average only 7.4 pounds of weight loss and σ is known to be 2.25, should the company be asked to retract its claim?

b. A local bank says that the mean wait time for customers is less than three minutes. Based on a sample of 125 customers with an average wait time of 2.83 minutes (and a σ of 1.16 minutes), do you believe the bank's claim?

c. A study claims that employees who smoke cigarettes average 30 minutes of smoking breaks each day with a standard deviation of eight minutes. To help reduce the time wasted on smoking breaks, management has implemented an anti-smoking campaign, offering employees who wish to quit smoking the necessary help and resources. A sample of 110 smokers taken four months after the initiation of the campaign revealed a sample mean of 25 minutes of smoking breaks each day. Was the campaign effective in reducing the time wasted on smoking breaks? What is the highest level of significance for your conclusion?

END-OF-CHAPTER PRACTICE

1. For questions 1.a. through 1.c., state the null and alternative hypotheses and explain under what situation a type I error would be committed.

    a. You are testing whether the average number of cups of coffee that people drink daily has increased from 2.6 cups since 2003.

    b. Architects at a firm claim their salaries are less than the State average of $48,520.

    c. You believe that the mean weight of adult males in the U.S. is not 195 lbs, as a just-published study claims.

2. For questions 2.a. through 2.c.: first calculate the test statistic; then calculate the p-value; and, finally, interpret the test results.

    a. $H_0: \mu \geq 100$; $H_1: \mu < 100$; $\bar{X} = 97$; $\alpha = 0.05$; $\sigma = 20$; $n = 81$

    b. $H_0: \mu \leq 15$; $H_1: \mu > 15$; $\bar{X} = 18$; $\alpha = 0.01$; $\sigma = 5$; $n = 16$

    c. $H_0: \mu = 50$; $H_1: \mu \neq 50$; $\bar{X} = 47$; $\alpha = 0.05$; $\sigma = 15$; $n = 100$

3. For questions 3.a. through 3.f., calculate the test statistic and p-value for each test as one of the parameters changes.

    a. $H_0: \mu \leq 50$; $H_1: \mu > 50$; $\sigma = 5$; $n = 9$; when $\bar{X} = 51$, $\bar{X} = 52$ and $\bar{X} = 54$

b. $H_0: \mu \geq 15$; $H_1: \mu < 15$; $\sigma = 2$; $n = 25$; when $\bar{X} = 14.5$, $\bar{X} = 14.4$ and $\bar{X} = 14.3$

c. $H_0: \mu = 100$; $H_1: \mu \neq 100$; $\sigma = 9$; $\bar{X} = 98$; when $n = 20$, $n = 50$ and $n = 100$

d. $H_0: \mu \leq 70$; $H_1: \mu > 70$; $\sigma = 20$; $\bar{X} = 80$; when $n = 64$, $n = 25$ and $n = 9$

e. $H_0: \mu \geq 200$; $H_1: \mu < 200$; $\bar{X} = 190$; $n = 9$; when $\sigma = 10$, $\sigma = 30$ and $\sigma = 50$

f. $H_0: \mu \leq 1000$; $H_1: \mu > 1000$; $\bar{X} = 1190$; $n = 100$; when $\sigma = 100$, $\sigma = 50$ and $\sigma = 25$

4. A young baseball player is trying to provide evidence that he hits very well in important situations, even though his overall batting average is low. The player claims his clutch hitting average is .300, but one scout is skeptical of this number. The scout attends some games and takes a sample of the player's hitting in clutch situations. In the data below, 1 represents a hit and 0 represents an out:

   0 1 0 0 0 0 0 0 1 1 0 0 0 1 0 0 1 0 0 0 0 1 0 0 0
   (Average: 6/25 = 0.240)

   If the population is normally distributed with a standard deviation of 0.38, can the scout conclude at the 5% significance level that the player hits less than .300 in clutch situations?

5. A coffee shop owner is worried about a slowdown in business because a new shop opened up one block away. Before the new shop opened, the owner's customer traffic (customers per hour) was normally distributed with a standard deviation of three and a mean of fifteen. The owner randomly surveyed sixteen hours and recorded a mean of thirteen customers per hour. Can he conclude at the 1% significance level that his shop is in trouble?

6. The United States government claims DUI arrests average 22,096 per state per year. A sample of size n=10 states finds the mean to be 44,002. If DUI arrests are normally distributed with a standard deviation of 28,584, can it be concluded at the 1% significance level that DUI arrests are higher than the government claims?

7. In 2011 *Twitter* was already an extremely popular social networking and communication tool, reaching a mean of 140 million tweets per day in March

2011, with a standard deviation of 29 million.[77] (To put this in perspective, the *Twitter*-world types the equivalent of 8,000 copies of Leo Tolstoy's *War and Peace* daily). To see if there was more *Twitter* traffic during the 2012 *Summer Olympic Games*, a nine-day random sample was taken and the results were a mean of 156 million tweets daily. Assuming that tweets are normally distributed, is there enough evidence at the 5% significance level to indicate that people tweeted more during the *Olympics*?

8. State the *final conclusion* you would reach with each of the following hypothesis testing scenarios:

    a. Original research hypothesis: The mean of the statistics quiz is less than 75. Test result: Reject the null hypothesis.

    b. Original research hypothesis: The mean amount of money spent by college freshmen on soda per day is at least $3.50. Test result: Do not reject the null hypothesis.

---

[77] Source: http://blog.twitter.com/2011/03/numbers.html

CHAPTER 8: Additional Concepts in Hypothesis Testing

The previous chapter introduced the logic and process of hypothesis testing. We discussed the notion of a hypothesis, introduced the central limit theorem and the concept of a sampling distribution, and described how to carry out a hypothesis test for the population mean. We also talked about the two types of errors that might occur in hypothesis testing: type I and type II errors.

In our discussion of hypothesis testing in Chapter 7, we assumed that sigma (σ), the standard deviation of the population, was known and we made use of this information in performing a hypothesis test. In this chapter, we relax this assumption and instead work with s, the sample standard deviation. Relaxing this assumption about knowing the value of σ is actually quite realistic as we frequently know very little about a population's σ, especially in the early stages of a data analysis project. Hence, while we hypothesize about the true value of the mean (μ) of some variable of interest, we conduct our test based solely on sample data that we collect. In other words, we use the mean and standard deviation obtained from our sample in conducting our hypothesis test.

Because we will be working with the sample standard deviation, as opposed to the population standard deviation, we no longer use the standard normal distribution. Instead, we use the t-distribution in conducting hypothesis tests. Recall from Chapter 6 that the t-distribution is very similar to the standard normal distribution, with the exception that its shape depends on the number of degrees of freedom associated with the analysis being performed. Because we are working with sample-estimated values (specifically, the sample standard deviation), we lose a degree of freedom – we thus work with a t-distribution with n-1 degrees of freedom.

This chapter also introduces two new techniques for conducting hypothesis testing. In the previous chapter, we used the *p-value approach* to hypothesis testing. With the p-value approach, we examine the *probability* (i.e., the p-value) of our test statistic, comparing this p-value against a desired level of significance (i.e., α). Here, we learn a related technique, referred to as the *critical value approach*, which involves comparing our test statistic to a *critical value* defining a *rejection region* (again based on α) in our population distribution. In addition, we also introduce a new concept called a *confidence interval*, which enables us to obtain an estimate of the population mean based on our sample mean. We also describe how confidence intervals can be used in hypothesis testing. This chapter covers the following topics:

- The t-Distribution
- Hypothesis Test of the Population Mean when the Population Standard Deviation is Unknown
- Confidence Interval for a Single Population Mean
- Confidence Interval for the Population Mean when Sigma is Known
- Confidence Interval for the Population Mean when Sigma is Unknown

UNIT 1

The t-Distribution

In this unit we cover how to conduct hypothesis tests of a single population mean when σ, the population standard deviation, is unknown. Because we will be working with the sample standard deviation, s, rather than the population standard deviation, σ, we must use the t-distribution in conducting these hypothesis tests. We begin with a brief overview of the t-distribution. Then, we describe how the hypothesis test is conducted using the p-value approach to hypothesis testing that was covered in Chapter 7.

Again, the t-distribution is similar to the normal distribution, with its precise shape depending on the sample size – specifically, the number of degrees of freedom (here, df=n–1). As the number of degrees of freedom increases, the shape of the t-distribution gets closer and closer to that of the normal distribution.

We can use Excel to locate specific values in the t-distribution, or we can use a slightly older tool known as the t-distribution table ('t-table'). We explain both processes here, but will proceed with using Excel throughout the chapter.

In Excel, the function =T.INV(α, degrees of freedom) can be used to obtain values from the t-distribution. The function returns a value t*, such that:

$$P(t \leq -t^*) = \alpha$$

For example, =T.INV(0.05,5) returns a value of -2.015. In other words, in the t-distribution with 5 degrees of freedom (df=5), the probability of finding a value lower than -2.015, is 0.05 (α). The value of -2.015 creates a 5% tail at the lower end of the distribution. Another Excel function is =T.DIST(x, degrees of freedom, cumulative), which returns the probability P(t≤x). For example, =T.DIST(-2.015, 5,1) will return the value 0.05. As with the normal distribution Excel functions, we always select the value of '1' (or 'true') for the cumulative option to obtain an area rather than a specific point.

To better understand these values, consider Figure 8-1, which shows a t-distribution graph with five degrees of freedom. The area marked in blue is the 5%

lower-tail, with the t value corresponding to the value of -2.015 we obtained from the above formula.

Figure 8-1
A t-Distribution Illustration (df=5)

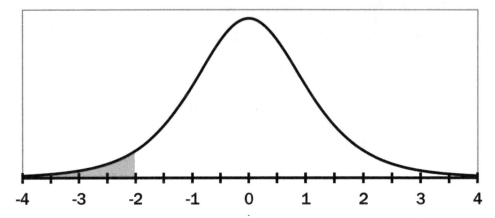

We can use these Excel functions to find any value we wish in the t-distribution. For example:

- To find the lower-tail corresponding to a 10% level of significance and nine degrees of freedom, we type =T.INV(0.1,9) to obtain -1.3830.

- To find the probability of a t value lower than -1, given 24 degrees of freedom, we type =T.DIST(-1,24,1) to obtain 0.1636.

- And, of course, since the distribution is symmetric, we can find the probability of a value being *greater* than 1 by typing =1-T.DIST(1,24,1) and the value that creates an *upper* 10% tail by typing =T.INV(0.9,9).

We will continue to practice these Excel functions throughout the chapter. Now we turn to explain another tool that can be used when working with the t-distribution: the t-distribution table.

Table 8-1 provides a partial view of the t-table and the full t-table is provided in this chapter's Excel file as the worksheet titled 't-Distribution Table'. The top row of the t-table provides specific significance level (α) values, with each α value corresponding to a worksheet column. The notation $t_{0.05}$ thus indicates that this column can be used to find the probability of a specific value of t (denoted as t*) such that:

$$P(t \geq t^*) = \alpha$$

and, with $\alpha=0.05$, this expression becomes:

$$P(t \geq t^*) = 0.05$$

In other words, the t-table provides a t value that defines an area at the right-tail of the distribution corresponding to a probability of α. Each row in the table corresponds to a different number of degrees of freedom (in other words, the first column of the worksheet contains the number of degrees of freedom associated with each row of the worksheet). For example, given five degrees of freedom and an α of 0.05, the table returns a value of 2.015, meaning that:

$$P(t \geq 2.015) = 0.05.$$

Table 8-1 highlights how the value 2.015 was located.

Table 8-1
A t-Distribution Table

df	$t_{0.1}$	$t_{0.05}$	$t_{0.025}$	$t_{0.01}$	$t_{0.005}$
1	3.078	6.314	12.706	31.821	63.657
2	1.886	2.920	4.303	6.965	9.925
3	1.638	2.353	3.182	4.541	5.841
4	1.533	2.132	2.776	3.747	4.604
5	1.476	2.015	2.571	3.365	4.032
6	1.440	1.943	2.447	3.143	3.707
...	...	...	...	...	...
∞	1.282	1.645	1.960	2.326	2.576

Because the t-distribution is symmetric, we use negative values of those shown in Table 8-1 when looking at the distribution's left-tail. This is expressed as:

$$P(t \leq -t^*) = \alpha$$

with an example being:

$$P(t \leq -2.015) = 0.05$$

You can also observe from Table 8-1 that, as the sample size increases, the values in the t-table approach the Z score values associated with the standard normal distribution (e.g., compare the bottom row of Table 8-1 with the values found in the standard normal table for similar α values).

As we progress with the examples in this chapter, you will gradually gain a level of comfort working with the t-table and with Excel's t-distribution functions. Now, armed with a basic understanding of the t-distribution, we can turn our attention back to the process of hypothesis testing of the population mean when the population standard deviation is unknown.

Hypothesis Test of the Population Mean when the Population Standard Deviation is Unknown

We begin by reviewing the p-value approach (introduced in Chapter 7) for conducting a hypothesis test of the population mean, but *now* for the case where the *population standard deviation is unknown*. Then, we describe how to carry out this same test using the critical value approach.

The P-Value Approach

Consider the following example. According to one source[78], 71% of college students who graduated in 2015 had outstanding student loans, carrying an average debt of $35,000. You believe that this number might be higher than the average debt of students graduating from your college and decide to collect data to

---

[78] Source: http://blogs.wsj.com/economics/2015/05/08/congratulations-class-of-2015-youre-the-most-indebted-ever-for-now/

test your belief (using α=0.05). You survey 40 recent graduates from your college and ask about their debt upon graduation. You find that the average debt with your sample is $32,743.85 with a sample standard deviation of $6,313.65. These data are provided in Column A of the worksheet titled 'College Debt Example' in this chapter's Excel file.

As you now know, the first step in any hypothesis test is to correctly formulate the hypotheses. The parameter of interest in this case is the mean and we would like to find out whether the mean debt upon graduation for students at your college is lower than the national average of $35,000. Hence, the hypotheses are:

$$H_0: \mu \geq 35,000$$

$$H_1: \mu < 35,000$$

where μ is the average debt upon graduation of students at your college.

Following the procedure outlined in Chapter 7, our second step involves computing the test statistic and corresponding p-value. This test statistic reflects the distance (in number of standard errors) between the sample mean and the hypothesized mean. The formula used for this calculation is:

$$t = \frac{\bar{X} - \mu}{\frac{s}{\sqrt{n}}}$$

where $\bar{X}$ is the sample mean, μ is the hypothesized mean, s is the sample standard deviation, and n is the sample size. In this calculation, the numerator is the distance that the sample mean is from the hypothesized mean and the denominator is the standard error of the sample mean.

The values for $\bar{X}$ and s are computed from the collected data. This example's Excel file provides the mean and standard deviation of the sampled data in cells A43 and A44, respectively. The test statistic is calculated in cell B48. You may understand the nature of this test statistic a little better if you can take some time to change the values in the Excel worksheet and see how they impact the resulting test statistic. For example, what happens when you increase the sample size by adding additional data values? Or, what happens when you reduce the sample standard deviation by changing some of the more extreme values in the sample so that they are closer in value to the sample mean? Remember that the denominator in the formula for calculating the test statistic is the standard error of the mean:

$$\frac{s}{\sqrt{n}}$$

Because of this, larger values of s or smaller values of n will increase the value of this denominator, resulting in smaller absolute values of the test statistic – meaning that the sample mean would be closer (in terms of number of standard errors) to the hypothesized population mean.

Returning to the second step of our hypothesis testing procedure, knowing that $\bar{X}$=32,743.85, that s=6,313.65, and that µ=35,000 (from the null hypothesis), we compute the test statistic:

$$t = \frac{\bar{X} - \mu}{\frac{s}{\sqrt{n}}} = \frac{32,744 - 35,000}{\frac{6,314}{\sqrt{40}}} = -2.26$$

We can now find the p-value that corresponds to this computed test statistic. Recall from Chapter 7 that this derived p-value refers to the probability of finding a sample mean as extreme as the one we have found, or for the sampled data:

$$P(t \leq -2.26) = \text{p-value}$$

Using Excel we type: =T.DIST(-2.26,39,1) to obtain a p-value of 0.0147.

We now execute the third step of our hypothesis testing procedure: identifying the decision rule to use in order to reach a conclusion. As presented in Chapter 7, the decision rule for the p-value approach is: reject the null hypothesis if the derived p-value is less than α.

Step four of the p-value approach applies this decision rule. As we are conducting the test at $\alpha=0.05$, our decision rule is to reject the null hypothesis if the derived p-value is less than 0.05. Since 0.0147 is less than 0.05, we reject the null hypothesis.

The last step in hypothesis testing always involves interpreting the test result in the context of the question being asked. We consequently conclude, at the 5% level of significance, that the mean college debt at your school is lower than the national average of $35,000 (the null hypothesis).

The Critical Value Approach

An equivalent approach to conducting hypothesis tests relies on a different decision rule based on whether or not the sample mean lays greater than a threshold distance (defined in terms of number of standard errors) beyond the hypothesized population mean. This threshold distance is known as the *critical value*. Deciding to use either the p-value approach or the critical value approach is largely a matter of personal preference, though the availability of needed resources (such as statistical software) can be a determining factor. Regardless of which approach is followed, the conclusion reached will be the same. We now turn our attention to describing the procedure followed in carrying out the critical value approach to hypothesis testing.

We begin with the same example used above in describing the p-value approach: determining whether the mean college loan debt upon graduation for students at your college is lower than the national average of $35,000. The hypotheses were:

$$H_0: \mu \geq 35,000$$

$$H_1: \mu < 35,000$$

where μ is the average debt upon graduation. We used $\alpha=0.05$ as the specified level of significance in conducting the test.

Our sample data revealed a mean debt of $32,743.85 for students at your college. We also computed a test statistic using the sample mean and standard deviation (given in the worksheet 'College Debt Example'), with the computed value of the test statistic being -2.26.

At this stage we need to bring in our new decision rule, which has not yet been specified. We know, from our prior discussions about the logic of hypothesis testing, that what is needed is some boundary point beyond which the null hypothesis would be rejected. Maintaining the logic developed in Chapter 7 of rejecting the null hypothesis when the sampled parameter (e.g., the sample mean) is found to be in an extreme tail of the hypothesized distribution, we define the needed boundary point as a *threshold distance* – that is, the distance beyond which we would reject the null hypothesis. Such a threshold distance is called the **critical value**.

The critical value is computed based on the desired level of significance for the test such that:

$$P(t \geq |critical\ value|) = \alpha$$

We can now state the decision rule for the critical value approach: reject the null hypothesis if the test statistic, in absolute value, is greater than the critical value. That is, reject the null hypothesis if the distance (in number of standard errors) between the sample mean and the population mean is greater (in absolute value) than the derived critical value.

Figure 8-2 shows a lower-tail critical value in the t-distribution. This value was calculated based on a desired significance level (α) and a given number of degrees of freedom. In our college debt example, using 5% significance level and given 39 degrees of freedom, the critical value would be calculated using the Excel function =T.INV(0.05,39), which returns a value of -1.6849. Also at the lower-tail of the distribution, beyond the critical value, is an area highlighted in blue. This area is called the **rejection region** (thinking in terms of a rejection region may help you to understand the logic of the critical value approach to hypothesis testing). Remember that the test statistic was computed by converting the sample mean into a standardized score representing the number of standard errors by which the sample mean lays below (or above) the hypothesized mean. If this test statistic falls into the rejection region, we would conclude that the sample mean is indeed significantly lower (or higher) than the hypothesized mean and we would reject the null hypothesis. Thus, we restate the critical value decision rule to make it more intuitive: reject the null hypothesis if the test statistic falls in the rejection region defined by the critical value.

Figure 8-2
Critical Value and Rejection Region: Lower-Tail Test

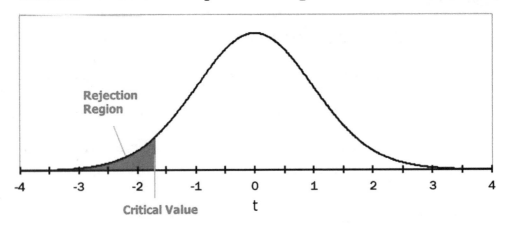

Let's now return to the college debt example. As previously mentioned, we are conducting a lower-tail test at the 5% level of significance and therefore the derived critical value for this example is computed as:

$$P(t \leq \textit{critical value}) = 0.05$$

In other words, we wish to find a value from the t-distribution that defines a 5% tail at the lower side of the distribution. Earlier, we used the Excel function =T.INV(0.05,39) function to obtain a critical value of -1.684,[79] which is a value at the left-hand tail of the t-distribution such that:

$$P(t \leq \textit{-1.684}) = 0.05$$

Now that we have obtained a critical value (-1.684), we formulate our decision rule as: reject the null hypothesis if the test statistic falls in the rejection region defined by the critical value. Comparing the test statistic (-2.26) to the

---

[79] To find this critical value by using the t-table, we first look up 39 degrees of freedom in the left-most column (df=40-1), and α=0.05 in the top row. The t-table in this chapter's Excel file was created to allow printing on a single page. For this reason, it has a more limited number of rows, which means it does not include every possible value of degrees of freedom. Since there is no row for 39 degrees of freedom, we can simply approximate to the nearest degrees of freedom, or df=40, in order to locate the value sought: -1.684 (remember that this is a lower-tail test and that the t-distribution is symmetrical).

critical value (-1.684), we find that the test statistic does, indeed, fall within the rejection region (i.e., -2.26<-1.684) and we reject the null hypothesis. This is shown in Figure 8-3. By rejecting the null hypothesis, we thus conclude, at the 5% level of significance, that the average debt of students graduating from your college is lower than the hypothesized national average of $35,000.

Figure 8-3
Comparing the Critical Value and Test Statistic:
Lower-Tail Test

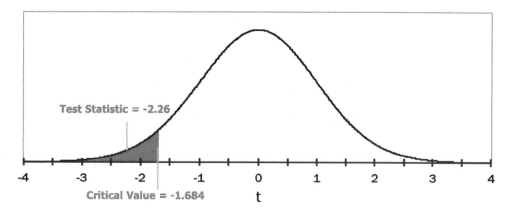

This college debt example involved a lower-tail test: the alternative hypothesis was about the mean being *lower* than some hypothesized value and, therefore, we defined our rejection region at the lower-tail of the distribution. In carrying out the critical value approach to hypothesis testing, we essentially followed the same five steps introduced in Chapter 7 for the p-value approach: (1) we formulated the hypotheses; (2) we computed our test statistic; (3) we identified a specific decision rule based on the critical value derived from the desired level of significance (reject the null hypothesis if the test statistic falls within a rejection region defined by the critical value); (4) we applied the decision rule, comparing the test statistic to the critical value to draw a conclusion; and, (5) we interpreted the conclusion with respect to the original question. Let's go through these steps again, but this time for a two-tail test, using a different example.

This two-tail test example follows steps similar to the lower-tail college debt example with only a minor change to how we determine the critical value. Consider a situation where your family had purchased a home in 1992 for $225,000, which was the average price for a home in 1992. Now your parents are thinking of selling the family home. To help your parents with the decision, you take a sample of ten homes sold in your neighborhood during the past six months and find the following sales prices:

$190,000, $195,500, $196,000, $200,000, $220,000,

$227,000, $228,000, $234,000, $237,400, $240,000

Following the steps for the critical value approach, you test whether the average of these recent selling prices has changed from the 1992 average selling price of $225,000. (Assume that selling prices are normally distributed and use $\alpha=0.05$).

Step 1: Formulate the Hypotheses

The parameter of interest is the mean and we would like to find out whether the mean selling price has changed (up or down) from the original price of $225,000. Hence, the hypotheses are:

$$H_0: \mu = 225,000$$

$$H_1: \mu \neq 225,000$$

where μ is the average selling price of homes.

Step 2: Compute the Test Statistic

The question does not provide either the sample mean ($\bar{X}$) or the sample standard deviation (s). Thus, these values need to be computed from the collected recent selling prices. Using these data, you compute: $\bar{X}$=$216,790 and s=$19,393. Now, you can calculate the test statistic:

$$t = \frac{\bar{X} - \mu}{\frac{s}{\sqrt{n}}} = \frac{216{,}790 - 225{,}000}{\frac{19{,}393}{\sqrt{10}}} = -1.339$$

Step 3: Formulate an Appropriate Decision Rule

Recognizing that this is a two-tail test, the specified decision rule is: reject the null hypothesis if the test statistic falls into either the lower-tail or upper-tail rejection regions defined by the critical value. But, in order to execute this decision rule, you need to find the critical value to be used with this hypothesis test.

As described earlier, you are conducting the test at a 5% level of significance. You need to find the critical value in the t-table such that:

$$P(t \geq |critical\ value|) = 0.025$$

Two important points about this expression are: (1) the absolute value is used in determining the critical value, since this is a two-tail test and critical values are needed for both tails of the distribution; and, (2) α/2 is used for the same reason. This is illustrated in Figure 8-4.

Figure 8-4
Two-Tail Critical Values and Rejection Regions

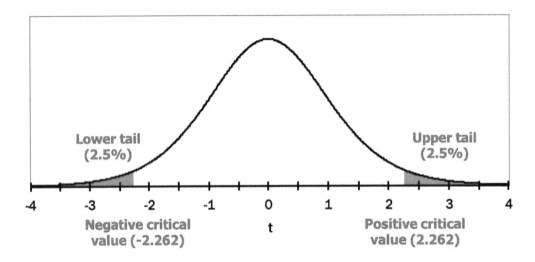

To locate this critical value, we can use the t-table (refer to this chapter's Excel file for this table) to identify the intersection of nine degrees of freedom (in the left-most column) and $\alpha=0.025$ (in the top row). Alternatively, we can use Excel to find the critical value by typing =T.INV(0.025,9). Using either method, the critical value is: ±2.262. Once these two critical values have been identified, you can now restate the decision rule as: reject the null hypothesis if the test statistic is either greater than 2.262 or less than -2.262.

Step 4: Apply the Decision Rule

The decision rule states that you should reject the null hypothesis if the computed test statistic falls within either the upper or lower rejection regions. Comparing the test statistic to these two critical values, you find:

$$-2.262 < -1.339 < 2.262$$

This is illustrated in Figure 8-5. Thus, at the 5% level of significance, you do not reject the null hypothesis.

Figure 8-5
Comparing the Critical Value and Test Statistic: Two-Tail Test

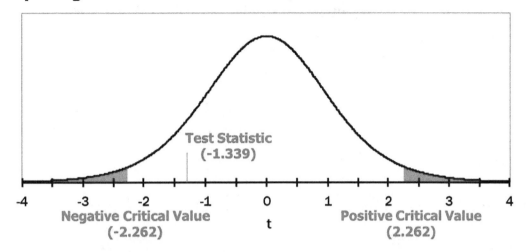

Step 5: Draw and Interpret a Conclusion

This final step involves interpreting your results in the context of the question being asked. By not rejecting the null hypothesis, you conclude, at the 5% level of significance, that the average price of homes in your family's neighborhood has not changed significantly since 1992.

Unit 1 Summary

- In conducting a hypothesis test of a single population's mean when σ (the population standard deviation) is unknown, the t-distribution with n-1 degrees of freedom is used.

- The overall process of the test is summarized in Figure 8-6. The figure presents, side-by-side, the p-value approach, as described in Chapter 7, and the critical value approach introduced in this chapter:

Figure 8-6
Process of Hypothesis Testing:
P-Value Approach vs. Critical Value Approach

P-Value Approach	Critical Value Approach
Formulate the Hypotheses	**Formulate the Hypotheses**
• For an upper-tail, lower-tail or two-tail test	• For an upper-tail, lower-tail or two-tail test
Compute the test statistic and p-value	**Compute the test statistic and p-value**
• In this chapter, we computed these values from the t-distribution	• In this chapter, we computed these values from the t-distribution
Formulate the decision rule	**Formulate the decision rule**
• Reject if p-value is less than α	• Find the critical value based on α • Reject if test statistic falls in the rejection region
Apply the decision rule	**Apply the decision rule**
• Compare the derived p-value to α	• Compare the test statistic to the critical value
Draw and interpret your conclusion	**Draw and interpret your conclusion**
• Decide whether or not to reject the null hypothesis and then answer the original research question	• Decide whether or not to reject the null hypothesis and then answer the original research question

- The ***critical value*** is computed based on our desired level of significance, such that:

$$P(t \geq |critical\ value|) = \alpha$$

In other words, the probability of a value falling farther away from the mean than the critical value equals α. For a two-tail test, we would use $\alpha/2$ for each of the tails.

- The critical value defines a ***rejection region*** at the extreme tails (or ends) of the distribution. The probability of a sample mean falling within the rejection region is α. This is illustrated in Figure 8-7 for a lower-tail test.

Figure 8-7
Illustration of the Critical Value and Rejection Region

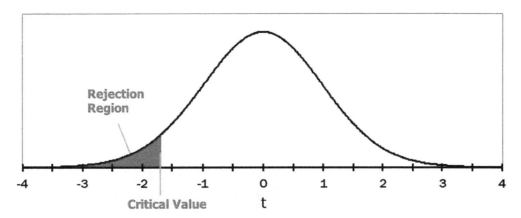

- The *test statistic* used with the critical value approach is calculated as:

$$t = \frac{\bar{X} - \mu}{\frac{s}{\sqrt{n}}}$$

- Use the *decision rules* in Table 8-2 to decide whether or not to reject the null hypothesis.

Table 8-2
Critical Value Approach Decision Rules for a Hypothesis Test of a Single Population Mean

Type of Test	Rejection Rule
Lower-tail test H_0: $\mu \geq$ value H_1: $\mu <$ value	Reject the null hypothesis if: test statistic < critical value
Upper-tail test H_0: $\mu \leq$ value H_1: $\mu >$ value	Reject the null hypothesis if: test statistic > critical value
Two-tail test H_0: $\mu =$ value H_1: $\mu \neq$ value	Reject the null hypothesis if: test statistic < lower critical value Or if: test statistic > upper critical value

Unit 1 Exercises

1. For each of the following cases: (1) calculate the test statistic; (2) find the critical value; and (3) state your rejection rule:

   a. n = 50; $\bar{X}$ = 175; s = 22; H_1: $\mu \neq$ 180; α = 5%

b. $n = 150$; $\bar{X} = 155$; $s = 30$; H_1: $\mu \neq 160$; $\alpha = 10\%$

c. $n = 500$; $\bar{X} = 400$; $s = 90$; H_1: $\mu < 405$; $\alpha = 10\%$

d. $n = 225$; $\bar{X} = 1511.4$; $s = 35.7$; H_1: $\mu > 1502.5$; $\alpha = 1\%$

2. Test the claim that the mean starting salary for a business school graduate is more than $42,500, given that: $n = 500$; $\bar{X} = \$43,000$; $s = \$3,000$; and, $\alpha = 0.025$.

3. Test the claim that the average number of customers per day at a bank's drive-through window is at least twenty, given that: $n = 50$; $\bar{X} = 19.75$; $s = 1.3$; and, $\alpha = 0.05$.

4. Test the claim that the mean time spent on homework by students at your college is less than three hours per week, given that: $n = 135$; $\bar{X} = 2.9$; $s = 1.22$; and, $\alpha = 0.01$.

5. A seller of a speed-typing software package claims that customers average an increase of 25 words per minute after using the software package for one month. If the next 40 customers average an increase of 26.3 words per minute with a standard deviation of 4.21, is the seller's claim true? (Use $\alpha=0.05$.)

6. A local juice company advertises that its orange juice is a healthier, more nutritious option than its leading competitor, as the local company's juice has "less than 100 calories per bottle!". To test this claim, a scientist measured the number of calories in 25 juice bottles. He found that the average caloric content was 98.3 with a standard deviation of five calories. Can the scientist's measurements support the advertisement if the test is carried out at the 5% level of significance? What if the test uses a significance level of 10%?

7. Without directly calculating it, what can you say about the p-value of the test in Question 6?

UNIT 2

Confidence Interval for a Single Population Mean

In the previous section, we discussed the process of *testing* our population parameter against a hypothesized value. Testing requires that we have a specific value of the parameter in mind, which we formulate into a hypothesis. We can also use inferential statistics to *estimate* the population parameter, learning more about its possible values. An estimate of the population mean that you are already familiar with is the *sample mean*. The sample mean is considered a **point estimator**, defined as a single-value estimate computed from sampled data.

We expect point estimators to be *unbiased* and *consistent*. Suppose that we are interested in estimating (that is, learning more about) a population mean using the sample mean as our point estimator. Suppose further that we repeatedly sample from the population, calculating and recording the sample mean as we collect each new sample. If the average of all sample means equals the population mean (the parameter of interest), then we say that our sample mean is an **unbiased** estimator of the population mean. That is, we have not systematically over- or under-estimated the population mean. In other words, the expected value of an unbiased estimator is the parameter itself. The second expectation we have about the point estimator is that it is **consistent**. A point estimator is said to be consistent if, as we increase the sample size, the values of the estimator converge to the parameter. In other words, if we look at the sampling distribution of our point estimator, we expect it to become narrower as we increase the sample size.

A point estimator, however, even if it is unbiased and consistent, is still subject to *sampling error*. Recall from Chapter 1 that a sampling error is defined as the difference between the population parameter and the sample statistic. In fact, sampling error is why, in the hypothesis tests described in Unit 1, we expect a sample mean to deviate somewhat from a true hypothesized mean and why we define rejection regions at the tails of a population's distribution.

An alternative to using a point estimator for estimating a population parameter that accounts for sampling error is to use an *interval estimator*. An **interval estimator** is an interval that is constructed around the point estimator. In other words, rather than saying "based on our sample, we believe that the mean college debt is $32,743.85", we would say "based on our sample, we believe that the mean debt is between $31,061 and $34,425". But, we cannot state a belief about an estimate without providing some indication of the strength of this belief. Thus, a **confidence interval** is an interval estimator associated with a specific **confidence level** that indicates the strength of our belief that the interval, indeed, does include the true value of the parameter. This *confidence level* is stated as a percent (e.g., a 90% confidence interval or a 95% confidence interval) and is determined by the researcher. More formally, the confidence level is defined as $1-\alpha$, where α is the same significance level that was introduced in the discussions of hypothesis testing. If we construct a 95% confidence interval for mean college debt, for example, we can now say, "We are 95% confident that the interval [$31,061 ... $34,425] captures the true mean debt of college graduates". In the remainder of this unit, we first explain how to construct such confidence intervals for estimating the population mean when sigma (σ) is known and then we explain how to construct such confidence intervals for estimating the population mean when sigma (σ) is unknown. Then, we apply what has been learned about confidence intervals to hypothesis testing.

Confidence Interval for the Population Mean when Sigma is Known

Recall from Chapter 7 that knowing the true population standard deviation (σ) allows us to work with the standard normal (or, Z) distribution. We start with the same example used previously in Chapter 7: a consumer group claims that wireless companies are worsening their customer service, with the time customers are kept on hold when contacting the company reaching an average of 4.4 minutes. Suppose you know that the population's standard deviation of hold time is 54 seconds (or 0.9 minutes). You are a lobbyist for wireless providers and would like to estimate the true average time customers are kept on hold. To accomplish this, you collect data on 50 random customer service calls and find that the sample mean time a customer is kept on hold is 4.0 minutes.

This sample mean of 4.0 is now your best *point estimator* of the true (and unknown) population mean.[80] In order to obtain a higher level of confidence in your estimate of the population mean, you decide to construct a *95% confidence interval*. In other words, you would like to stretch your point estimator into an interval such that, for 95% of the samples obtained from this population (i.e., 95 out of 100 samples), the true population mean would be captured somewhere in this interval.

Conceptually, in constructing this interval you would start from $\bar{X}$ (the sample mean) and add and subtract a certain margin to it:

$$\bar{X} \pm margin$$

Because the width of the sampling distribution of the mean is determined by its standard error ($\frac{\sigma}{\sqrt{n}}$), we also use the standard error in determining the margin of a confidence interval. Thus, our interval is determined as our sample mean ($\bar{X}$) plus

---

[80] Remember that the value of '4.4' is the one *claimed* by the consumer group and is not necessarily the true mean. This is what you set out to test in the first place. The fact is that no one really knows the value of the true mean, and the only way to determine this true mean is to conduct a census and compute the mean based on every single member of the population. Without such a census, the best we can do is estimate the mean with some level of confidence.

and minus a certain number of standard errors. The exact number of standard errors that we add and subtract depends on our specified confidence level.

Intuitively, in order for someone to have greater confidence in an estimated confidence interval, the estimated interval would need to be wider. Taken to an extreme, in order to be 100% confident, the estimated confidence interval would need to include every possible value in the population. As the specified level of confidence is reduced, the interval narrows. Putting together all of the above, we compute the confidence interval for the population mean, when σ is known, using the following formula:

$$\bar{X} \pm Z_{\alpha/2} \frac{\sigma}{\sqrt{n}}$$

where $\bar{X}$ is the sample mean (the point estimator), $\frac{\sigma}{\sqrt{n}}$ is the standard error of the mean, and $Z_{\alpha/2}$ is the number of standard errors we need to add to and subtract from the sample mean in order to obtain an interval estimate with a 1-α confidence level. To better understand this last part, recall, using what we learned from Chapter 6, that we can define an interval under the normal distribution curve that captures 1-α percent of the area under the distribution curve as:

$$P(-Z_{\alpha/2} < Z < Z_{\alpha/2}) = 1 - \alpha$$

where $\pm Z_{\alpha/2}$ are the cutoff values at both tails of the distribution.

Returning to our customer hold-time example, we know: $\bar{X}=4.0$, σ=0.9, n=50, and 1-α=0.95 (since we are looking for a 95% confidence interval). Looking up the corresponding Z values in Excel (or in the standard normal distribution table), we find:

$$P(-1.96<Z<1.96) = 0.95$$

Now we can compute:

$$\bar{X} \pm Z_{\alpha/2}\frac{\sigma}{\sqrt{n}} \Rightarrow 4.0 \pm 1.96\frac{0.9}{\sqrt{50}} \Rightarrow 4.0 \pm 0.25$$

We thus obtain a **lower confidence limit (LCL)**, which is the lower bound of our interval, of 3.75, and an **upper confidence limit (UCL)**, which is the upper bound of our interval, of 4.25. Hence, our 95% confidence interval of the mean hold time of customer service calls is [3.75 ... 4.25]; and, we are able to state, with a confidence level of 95%, that this interval captures the true population mean.

Let us take a minute to consider the meaning of the above result. If we are 95% confident that the interval [3.75 ... 4.25] captures the true population mean of customer service hold time, how likely are we to believe the claim of the consumer group that the true mean call time is 4.4 minutes (as stated in the original question)? Well, at the 95% level of confidence, we are not likely to believe this claim, because our interval does *not* include the value 4.4. The upper confidence limit of our interval is 4.25, which is less than 4.4. This outcome demonstrates the important link regarding confidence intervals and hypothesis tests. In fact, determining confidence intervals is yet another approach to take in conducting tests of hypotheses, as summarized in Figure 8-8.

Figure 8-8
Hypothesis Testing Using Confidence Intervals

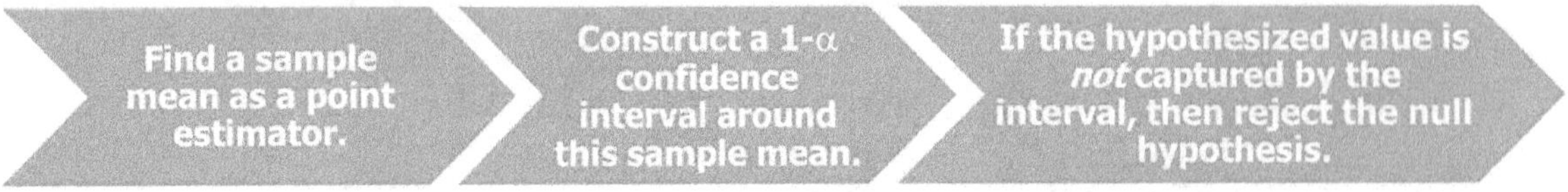

> Find a sample mean as a point estimator. → Construct a 1-α confidence interval around this sample mean. → If the hypothesized value is *not* captured by the interval, then reject the null hypothesis.

Confidence Interval for the Population Mean when Sigma is Unknown

We are going to walk through another example of computing a confidence interval for the mean, but this time sigma (σ), the population standard deviation, is *not* known. Recall that when σ is unknown, we work with the t-distribution rather than the standard normal distribution. The steps of constructing the confidence interval are exactly the same as above with one difference: rather than looking up a value in the Z distribution, we will look up a value in the t-distribution. Hence, the formula we use to calculate the confidence interval is:

$$\bar{X} \pm t_{\alpha/2} \frac{s}{\sqrt{n}}$$

where $\bar{X}$ is the sample mean (the point estimator), $\frac{s}{\sqrt{n}}$ is the standard error of the mean, and $t_{\alpha/2}$ is the number of standard errors we need to add to and subtract from the sample mean.

Recall the house-selling example from Unit 1 of this chapter: your family purchased a home in 1992 for $225,000, which was the average price of a home in your parent's neighborhood in 1992. Now your parents are thinking of selling the family home. To help your parents with the decision, you take a sample of ten homes sold in this same neighborhood during the past six months and find the following sales prices:

$190,000, $195,500, $196,000, $200,000, $220,000,

$227,000, $228,000, $234,000, $237,400, $240,000

Now, you will estimate, at a 99% level of confidence, the current mean selling price of homes in the neighborhood (note that we have changed our confidence level in this example from 95% to 99% for practice purposes).

From the data provided above, you compute: $\bar{X}$=$216,790 and s=$19,393. You also know that n=10 and 1-α=0.99 (since we are looking for a 99% confidence interval). Next, you need to look up the corresponding t values (with α/2=0.005 and df=9) in Excel (or in the t-table), finding:

$$P(-3.25 < t < 3.25) = 0.99$$

Now, you compute:

$$\bar{X} \pm t_{\alpha/2} \frac{s}{\sqrt{n}} \Rightarrow 216790 \pm 3.25 \frac{19393}{\sqrt{10}} \Rightarrow 216790 \pm 19931$$

to obtain a *lower confidence limit (LCL)* of 196,859 and an *upper confidence limit (UCL)* of 236,721. Hence, the 99% confidence interval of the mean selling price of homes is [$196,859 ... $236,721]; and, you can be 99% confident that this interval captures that true population mean.

The original question (in Unit 1 of this chapter) involved determining whether or not the average housing price in the neighborhood has changed from the 1992 average price of $225,000. Looking at the calculated confidence interval, you see that the value of $225,000 is, in fact, captured by this interval. Hence, at the 99% level of confidence, you have no reason to conclude that the average selling price has changed. Although this conclusion is consistent with the conclusion reached in Unit 1 using the critical value approach to hypothesis testing, the analysis in Unit 1 was conducted at the 5% significance level, whereas the just-calculated confidence interval was constructed using a 99% confidence level. In order to validly compare the conclusions reached through these two approaches, we would need to either construct a 95% confidence interval of the mean or conduct the hypothesis test using a significance level of 1%. You can try changing these values and see whether the results still match. If you do redo the analysis to more validly compare the approaches, you will find that the outcomes do match. *Regardless of the approach used to conduct our test of hypotheses (i.e., the p-value approach, the critical value approach or the confidence interval approach), the conclusion reached should always be the same.*

Unit 2 Summary

- A **point estimator** is a single, sample-based, value from which we aim to learn about the population parameter.

- An **unbiased** estimator is one whose expected value equals the population parameter. A **consistent** estimator is one that converges to the true parameter as the sample size increases.

- An **interval estimator** is an interval that is constructed around the point estimator in order to obtain a better estimate of the population parameter. A **confidence interval** is an interval estimator for which we can hold a specific level of confidence that the interval, indeed, includes the true value of the parameter. Formally, the **confidence level** is defined as 1-α, where α is the same significance level introduced in hypothesis testing discussions.

- The formula to use when constructing a 1-α confidence interval for the mean of a population with a *known standard deviation* is:

$$\bar{X} \pm Z_{\alpha/2} \frac{\sigma}{\sqrt{n}}$$

- The formula to use when constructing a 1-α confidence interval for the mean of the population with an *unknown standard deviation* is:

$$\bar{X} \pm t_{\alpha/2} \frac{s}{\sqrt{n}}$$

- The **lower confidence limit (LCL)** is the lower bound of the confidence interval; whereas the **upper confidence limit (UCL)** is the upper bound of the confidence interval.

- Confidence intervals can be used to test hypotheses regarding the population mean, as shown in Figure 8-9:

Figure 8-9
Hypothesis Testing Using Confidence Intervals

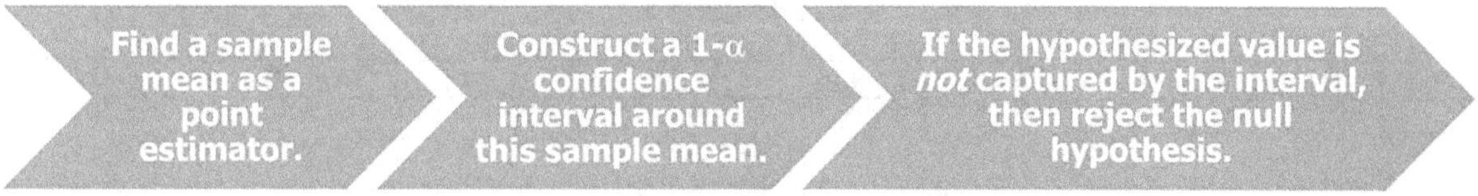

Find a sample mean as a point estimator. → Construct a 1-α confidence interval around this sample mean. → If the hypothesized value is *not* captured by the interval, then reject the null hypothesis.

Unit 2 Exercises

1. Estimate μ in each of the following cases with 95% confidence.

    a. $\bar{X} = 500$; s = 100; n = 100

    b. $\bar{X} = 500$; s = 100; n = 225

    c. $\bar{X} = 500$; s = 100; n = 400

    d. What is the effect of increasing the sample size on a confidence interval?

2. Estimate μ in each of the following cases with 95% confidence.

    a. $\bar{X} = 500$; s = 3; n = 100

    b. $\bar{X} = 500$; s = 5; n = 100

    c. $\bar{X} = 500$; s = 8; n = 100

    d. What is the effect of increasing the sample standard deviation on a confidence interval?

3. Estimate μ when $\bar{X} = 36$, s = 12, and n = 100 for each confidence level.

    a. With a 90% confidence level

    b. With a 95% confidence level

c. With a 99% confidence level

d. What is the effect of increasing the confidence level on a confidence interval?

4. Given the sample of NASA budgets provided in Table 8-e1 (in constant 2007 $US millions), estimate the population mean with 90% confidence.[81]

Table 8-e1

| 33,514 | 11,131 | 19,686 | 14,926 | 17,186 | 21,376 | 6,360 | 12,221 |

5. Suppose a sample of 100 master's degrees yielded an average cost of $40,000 and a standard deviation of $5,000. Estimate the mean cost with 95% confidence.

6. You just read a study examining whether or not it was a good financial investment to obtain a master's degree. Comparing 25 business majors with master's degrees with 25 business majors without master's degrees, the study found that the average increase in lifetime income with a master's degree was $375,000 with a standard deviation of $30,000. Estimate the mean increase with 95% confidence.

7. The same study described in Question 6 also compared 25 liberal arts majors with master's degrees against 25 liberal arts majors without master's degrees. The study observed that the average increase in lifetime income with a master's degree to be $15,000 with a standard deviation of $1,500. Estimate the mean increase with 95% confidence.

8. *Coca-Cola* is a globally recognized soft drink brand widely consumed around the world. A research project has surveyed the annual liters of *Coke* consumed per person in the U.S., Mexico and France. Given the data in Table 8-e2[82], estimate the population means for each country with 99% confidence.

Table 8-e2

	n	σ	$\bar{x}$
U.S.	200	15 L	170 L
Mexico	200	25 L	225 L
France	200	6 L	22.7 L

---

[81] Source: http://en.wikipedia.org/wiki/Budget_of_NASA#Annual_budget.2C_1958-2012
[82] Source: http://thumbnails.visually.netdna-cdn.com/behind-cocacola--datavisualisation_5157176868dfc.png

9. A random anonymous office survey of 30 employees finds that people waste an average of 150 minutes per day on non-work related activities. If the population standard deviation is known to be 25 minutes, estimate the population mean with 90% confidence.

10. Consider a sample of 100 women's *Facebook* pages, having an average of 394 posts with a standard deviation of 55 posts. Estimate the population mean for the number of women's *Facebook* posts with 95% confidence.

11. The same sample of men's *Facebook* posts finds a mean of 254 posts and a standard deviation of 40 posts. Estimate the population mean with 95% confidence.

12. Reports from the journal *Marine Policy* estimate between 63 million and 273 million sharks are killed each year, many of them illegally.[83] In some Asian cultures, soup made with shark fin is considered a delicacy and is highly sought after. By contrast, shark attacks on humans are uncommon and fatal incidents are extremely rare (despite being highly publicized by news media and in Hollywood movies). Table 8-e3 provides a sample of the total annual *number* of shark attack incidents. Table 8-e4 provides a sample of the total annual *fatal* shark attack incidents. Estimate the population mean for each sample with 99% confidence.

Table 8-e3

| 76 | 58 | 71 | 53 | 66 | 80 |

Table 8-e4

| 4 | 4 | 1 | 7 | 3 | 6 |

13. In late August 2012, Hurricane Isaac caused an estimated $2.34 billion worth of damage in the Caribbean and the Gulf Coast of the United States. The storm reached wind speeds of 130 km/h (80 mph).[84] Given the random sample of the hurricane's wind speeds provided in Table 8-e5, estimate the overall average wind speed with 95% confidence.

Table 8-e5

| 130 | 90 | 100 | 105 |
| 85 | 115 | 80 | 75 |

---

[83] Sources: http://www.flmnh.ufl.edu/fish/sharks/statistics/statsw.htm,
http://www.huffingtonpost.com/2013/03/06/100-million-sharks-killed-every-year_n_2813806.html
[84] Source: http://en.wikipedia.org/wiki/Hurricane_Isaac_%282012%29

14. Jupiter is the largest planet in the solar system. Its volume is about 1,321 Earths. Suppose 60 estimates of Jupiter's radius taken throughout the planet's rotational cycle yield an average of 69,000 km with a standard deviation of 7,000 km.[85] Estimate Jupiter's radius with 99% confidence.

END-OF-CHAPTER PRACTICE

1. A recent newspaper article claims that university students average 4.5 hours of sleep each night during their exam periods, a very unhealthy circumstance. To test this claim, you survey the fifteen students in your class and find the average amount of sleep each night during a recent exam period to be five hours with a standard deviation of 1.25 hours. At the 5% level of significance, do you have enough evidence to refute the newspaper's claim?

2. Your faculty advisor claims that the average GMAT score required to be accepted into top university MBA programs is at least 663. To test this, you decided to research 27 MBA programs across the United States and found the average GMAT score for accepted applicants to be 634 with a standard deviation of 65. Does your research support or refute your advisor's claim at the 1% level of significance?

3. According to a health magazine, an average of eight glasses of water should be drunk each day to maintain a healthy diet. If readers drink less than the encouraged amount, the magazine would like to launch a campaign to promote the drinking of water on a regular basis. To test if readers are actually drinking this average amount, a random sample of twenty readers was taken, with a sample mean of 6.9 glasses of water and a standard deviation of 2.46 glasses. At the 5% level of significance, should the magazine launch their campaign?

4. A tire company claims that their tires will last at least 35,000 miles. To test this claim, a consumer advocate tests a sample of 50 tires. The test results showed that the tires lasted an average of 33,950 miles with a standard deviation of 2,050 miles. Does this test support or refute the tire company's claims? (Use $\alpha=0.01$.)

5. According to the 2015 *American Time Use Survey*, "On an average day, adults age 75 and over spent 7.8 hours engaged in leisure activities--more than any other age group; 35- to 44-year-olds spent 4.0 hours engaged in leisure and sports activities--less than other age groups".[86] A sample of ten random individuals within the ages of 35 to 44 was collected, and the average time they spent on leisure and sports activities each day was recorded. The sample mean was 4.5 hours with a standard deviation of 45

---

[85] Source: http://solarsystem.nasa.gov/planets/profile.cfm?Object=Jupiter&Display=Facts
[86] Source: http://www.bls.gov/news.release/atus.nr0.htm

minutes. Does this sample support the survey claim about 35- to 44-year-olds? (Use α=10%.)

6. Tony claims that compared to other students in his school he saves an average of $7.50 per month on public transit, because he understands all the available ways to use transit transfers. His friend does not believe this claim and collects a random sample of Tony's monthly spending (he takes public transit 22 days in a month). The sample shows that Tony saves an average of $6.45 compared to other students (with a sample standard deviation of $0.89).

   a. Construct a 99% confidence interval for Tony's savings.

   b. Can Tony's claim be refuted at the 1% level of significance? Explain.?

7. A student has been spending exactly four hours on every assignment she has gotten this year. She took a random sample of sixteen of her assignments and their word count. She found the average word count in this sample to be 1,140 words, with a standard deviation of 397 words.

   a. Construct a 95% confidence interval for the mean number of words she writes.

   b. Can she conclude at a 5% significance level that the mean number of words she writes in four hours is 1,100?

8. Joe wants to know if he receives the same number of emails as his co-workers, who claim to get seventeen emails daily with a known population standard deviation of five emails. He takes a random sample of twenty days and records how many emails he receives. His sample average is 17.9 emails per day.

   a. Construct a 95% interval for the mean number of daily emails.

   b. Can Joe conclude at the 5% significance level that he gets the same number of emails per day as claimed by his co-workers?

The data for Questions 9 through 17 are provided in this chapter's accompanying Excel file. For each question, first carefully decide whether you should use the t-distribution or the Z distribution. If it is not stated in a question, assume that the data set follows a normal distribution. Then, conduct your test using the approach of your choice (p-value, critical value or confidence interval).

9. A computer science teacher believes that the average student types at least

33 words per minute (wpm). To test her belief, she randomly samples 30 of her students and measures their typing speeds in wpm. The data are provided in the worksheet titled 'Typing Speed'. Can the teacher infer at the 5% significance level that she is correct?

10. A machine that produces quarters at the *United States Mint* is set so that the average thickness is 1.75 mm with a standard deviation of 0.03555 mm. A sample of fifteen coins was measured for their thickness, with the results listed in the worksheet titled 'US Mint'. Can we conclude at a 1% significance level that the mean thickness is not 1.75 mm (and, hence, that the machine needs to be adjusted)?

11. A small boutique clothing store estimates its daily sales by assuming that the average dollar sale will be $55 per customer. A new manager tests this assumption by conducting a random survey of 65 customers as they complete a purchase, collecting the dollar sale amount associated with this purchase. The data are provided in the worksheet titled 'Daily Sales'. Can the manager infer at a 5% significance level that the current assumption about the average dollar sale amount is wrong?

12. Tenants living in a 32-story apartment building had been complaining about the elevator wait times. Previously conducted surveys have shown that the average wait time is five minutes, with a standard deviation of three minutes. Tenants' complaints are that the wait time is actually longer than five minutes. The apartment superintendent conducts a random survey of 89 tenants to test if the tenants' claim is true. The data are provided in the worksheet titled 'Elevator Wait Time'. Conduct a test to determine if the mean wait time for the tenants is indeed longer than five minutes, using the 5% significance level.

13. A museum tour guide claims that each gallery in the museum contains at least twenty ancient artifacts. To test the claim, a tourist conducts a random survey of thirteen galleries and counts the number of ancient artifacts each gallery contains. The data are provided in the worksheet titled 'Museum'. Can the tourist conclude at a 10% significance level that the museum guide's claim is true?

14. A famous musician claims that he sings for at least 50 minutes during each of his live concerts. To test the claim, a news reporter conducted a random survey of ten concerts the musician has performed and tested for the number of minutes the musician sang live. The data are provided in the worksheet titled 'Concerts'. Can the news reporter conclude at a 5% significance level that the musician's claim is false?

15. A ranked chess player claims that she wins her chess matches in under 30 minutes. A random sample of 23 of her chess matches was conducted to test for the number of minutes the match lasted. The data are provided in the worksheet titled 'Chess'. Can we conclude with a 5% significance level

that the chess player's claim is true?

16. A primary school teacher expects his students to be able to read at least four pages in a book in ten minutes. A random sample of eighteen of his primary school students was drawn to test the number of pages in a book the students could read in ten minutes. The data are provided in the worksheet titled 'Reading'. Conduct a test to determine whether there is enough evidence for the teacher to conclude that the average student can read at least four pages in a book in ten minutes. (Use $\alpha=0.01$.)

17. A study by the *BBC* looked at the wages earned monthly in Purchasing Power Parity dollars for workers in all of the world's countries. A sample of the mean wages earned by workers across sixteen randomly selected European countries has been obtained. The data are provided in the worksheet titled 'Wages'. If wages per country across the world are normally distributed with a mean of $1,538 and a standard deviation of $993, do the European countries have wage amounts that match exactly with the world average at the 5% significance level?

CHAPTER 9: Hypothesis Testing for a Population Proportion and Variance

Chapters 7 and 8 have described various ways of conducting hypothesis tests for a population's mean and of determining confidence intervals for a population's mean. Our questions about the world around us, however, are not limited to a population's mean. For example, we may wish to learn about a store's proportion of satisfied customers or about the standard deviation of product defects within a specific manufacturing process. In this chapter, we expand our exploration of hypothesis testing to cover a population's proportion and variance.

Regardless of the population parameter for which a hypothesis test is being performed, the basic logic of the testing procedure remains the same: (1) formulate the hypotheses, (2) compute the test statistic and/or p-value, (3) identify a decision rule and compute, if needed, a critical value, (4) apply the decision rule, and (5) draw and interpret your conclusion. We continue to follow this process in Chapter 9 and also compute confidence intervals for the population proportion and variance. This chapter covers the following topics:

- Test of a Single Population Proportion
- Sampling Distribution of Proportions
- Conducting a Hypothesis Test of a Single Population Proportion
- Confidence Interval for a Single Population Proportion
- Test of a Single Population Variance
- The χ^2 Distribution
- Conducting a Hypothesis Test of a Single Population Variance
- Confidence Interval for a Single Population Variance

UNIT 1

Test of a Single Population Proportion

Proportions are computed as the number of successes in n trials. For example, when we say that 60% of the students in a particular class are female, we are (1) defining a success as a student being female, (2) adding up the number of females (successes) in the class, (3) adding up the total number of students (trials) in the class, and (4) dividing the number of female students (successes) by the total number of students (trials) to find this proportion.

The procedure followed in testing proportions is similar to that described in the previous two chapters. We have a certain expectation about the population's proportion (which we denote as p) and we test this expectation using sample data. We illustrate the test making use of *Angry Birds*, a game that quickly became a global phenomenon after its release in December 2009. According to one source[87], 45% of those who played *Angry Birds* have purchased a version of the game (as opposed to downloading and playing the free version). These data are interesting because they can tell us about the willingness of online-game users to pay for their apps (be it *Angry Birds*, *Candy Crush*, *Minecraft*, or *Pokemon Go*). Suppose that we wish to test the accuracy of the claim that 45% of *Angry Birds* users purchased the paid version of the game, and we wish to conduct the test at the 10% level of significance. To test this claim, we first formulate the following null hypothesis about the proportion of people who purchased a version of the game:

$$H_0: p = 0.45$$

and our alternative hypothesis will therefore be:

---

[87] Source: http://aytm.com/blog/research-junction/angry-birds-addiction/

$$H_1: p \neq 0.45$$

We next collect data from a sample of past *Angry Bird* players, asking each whether or not he or she had purchased a version of the game. Out of 250 users surveyed, 95 responded that they had purchased a version of the game. Our sample proportion, which we denote as $\hat{p}$ (p-hat) can be computed as:

$$\hat{p} = \frac{95}{250} = 0.38$$

which is the number of successes (i.e., people who have purchased the game) divided by the sample size.

We are again faced with some familiar questions: does our sample provide sufficient evidence to reject the null hypothesis and, if so, at what level of significance? From previous chapters, you now know that just noting that the sample proportion (0.38) is different from the hypothesized proportion (0.45) is insufficient evidence. In order to determine if the null hypothesis can be rejected, we need to go through each of the steps of hypothesis testing. But, just as we rely on knowledge of the sampling distribution of the mean to carry out hypothesis tests regarding the population mean, we now rely on knowledge of the sampling distribution of proportions when carrying out hypothesis tests about the population proportion. Hence, we take a brief detour to learn about the *sampling distribution of proportions*.

Sampling Distribution of Proportions

The population proportion (p) represents the proportion of a population that has a specific attribute of interest (e.g., people who have purchased *Angry Birds*, satisfied customers, female students, democratic voters, etc.). We can also compute a sample proportion ($\hat{p}$), as the number of observations holding the attribute of interest (or, the number of successes) in a sample of size n.

For example, if we hypothesize that 45% of *Angry Birds* players have purchased a version of the game, then we believe that p=0.45. If we sample 250 Angry Birds players and find that 95 have purchased the paid version (x=95), the sample proportion is:

$$\hat{p} = \frac{x}{n} = \frac{95}{250} = 0.38$$

Remember that the above sample proportion just represents a single sample taken out of the population of *Angry Birds* players. We can take another sample of 250 players and are likely to find, say, a sample proportion of 0.40, or 0.46, and so on. Just like we did in Chapter 7, we can repeat the sampling process out of this population to create the **sampling distribution of proportions**, which describes the probabilities attached to all possible sample proportions (for samples of size n) that are repeatedly taken from the same population.

Applying the same *central limit theorem* introduced in Chapter 7 and given that our sample size is sufficiently large (i.e., when n*p≥5 *and* n*(1-p)≥5, with n being the sample size and p the population proportion), the sampling distribution of proportions is:

$$\hat{p} \sim N\left(p, \sqrt{\frac{p(1-p)}{n}}\right)$$

meaning that the sample proportion ($\hat{p}$) of samples of size n taken from the same population follows a normal distribution with a mean of p (the population proportion) and a standard deviation of $\sqrt{\frac{p(1-p)}{n}}$. This standard deviation is called the *standard error of the proportion*.

Let us consider a new example to understand this sampling distribution better before continuing with our hypothesis testing. *Wikipedia* has become one of the most popular sites on the Internet. The *Wikipedia* (English language version) contains over four and a half million articles,[88] with its most popular subject areas being: culture & arts (30%), biographies (15%) and geography (14%). Suppose we took a simple random sample of 60 articles. What is the probability that eighteen or more of these articles are biographies? In other words, assuming that the true population proportion (p) of biographies is 15%, what is the probability of selecting a sample of size n=60 in which the proportion of biographies ($\hat{p}$) will be 18/60=0.3 or higher? Or:

$$P(\hat{p} \geq 0.3) = ?$$

In order to compute this probability, we need to apply the sampling distribution of proportions. To do so, we first check to make sure that the sample is large enough, i.e., that *both* n*p and n*(1-p) are greater than five. As n*p=60*0.15=9 and n*(1-p)=51 for our example, we can proceed, knowing that:

$$\hat{p} \sim N\left(p, \sqrt{\frac{p(1-p)}{n}}\right)$$

Now, using the standard normal distribution (and converting our probability expression into Z scores), we compute:

---

[88] Source (data from December 2014): http://en.wikipedia.org/wiki/Wikipedia:Size_of_Wikipedia

$$P(\hat{p} \geq 0.3) = P\left(Z \geq \frac{\hat{p} - p}{\sqrt{\frac{P(1-p)}{n}}}\right) = P\left(Z \geq \frac{0.3 - 0.15}{\sqrt{\frac{0.15 * 0.85}{60}}}\right) = P(Z \geq 3.254) = 0.0006$$

That is, there is a very small chance (only 0.0006) of selecting a sample of 60 articles in which eighteen or more of these articles are biographies. This makes sense if you consider that we believe the true proportion of biographies to be 15%, which is much lower (in terms of standard errors) than our sought value of 30%. This is why we obtained such a high Z score and a correspondingly low probability.

Conducting a Hypothesis Test of a Single Population Proportion

Let us return to the *Angry Birds* example. It has been claimed that only 45% of players purchased a version of the game. To test the correctness of this number, we asked 250 *Angry Birds* players if they had purchased a version of the game. 95 of those surveyed indicated that they had paid for the game, with the remaining 155 players surveyed indicating that they had not purchased a version of the game. Use $\alpha=10\%$ to test whether the original claim is true.

We first describe the *critical value approach* for this hypothesis test.

Step 1: Formulate Hypotheses

The parameter of interest in this example is the population proportion of the people purchasing a version of the game (as opposed to downloading and playing the free version). Our hypotheses are:

$$H_0: p = 0.45$$

$$H_1: p \neq 0.45$$

where p is the hypothesized population proportion.

Step 2: Compute the Test Statistic

We know that the sampling distribution of proportions is approximately normally distributed, given a sufficiently large sample. We thus need to check to see if our sample is large enough by examining whether both n*p and n*(1-p) are at least five. Using n=250 and a p of 45%, we find both conditions to hold:

$$n*p = 250*0.45 = 112.5$$

$$n*(1-p) = 250*0.55 = 137.5$$

Thus, we can rely on the central limit theorem to know:

$$\hat{p} \sim N\left(p, \sqrt{\frac{p(1-p)}{n}}\right)$$

As in the previous tests we learned, the test statistic is a measure of the distance (in standard errors) of our sample proportion from the hypothesized proportion, which means that our test statistic is computed as the sampled value minus the hypothesized value divided by the standard error of the estimate. Hence, our test statistic is

$$Z = \frac{(\hat{p} - p)}{\sqrt{\frac{p(1-p)}{n}}}$$

where $\hat{p}$ is the sample proportion, p is the hypothesized proportion, and n is the sample size. From the information provided in the example, we know that p is 0.45 and n=250. We can compute $\hat{p}$ as: $\hat{p}$=95/250=0.38, which means that 95 out of the 250 survey respondents, or 38%, said they purchased a version of the game. Calculating our test statistic yields:

$$Z = \frac{(0.38 - 0.45)}{\sqrt{\frac{0.45(1 - 0.45)}{250}}} = -2.225$$

Step 3: Formulate the Decision Rule

This is a two-tail test and we will reject the null hypothesis if our test statistic is much higher or much lower than the hypothesized value. As we are looking at two rejection regions (one in the distribution's lower-tail and one in the distribution's upper-tail), our decision rule is: reject the null hypothesis if the test statistic falls in either rejection region. In order to define the two rejection regions, we need to determine the associated critical values, such that:

P(Z≥|critical value|) = α/2

At this point in our study of statistics, you should be able to figure out that this critical value is ±1.645, given that the significance level for the test is 10%. Our decision rule becomes: reject the null hypothesis if the test statistic falls in either the upper or lower rejection region defined by ±1.645.

Step 4: Apply the Decision Rule

Noting that the test statistic, -2.225, is to the left of our lower-tail critical value (-1.645), as shown in Figure 9-1, we reject the null hypothesis.

Figure 9-1
Comparing the Critical Value and Test Statistic
for the *Angry Birds* Example

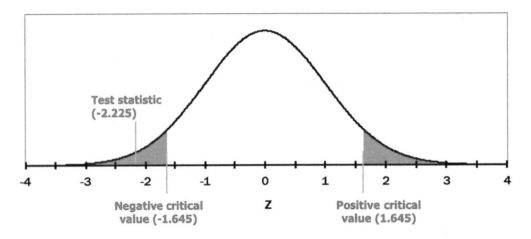

Step 5: Draw a Conclusion

By rejecting the null hypothesis, we conclude at the 10% level of significance that the claim that 45% of *Angry Birds* players purchased a version of the game is too high. Based on the hypothesis test using data from our survey, the proportion of players purchasing *Angry Birds* should be lower than 45%.

We could have just as easily tested our hypothesis by using the *p-value approach* to hypothesis testing. Specifically, recall that the p-value is the probability of finding a test statistic as extreme as the one found in our sample. In the critical value approach to hypothesis testing, the value of the test statistic was found to be -2.225. Because this is a two-tail test, the p-value should capture both tails of the distribution; therefore, it is computed as: $2*P(Z \leq -2.225)$. We can use the standard normal table or Excel to obtain this p-value. Using Excel and typing =NORM.S.DIST(-2.225,1), a value of 0.013 is obtained. Because this is a two-tail test, the p-value is actually: $2*0.013=0.026$. Finally, we compare this p-value with the specified α, which in this example is 0.1. Because the p-value of 0.026 is lower than α, we reject the null hypothesis and conclude that fewer than 45% of *Angry Birds* players purchased a version of the game. As would be expected, this is the same conclusion reached using the critical value approach to hypothesis testing.

Confidence Interval for a Single Population Proportion

Similar to the sample mean, the sample proportion is a *point estimator* of the population proportion. Lacking any other knowledge, the sample proportion is our best estimate of the true population proportion, the parameter of interest. Similar to Chapter 8, we can construct a *confidence interval* around the sample proportion to more confidently estimate the value of the true population proportion. Applying what you learned about computing confidence intervals in Chapter 8, the formula to use is:

$$\hat{p} \pm Z_{\alpha/2} \sqrt{\frac{\hat{p}(1-\hat{p})}{n}}$$

where $\hat{p}$ is the sample proportion, n is the sample size, and $Z_{\alpha/2}$ is the cutoff value for a 1-α confidence interval. Using our *Angry Birds* example, if we wish to construct a 95% confidence interval for the population proportion based on our sample proportion of 38% (i.e., 95 out of the 250 people surveyed indicated that they had purchased a version of the game), we can compute (filling in the example's values in the above formula):

$$0.38 \pm 1.96 \sqrt{\frac{0.38(1-0.38)}{250}}$$

to find a *lower confidence limit* of 32% and an *upper confidence limit* of 44%. We can further see that the hypothesized value of 45% is *not* included in our confidence interval; therefore, at a 95% confidence level, we can state that the true proportion of those who purchased the game is lower than 45%.

Let us go through this example again, but now use a 90% confidence interval, as we have initially conducted our hypothesis tests using the critical value and p-value approaches using $\alpha=10\%$. Before we actually compute this confidence interval, think about the change you expect to see in the width of the interval. Will it be *narrower* or *wider*?

The 90% confidence interval becomes:

$$0.38 \pm 1.645 \sqrt{\frac{0.38(1-0.38)}{250}}$$

(Note that, here, the value of $Z_{\alpha/2}$ is identical to the critical value we had previously computed in conducting the hypothesis test.) Using this mathematical expression, we compute a *lower confidence limit* of 33% and an *upper confidence limit* of 43%. This 90% confidence interval is narrower than the 95% confidence interval previously determined. Is this what you expected? Recall that the width of a confidence interval depends on the confidence level such that, the more confident we wish to be, the wider the interval. In this case, we have reduced our confidence level from 95% to 90% and we obtained a narrower interval.

Again, we see that the hypothesized value of 45% is *not* included in our confidence interval. Therefore, at a 90% confidence level, we can state that the true proportion of those who had purchased the *Angry Birds* game is lower than 45%.

Unit 1 Summary

- The hypothesis test of a single population's proportion follows the same basic five-step process used in Chapters 7 and 8.

- The **sampling distribution of proportions** describes the probabilities attached to all possible sample proportions that are repeatedly taken from the same population. For large enough samples, i.e., when *both* n∗p and n∗(1-p) are at least five, the sampling distribution of proportions can be defined as:

$$\hat{p} \sim N(p, \sqrt{\frac{p(1-p)}{n}})$$

- Use the following formula to compute the test statistic:

$$Z = \frac{(\hat{p} - p)}{\sqrt{\frac{p(1-p)}{n}}}$$

- The critical value is a Z score value obtained using α from the standard normal distribution table.

- Use the rules in Table 9-1 to help you decide whether or not to reject the null hypothesis.

Table 9-1
Critical Value Approach Decision Rules for a Hypothesis Test of a Single Population Proportion

Type of Test	Decision Rule
Lower-tail test H_0: p ≥ value H_1: p < value	Reject the null hypothesis if: test statistic < lower critical value
Upper-tail test H_0: p ≤ value H_1: p > value	Reject the null hypothesis if: test statistic > upper critical value
Two-tail test H_0: p = value H_1: p ≠ value	Reject the null hypothesis if: test statistic < lower critical value Or if: test statistic > upper critical value

- The confidence interval of the sample proportion is computed as:

$$\hat{p} \pm Z_{\alpha/2} \sqrt{\frac{\hat{p}(1-\hat{p})}{n}}$$

Unit 1 Exercises

1. For each of the following cases: (1) find the test statistic; (2) find the critical value(s); and (3) state your rejection rule:

   a. $n = 125$; $x = 55$; H_1: $p < 0.5$; $\alpha = 5\%$

   b. $n = 83$; $x = 21$; H_1: $p < 0.3$; $\alpha = 10\%$

   c. $n = 210$; $x = 150$; H_1: $p \neq 0.65$; $\alpha = 10\%$

   d. $n = 35$; $x = 30$; H_1: $p > 0.85$; $\alpha = 5\%$

   e. $n = 108$; $x = 8$; H_1: $p > 0.01$; $\alpha = 1\%$

2. The U.S. wine industry has a market value of $35 billion, with California producing 90% of all U.S. wine.[89]

   a. In a sample of 200 bottles of U.S.-produced wine, what is the probability of more than 75% of the bottles originating from California?

   b. Repeat Question 2.a. with a sample size of 100 and a sample size of 50.

   c. What happens to the probability as the sample size decreases?

3. Construct the confidence interval for a population proportion in each of these cases:

   a. $n = 19$; $x = 9$; $1-\alpha = 90\%$

   b. $n = 41$; $x = 32$; $1-\alpha = 95\%$

   c. $n = 60$; $x = 55$; $1-\alpha = 92\%$

   d. $n = 67$; $x = 60$; $1-\alpha = 97\%$

   e. $n = 81$; $x = 10$; $1-\alpha = 99\%$

4. 30% of wealthy investors do not use financial advisors.[90] If a sample of 80 wealthy investors is taken, what is the probability that more than 40% of them do not use a financial advisor?

---

[89] Source: http://wineamerica.org/policy/by-the-numbers
[90] Source: http://visual.ly/how-will-you-attract-next-generation-investors

5. 35% of coffee drinkers prefer their coffee black.[91] If a sample of 50 coffee drinkers is taken, what is the probability that more than twenty preferred their coffee black?

6. A new restaurant that just opened claims to have a 95% customer satisfaction level. Surveying 54 random customers exiting the restaurant, you find that 49 of the customers give positive reviews. Using a 5% significance level, does the restaurant's customer satisfaction claim hold true?

7. A university business program states that at least 52% of its graduates receive offers for an internship or a summer job in their desired field of study in the summer between their junior and senior academic years. To test this claim, you survey three classes with only junior students, right before summer break. 50 students are surveyed in each class. You find that only 70 of these students have received internship or job offers in their desired fields of study. Does this refute the claims of the university business program? Use a 5% significance level.

8. In today's new marketing era, social media is a must-have for product promotion, product advertising and customer engagement. A research study claims the following percentages of companies monitoring specific social media activities[92]:

    - Changes of friends/followers: 65.5%
    - Site traffic: 59.5%
    - Mentions: 39%
    - New leads generated: 35.7%
    - Sales: 28.5%
    - Key influencers/reach of message: 25.7%
    - Duration of engagement: 21.6%

    Based on a random sample of 1,000 businesses, you collect the following data:

    - 640 track changes of friends/followers
    - 575 track site traffic
    - 335 track mentions
    - 310 track new leads generated
    - 250 track sales
    - 215 track key influencers/reach of message
    - 200 track duration of engagement

    Construct a 95% confidence interval for each of the above statistics. Which of the reported statistics are overstated, given your sample?

---

[91] Source: http://visual.ly/coffee-facts-0
[92] Source: http://visual.ly/new-world-marketing

9. You need to decide whether to use *UPS* or *FedEx* to transport animals from the island of Madagascar to the *New York City Zoo* in time for an exhibit. It has been shown that 91% of *UPS* packages are delivered on time.[93] Recently, you have been using *FedEx*, with 66 of the 75 packages delivered by *FedEx* being delivered on time. Using $\alpha=0.01$, test to see if *FedEx* has a lower on-time delivery percentage than the *UPS* figure of 91%.

10. The gift shop at *Harvard University* receives hundreds of tourists who purchase memorabilia daily. With most of the sales attributable to sweaters, the manager is thinking about setting up a sale on other items. However, he will only do so if sweaters exceed 40% of total sales. But, no one has yet set up a proper inventory system at the gift shop. As a result, the only way to estimate the true proportion of sweater sales is by manually recording items sold. Of the next 150 purchases, 68 are sweaters. Using a 5% significance level, advise the manager on whether or not to have a sale on all non-sweater items.

11. *Coachella* is an annual three-day music and arts festival in California and includes rock, indie, hip-hop, and electronic music performers. Although some acts are well-known, most are not mainstream. Your friend claims that over 50% of the performers have less than 100,000 *Facebook* likes. You randomly selected 40 acts, 22 of which have fewer than 100,000 likes. At the 1% significance level, was your friend right?

12. The phenomenon of a patient's perceived medical improvement following treatment with an inert substance is called the *placebo effect*. A university study claims 42% of headache patients respond positively after taking a placebo. A recent investigation administered 400 headache patients with placebos, with 152 positive responses. Was the university study too eager to report such a high rate, given your sample? (Use $\alpha=0.01$.)

13. America represents less than 5% of the world's population, but an astonishing 23% of the world's incarcerated population.[94] One problem is the percentage of felons who end up back in prison after their sentence is up, estimated at 52%. A study believing this percentage to be too low takes a random sample of 6,000 released felons, 3,345 of whom were again imprisoned. Test at the 5% significance level to see if the estimated rate is accurate.

14. A school system in the Yukon Territory in Canada claims that over 2% of their school days are snow days, i.e., days in which the school is closed due to excessive snow and/or ice. A random sample of 200 days in the past ten years finds that eight were snow days. Test this claim at $\alpha=0.05$.

---

[93] Source: http://visual.ly/battle-royale-fedex-vs-ups
[94] Source: http://en.wikipedia.org/wiki/United_States_incarceration_rate

15. 15% of video game sales are games rated M for mature.[95] You believe that within your group of friends, the proportion of M-rated games is higher than this. You take a random sample of the game collections of you and your friends, and you find that nine of the 50 games in this sample are rated M. Test the original claim at the 5% significance level.

16. Every year the *Academy Award* winners are chosen by the *Academy of Motion Picture Arts and Sciences* (*AMPAS*), a professional honorary organization whose members' identity is not made public. Recently, concerns about the small number of minority and female voters were raised by numerous sources[96] and the academy is making an effort to correct this bias. Suppose that the last estimate you were aware of was that females constituted 22% of *Oscar* voters. To see if public opinion has affected *AMPAS* membership, you look at a sample of 400 members whose identity is known and find that 95 of which are female. At $\alpha=0.01$, is this enough evidence to conclude female make up more than 22% of the total number of *AMPAS* members?

Unit 2

Test of a Single Population Variance

When an educational testing service designs a test, they need to ensure that the scores fall within a specific range and that the distribution of scores allows for differentiating among students' performance. A new test is designed so that the standard deviation in student scores would be *at least* 70 points, enabling proper differentiation among students. To determine whether this objective was achieved, the testing service gave the test to a random sample of 45 students and found that the sample mean and standard deviation were 673 and 48.5, respectively. At the 5% level of significance, what can you conclude about the standard deviation of the test?

Here is an example of a hypothesis test of yet another population parameter, the standard deviation. The test we review in this unit allows us to draw inferences

---

[95] Source: http://visual.ly/video-game-statistics
[96] For example: http://graphics.latimes.com/oscars-2016-voters/

about both the standard deviation and the variance of a population of interest. The test itself is defined as a test of a population variance due to the distribution we will work with, the χ^2 distribution introduced below. Because the standard deviation is computed as the square root of the variance, we can draw inferences about it from hypothesis tests done on the variance.

As with other tests, we hypothesize about the true value of the variance (σ^2) of some variable of interest vis-à-vis a predefined value (e.g., H_0: $\sigma^2=4,900$) and we follow the same five-step process to conduct the test. Also similar to the other hypothesis tests that we have previously covered, we need to work with the appropriate distribution for sample variances. In this case, the appropriate distribution is the χ^2 distribution. Thus, we again begin with a brief detour introducing the χ^2 distribution prior to demonstrating how to conduct the hypothesis test for a single population variance.

The χ^2 Distribution

Unlike the other distributions we have worked with so far in hypothesis testing, the χ^2 distribution is not symmetrical; instead, it is positively skewed. This fact is important in conducting hypothesis tests because it means that we need to obtain (1) different critical values in the distribution depending on whether we are conducting an upper-tail or a lower-tail test and (2) two different critical values for a two-tail test. This is illustrated in Figure 9-2.

Figure 9-2
Chi-Square Distribution

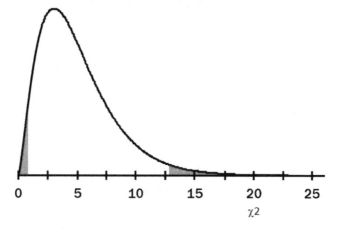

As with the t-distribution, the shape of the χ^2 distribution is also a function of the number of degrees of freedom (again defined as n-1 with n indicating the sample size). Finally, you can never have negative values in this distribution because neither the population variance nor the sample variance can take on negative values.

We can use either Excel or the χ^2-distribution table to find values in the χ^2 distribution. We explain first how to use the printable table (provided in this chapter's Excel file in the worksheet titled 'Chi Square Distribution Table'). But, our discussions of the hypothesis testing process will use associated Excel functions.

When looking up values in the χ^2-distribution table, you need to consider your degrees of freedom (n-1) and the desired significance level of the test (α). (The top row of the table provides different values of α and the first column of the table provides different numbers of degrees of freedom.) Looking at this χ^2-distribution table (or, χ^2-table), you should see that there is a division halfway through the top row. The first five headings provide α values for 90% or higher, whereas the next five headings provide α values for 10% or lower. This is intentional, as it reflects the non-symmetrical nature of the χ^2 distribution. The χ^2-table, thus, provides two sets of α values: one set for the lower-tail and one set for the upper-tail. The table is constructed in such a way that it always displays the right-tail corresponding to the probability given in the column heading. In other words, the table provides a value in the χ^2 distribution such that the probability of values falling to the right of this value is our pre-specified probability (i.e., α). (We have used this formatting of the χ^2-distribution table because this is the way the table normally appears in statistics books.) Therefore, if we are looking for an upper-tail *critical value*, we use the specified value of α to obtain a value from the table; and, if we are looking for the lower-tail *critical value*, we use $1-\alpha$. For example, say we are conducting an upper-tail test with $\alpha=0.05$ and $n=30$. We thus look up 29 degrees of freedom in the left-most column and $\alpha=0.05$ in the top row. The corresponding *critical value* of 42.6 is highlighted in Table 9-2. Likewise, if we are looking for the lower-tail critical value with $\alpha=0.05$ and $n=30$, we look up 29 degrees of freedom in the left-most column and $1-\alpha=0.95$ in the top row to obtain a *critical value* of 17.7 (also highlighted in the Table 9-2). Finally, if we were conducting a two-tail test, we would use $\alpha/2$ and $1-\alpha/2$ to obtain the needed two *critical values* from the χ^2-table.

Table 9-2
Finding Values in the χ^2 Table

df	$\chi^2_{.995}$	$\chi^2_{.990}$	$\chi^2_{.975}$	$\chi^2_{.950}$	$\chi^2_{.900}$	$\chi^2_{.100}$	$\chi^2_{.050}$	$\chi^2_{.025}$	$\chi^2_{.010}$	$\chi^2_{.005}$
...	...	...	...	...	...	...	...	...	...	...
26	11.2	12.2	13.8	15.4	17.3	35.6	38.9	41.9	45.6	48.3
27	11.8	12.9	14.6	16.2	18.1	36.7	40.1	43.2	47.0	49.6
28	12.5	13.6	15.3	16.9	18.9	37.9	41.3	44.5	48.3	51.0
29	13.1	14.3	16.0	**17.7**	19.8	39.1	**42.6**	45.7	49.6	52.3
30	13.8	15.0	16.8	18.5	20.6	40.3	43.8	47.0	50.9	53.7
40	20.7	22.2	24.4	26.5	29.1	51.8	55.8	59.3	63.7	66.8
...	...	...	...	...	...	...	...	...	...	...

In Excel, the same DIST and INV functions we have used with previous distributions are also available for the Chi-Square distribution. The function =CHISQ.INV(α,df) returns the *lower-tail* critical value. For example =CHISQ.INV(0.05,29) returns the lower-tail *critical value* of 17.7. Note that in Excel we use α (0.05) rather than 1-α (0.95), as we did in Table 9-2, to obtain the lower-tail value. We recommend that you choose either Excel or the table as you move forward with the material. Switching from one to the other may be confusing. To obtain the upper-tail *critical value* of 42.6, you use =CHISQ.INV.RT(0.05,29). With the latter function, the 'RT' indicates you wish to obtain a right-tailed critical value.

The function =CHISQ.DIST(test_statistic,df,cumulative) returns the p-value for a given test statistic. For example, =CHISQ.DIST(17.7,29,1) returns a value of 0.05, which is the probability of a value falling *below* 17.7 in our distribution. To find an upper-tail p-value, you use =CHISQ.DIST.RT(test_statistic,df). For example, =CHISQ.DIST.RT(42.6,29) returns a value of 0.05, meaning that the probability of a value being *greater than* 42.6 is 0.05.

Conducting a Hypothesis Test of a Single Population Variance

Let us return to this unit's opening example to go through the steps followed in conducting a hypothesis test of a single population variance. An educational testing service has designed an achievement test so that the standard deviation in student scores would be *at least* 70 points (or a variance of at least 4,900). To determine whether this objective was achieved, the testing service gave the test to a random sample of 45 students and found that the sample mean and standard deviation were 673 and 48.5, respectively. At the 5% level of significance, what can you conclude about the standard deviation of the test?

Step 1: Formulate the Hypotheses

The parameter of interest in this example is the standard deviation. However, we will be working with the variance to conduct the test (since the variance follows the chi-squared distribution), and we will return to the standard deviation in Step 5 when we draw our final conclusion. Our hypotheses are:

$$H_0: \sigma^2 \geq 4,900$$

$$H_1: \sigma^2 < 4,900$$

where σ^2 is the desired variance in test scores.

Step 2: Compute Your Test Statistic

The formula for computing the test statistic is shown below:

$$\chi^2 = \frac{(n-1)s^2}{\sigma^2}$$

where s^2 is the sample variance, σ^2 is the hypothesized variance, and n is the sample size. Using the data provided, we compute the value of our test statistic to be:

$$\chi^2 = \frac{(n-1)s^2}{\sigma^2} = \frac{(45-1)*48.5^2}{70^2} = 21.12$$

Step 3: Formulate the Decision Rule

We are conducting a lower-tail test, and we will use the critical value approach to hypothesis testing. Hence, the decision rule to be applied is: reject the null hypothesis if the test statistic falls in the lower-tail rejection region.

In order to find the corresponding critical value we can use Excel by typing =CHISQ.INV(0.05,44) to obtain the value of 29.79. Given this critical value, we restate the decision rule: reject the null hypothesis if the test statistic falls in the lower-tail rejection region defined by the critical value 29.79.

Step 4: Apply the Decision Rule

Comparing the test statistic (21.12) to this critical value (29.79), we find the test statistic to be to the left of the critical value and, hence, in the lower-tail rejection region. This is illustrated in Figure 9-3. Therefore, at the 5% significance level, we reject the null hypothesis.

**Figure 9-3
Comparing the Test Statistic and Critical Value
for the Education Testing Service Example**

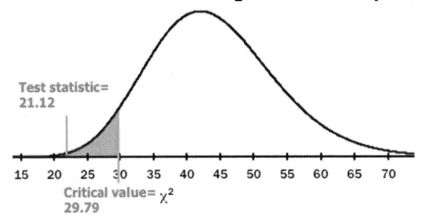

Step 5: Draw Your Conclusion

By rejecting the null hypothesis, we conclude that the variance of test takers' scores is lower than 4,900, meaning that the standard deviation is not 70 or higher, as desired by the educational testing service.

As pointed out regarding the t-distribution, the χ^2 distribution's dependence on degrees of freedom limits the usability of the χ^2-table to compute p-values. However, if Excel is available, we can use the =CHISQ.DIST() function or the =CHISQ.DIST.RT() function to obtain an exact p-value. Typing =CHISQ.DIST(21.12,44,1), we obtain a p-value of 0.0014. The decision rule with the p-value approach states that we should reject the null hypothesis if this p-value is less than α. Since 0.0014 is less than 0.05 (the significance level specified for this test), we reject the null hypothesis to reach the same conclusion as that obtained from applying the critical value approach to hypothesis testing. Thus, we again conclude that the standard deviation of test takers' scores is not 70 or higher, as desired by the educational testing service.

Now let us conduct a two-tail hypothesis test of a population variance. Consider an example where a cereal manufacturer wishes to test whether the population variance of the weight of empty cereal boxes is equal to 0.05 oz². A manufacturing engineer took a random sample of twenty boxes and found the standard deviation for the sample, s, to be 0.25 oz. Based on this sample, should the engineer question the weight of the boxes? Assume the population of box variances is normally distributed and conduct the test for $\alpha=0.05$.

Step 1: Formulate the Hypotheses

The parameter of interest is the population variance, σ^2, with the desired variance being 0.05 oz². This desired variance goes into our hypotheses:

$$H_0: \sigma^2 = 0.05 \text{ oz}^2$$

$$H_1: \sigma^2 \neq 0.05 \text{ oz}^2$$

where σ^2 is the variance of the weight of empty cereal boxes.

Step 2: Compute the Test Statistic

We now compute the test statistic:

$$\chi^2 = \frac{(n-1)s^2}{\sigma^2} = \frac{(20-1) * 0.25^2}{0.05} = 23.75$$

Note in particular that we used the square of the sample standard deviation (0.25²) in this mathematical expression. The denominator of 0.05 is already a squared value (it is the hypothesized variance) so we did not square it again.

Step 3: Formulate the Decision Rule

We are conducting a two-tail test and we will use the critical value approach to hypothesis testing. Hence, the decision rule to be applied is: reject the null hypothesis if the test statistic falls in either the lower-tail rejection region or the upper-tail rejection region.

Because this is a two-tail test, we need to find two critical values in the χ^2-table, one corresponding to $\alpha/2$ and the other to $1-\alpha/2$ (remember that in two-tail tests α is split over both tails and, hence, we divide it by two). To obtain these values, we will use Excel. We type =CHISQ.INV(0.025,19) to obtain a lower-tail value of 8.906 and we type =CHISQ.INV.RT(0.025,19) to obtain the upper-tail value of 32.852. Obtaining these two critical values (shown in Figure 9-4) enables us to restate the decision rule: reject the null hypothesis if the test statistic is less than 8.91 or is greater than 32.9.

Step 4: Apply the Decision Rule

Comparing the test statistic (23.75) to the critical values (also shown in Figure 9-4), we find the test statistic to be between the two critical values:

$$8.91 < 23.75 < 32.9$$

Therefore, we *do not* reject the null hypothesis.

Figure 9-4
Comparing the Test Statistic with the Two Critical Values for the Cereal Box Example

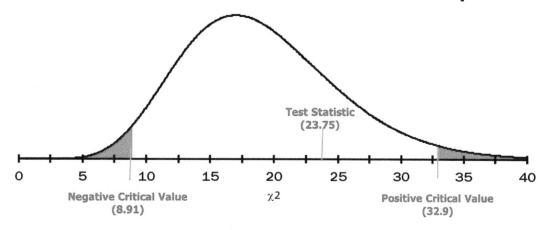

Step 5: Draw Your Conclusion

By not rejecting the null hypothesis, we conclude, at the 5% level of significance, that the variance in the weights of cereal boxes is not significantly different than 0.05 oz².

Given access to Excel, we could use the =CHISQ.DIST() function to obtain a p-value and then carry out the p-value approach to hypothesis testing. With this example we need to look for an upper-tail p-value, since our test statistic is located in the upper tail of the distribution. Thus, we use the computed test statistic (23.75) and type =CHISQ.DIST.RT(23.75,19) to obtain a value of 0.206. As this is a two-tail test, we double this value to obtain the p-value:

$$2*0.206 = 0.412$$

As this derived p-value is much higher than α, we do not reject the null hypothesis.

Confidence Interval for a Single Population Variance

We can estimate the population variance using a confidence interval, just like we did for the population mean and the population proportion. The key difference in how confidence intervals are computed for a population variance is directly associated to the non-symmetrical nature of the χ^2 distribution and the need to use distinct χ^2 values in computing the lower confidence limit (*LCL*) and upper confidence limit (*UCL*), as shown below:

$$LCL = \frac{(n-1)s^2}{\chi^2_{upper-tail}} \qquad UCL = \frac{(n-1)s^2}{\chi^2_{lower-tail}}$$

Note the denominators in the above formula correspond to the upper-tail and lower-tail critical value you would obtain form the chi-square distribution using α/2. We use the upper-tail value, which is the larger number, to obtain the lower confidence limit and the lower-tail value to obtain the upper confidence limit. Let us consider an example to see how confidence intervals are computed for the population variance.

A diabetes patient is concerned with the standard deviation of her daily glucose levels. High variation in glucose levels is likely to have health effects and may be risky to the patient. She would like to ensure that her glucose levels are relatively stable. She tracks her daily test results over a period of 25 days and obtains a standard deviation of 100 mg/dl.[97] Using the two formulas provided above and the =CHISQ.INV() function in Excel, she constructs a 95% confidence interval for the variance of her blood glucose levels:

$$LCL = \frac{(25-1)100^2}{39.364} = 6{,}096.94 \qquad UCL = \frac{(25-1)100^2}{12.401} = 19{,}353.28$$

In other words, at the 95% confidence level, she knows that the standard deviation of her daily glucose level is captured by the interval [78.083...139.12].[98]

---

[97] milligrams/deciliter, a standard glucose level measure.
[98] Using the square roots of the variances obtained in our *LCL* and *UCL* calculations.

Further, we can use this example to again demonstrate how hypothesis tests for a population variance can be used to draw conclusions about the population standard deviation. Assume that the reason she had calculated this confidence interval had to do with a concern for keeping the standard deviation of her glucose levels below 150 mg/dl. By computing the confidence interval, she can conclude, at the 95% confidence level, that the standard deviation does not exceed 150 mg/dl since this value is not included in the computed confidence interval.

Unit 2 Summary

- The hypothesis test of a single population variance follows the same five-step process we have been following for all hypothesis tests.

- The sample variance follows the χ^2 distribution with n-1 degrees of freedom.

- Use either α or 1-α and the number of degrees of freedom (n-1) to look up the *critical value* in the χ^2 table. In Excel, use the =CHISQ.INV(α,df) for the lower-tail critical value and =CHISQ.INV.RT(α,df) for the upper-tail critical value. Use $\alpha/2$ and 1-$\alpha/2$ for the two-tail tests.

- Use the following formula to compute the test statistic:

$$\chi^2 = \frac{(n-1)s^2}{\sigma^2}$$

- Use the rules in Table 9-3 to help you decide whether or not to reject the null hypothesis.

Table 9-3
Critical Value Approach Decision Rules for a Hypothesis Test of a Single Population Variance

Type of Test	Decision Rule
Lower-tail test H_0: $\sigma^2 \geq$ value H_1: $\sigma^2 <$ value	Reject the null hypothesis if: test statistic < lower critical value (using $1-\alpha$)
Upper-tail test H_0: $\sigma^2 \leq$ value H_1: $\sigma^2 >$ value	Reject the null hypothesis if: test statistic > upper critical value (using α)
Two-tail test H_0: $\sigma^2 =$ value H_1: $\sigma^2 \neq$ value	Reject the null hypothesis if: test statistic < lower critical value (using $1-\alpha/2$) Or if: test statistic > upper critical value (using $\alpha/2$)

- Use the following formulas to compute the confidence interval for the population variance:

$$LCL = \frac{(n-1)s^2}{\chi^2_{\alpha/2}} \qquad UCL = \frac{(n-1)s^2}{\chi^2_{1-\alpha/2}}$$

Unit 2 Exercises

1. For each of the following cases: (1) find the test statistic, (2) find the critical value(s), and (3) state your rejection rule:

   a. $n = 10$; $s^2 = 3.25$; H_1: $\sigma^2 \neq 4.03$; $\alpha = 5\%$

   b. $n = 35$; $s^2 = 1.12$; H_1: $\sigma^2 \neq 13.5$; $\alpha = 1\%$

   c. $n = 49$; $s^2 = 25.1$; H_1: $\sigma^2 < 25.8$; $\alpha = 1\%$

   d. $n = 100$; $s^2 = 220$; H_1: $\sigma^2 < 300$; $\alpha = 5\%$

   e. $n = 108$; $s^2 = 64.5$; H_1: $\sigma^2 > 61.8$; $\alpha = 5\%$

2. Construct a confidence interval of the population variance for each of the following cases:

   a. $n = 83$; $s = 5$; $\alpha = 10\%$

   b. $n = 81$; $s = 0.3$; $\alpha = 1\%$

   c. $n = 34$; $s = 1.6$; $\alpha = 5\%$

d. n = 10; s = 15; α = 10%

e. n = 27; s = 30; α = 5%

3. A stamp enthusiast believes his collection of rare stamps is unique not only because of the collection's worth, but also because the variation in stamp age is so high. He claims that the variance between the years that the stamps in his collection were created is over 200. To test his claim, a local magazine looked at a random sample of fifty stamps from his collection and recorded their year of print. The sample standard deviation is 14.307. Given this sample, test the enthusiast's claim at the 1% significance level.

4. The CFO of a chain of movie theaters wants to test the variation of ticket prices in his chain's theaters. He believes there should not be a significant difference in price between theaters and show times. The Board of Directors decides to take action if the standard deviation in ticket price exceeds $3. They sample eight movies and find that the sample standard deviation is 3.1472. Given this information, estimate, at the 95% confidence level, the variation in ticket prices. Should the Board take action?

5. The world record for solving a *Rubik's Cube* is an incredible 4.90 seconds and is held by an American named Lucas Etter.[99] To prepare for next year's competition, he wants to solve the cube both quickly and consistently. Last year, the variance between his tries was 0.245 seconds squared. He solves the cube ten times and records his times. His sample variance is 0.0982 seconds squared. Given this sample of attempts, has his consistency improved? (Use α=0.1.)

6. Jerry Seinfeld, the well-known comedian and actor, uses *Twitter* to connect with fans and share jokes. With over 1.5 million followers, his tweets attract lots of re-tweets and favorites. His publicity representative tells him his variance for re-tweets last month was 60,000. To check the consistency of his tweets' popularity this month, he takes a sample of the number of re-tweets on 200 random tweets. The sample variance is 52,902.9. Using α=0.05, has the variance decreased?

7. In baseball, a changeup pitch is used to make the pitch look like a fastball, but travel much slower, confusing the batter. One major league baseball pitcher wants to maximize the difference in speeds between his pitches, improving the variance to over 45. A sample of seven pitches reveals a variance of 48.476. Given this sample and using α=0.1, has he done so successfully?

---

[99] Source: https://www.worldcubeassociation.org/results/regions.php

8. Super giant slalom, or Super-G, is a skiing event where competitors ski 2.2 km down a slope with an average gradient of 27%. An Olympic skier is training for the next winter's competition and she would like to reduce the standard deviation between her runs to fewer than two seconds. A sample of seven runs shows a sample standard deviation of 1.679 seconds. At the 5% significance level, was she able to reduce her standard deviation to less than two seconds?

9. One employer looking to hire new graduates is interested in not only average grades, but also in the variance between grades. He will not hire candidates whose grade standard deviation is significantly higher than fifteen. He collected data on three students whose grades are shown in Table 9-e1. Construct a 95% confidence interval for the variance of grades for each of the three students. Which student (or students) should he consider hiring?

Table 9-e1

Student 1	Student 2	Student 3
90	72	86
82	78	90
75	83	69
89	91	91
71	93	80
74	78	80
86	84	75
74	74	79
84	92	93
93	87	89
84	70	70

END-OF-CHAPTER PRACTICE

1. George is concerned about his day-to-day heart rate fluctuation and is worried that the fluctuation is not normal. His friend (a second-year medical student) advises him to: first, measure his heart rate at the same time each day for a number of days; then, test the day-to-day variation to see if it is higher than ten; and, if it is, schedule an appointment with a doctor. The data are provided in the chapter's Excel file in the worksheet titled 'Heart Rate'. Given this sample, should George be concerned at the 1% significance level?

2. A political science student is concerned about the lack of student participation in government and believes that no more than half of all students have in any way participated in government. She randomly surveys 250 students at her university, asking each whether he or she has ever participated in their government in any way. Students that answered 'no' were recorded as 1 and students who answered 'yes' were recorded as 2. The data are provided

in the chapter's Excel file in the worksheet titled 'Government'. Based on the data she collected, can the political science student conclude at the 10% significance level that her belief is correct?

3. A recent press release by a respected national public opinion polling organization stated that more than 52% of voters are going to be casting their votes for Candidate A in the upcoming elections. To test this claim, a local journalist asked a random sample of 75 people whether they plan to vote for Candidate A. Their responses are provided in this chapter's Excel file in the worksheet titled 'Candidate A Voting'. Is there sufficient evidence to suggest that Candidate A has a solid chance of winning the election? Use a 1% significance level.

4. A middle school cafeteria assistant manager wants to see how many students finish all of the food that each has bought at the cafeteria. He believes that three-fourths of the students do not finish all of their food. To test this belief, he randomly samples 225 students. He marks a 1 if there was food left on a student's tray and a 0 if there was no food left on the tray. These data are provided in this chapter's Excel file in the worksheet titled 'Cafeteria'. Based on these data, can the assistant manager assume at the 10% significance level that his belief is correct?

5. The city's transit system has introduced new schedules for its buses on major routes and claims that less than 5% of the buses will be late in arriving at bus stops. You are a time-strapped university student. Before purchasing an annual bus pass, you do some research. You observe different bus stops and find that 145 out of the 150 arriving buses were, in fact, on time! Have the new schedules resulted in better on-time arrivals? Use a 10% significance level.

6. A steel rod manufacturing company is in the process of buying a new machine. To test out the capabilities of two different machines, the company tested the production of 25 steel rods on each machine. They found that Machine A produced rods with variance in the diameter of 6.2 mm^2 while Machine B's rods had diameter variance of 8.4 mm^2. Construct a 99% confidence interval for the variance of each the machines.

7. A professor keeps track of the scores students receive in the final exam for the Introduction to Finance course she teaches. She has computed the variance to be 420 over the past few years that she has taught this course. She just made significant changes to the way she structured and taught the course and she wants to determine if these changes were effective. To determine this, she tests to see if the variance in final exam grades has decreased, as a reduced variance would indicate that there were fewer extremely low grades on the final exam. Taking a random sample of fifteen students from her just-completed class, she found the variance to be 350. Can the professor infer that the changes in the course have helped her students? She is using a 5% significance level.

8. ABC Company has just released a new line of super camera batteries with longer battery lives. The company is promoting the new line as being perfect for extended trips, as fewer batteries will be required and a battery charger will not be needed. An independent testing firm examined the batteries and found them to last, on average, fifteen hours under normal usage, with a standard deviation of 3.4 hours. Before switching to these batteries for a major international shoot, a professional photographer tried out a random sample of twelve batteries. She found that the standard deviation in battery life was actually 4.5 hours. If this sample standard deviation proved indicative of battery performance, some of the new batteries may have shorter operating lives than might be expected given the testing firm's results; and, short battery operating lives could prove disastrous on a shoot in an isolated area. Should the photographer switch to these new batteries? Use a 5% significance level.

9. An economist believes that the unemployment percentage in Australia is at least 10%. She randomly samples 565,000 Australian citizens and finds that 509,159 are employed. Based on this information, can the economist conclude at the 1% significance level that her assumption about unemployment in Australia is correct?

10. A commercial for a new line of hair-care products claims that four out of five salon professionals support a new shampoo product. To test this claim, you visit 25 hairdressers in the area and find that eighteen are supporters of the new shampoo. Is this outcome consistent with the claim made in the commercial, at a 5% significance level?

11. In an effort to reduce a university's carbon footprint, an administrator encourages students to bring reusable water bottles. Her goal by the end of the semester is to have over 55% of the student body participating. A random sample of students was taken that showed 64 out of 100 are regularly using reusable bottles. Can it be concluded at the 5% significance level that the reusable water bottle initiative has been a success?

12. You read somewhere that one out of every thirteen people in the world has an active *Facebook* profile. You decide to test this claim with a random sample of 65 Americans. The sample shows that eleven respondents have an active *Facebook* account. Can it be concluded at the 1% significance level that Americans use *Facebook* more than the global average?

13. When the *iPhone 4S* came out, it included an exciting new feature called *Siri*. *Siri* lets the user talk to the phone, issuing commands about setting an alarm, searching the Internet, checking emails, etc. A colleague of yours claims she read somewhere that *Siri* has become very popular, with 55% of users satisfied with the system. To test this statistic, you take a sample of 75 *iPhone* users and find that only 32 are satisfied with how *Siri* operates. Test whether there is enough evidence at the 1% significance level to conclude less than 55% of users are satisfied with *Siri*.

14. A bookstore claims that 60% of its visitors are return customers. You sample 200 random customers entering the store and find that 130 of them had previously visited *and* bought items from the bookstore. Estimate the proportion of return customers entering the store at the 95% confidence level.

15. A new malaria vaccination is claimed by its manufacturer to reduce the risk of severe malaria in infants to less than 53%. This vaccine was given to 6,000 infants across malaria-prone regions of the world and 2,940 of these did not get the disease. Using a 5% significance level, is the vaccine as effective as its manufacturer claims?

16. A new restaurant aims to provide 'faster, more efficient service' and states that the variance in the time taken to serve a customer after the customer has been seated is only 3.7 minutes^2. This statistic was determined by measuring the variance in the time taken to serve ten customers. Estimate the variance of service time at the 90% confidence level.

17. Encouraging youth to vote has been a major issue in the United States with various media and marketing campaigns undertaken by both major political parties. Taking a random sample of 500 eligible-to-vote students at a state university, it was found that 150 of the students had voted in the last election. Estimate the proportion of college-age voters at the 99% confidence level.

CHAPTER 10: Hypothesis Testing of Parameters from Two Populations

One of the episodes of the ABC series *Shark Tank*[100] hosted a company called *eCreamery*. On this episode, *eCreamery*'s owners described how the company allows its customers to personalize and create their own unique ice cream flavors, selling its ice cream from a physical retail shop in Omaha, Nebraska and over the Internet. Internet purchases are shipped to the customer from the Omaha, Nebraska location. During the episode, a discussion developed around profitability and the 'sharks' (i.e., the investor panelists) recommended closing the brick and mortar (physical) store in favor of the Internet operation.

The above vignette is but one example of situations where researchers and organizations may wish to use statistics to compare specific attributes of two populations. In the case of *eCreamery*'s consideration of the desirability of closing their physical store, for example, the owners may wish to compare various statistics of their physical and online stores. They may, for example, wish to survey their online customers and their in-store customers in order to better understand and predict the future behaviors of each group. And, once these data are collected, the owners may compare the two groups on various parameters, such as the mean number of expected purchases, the variance in expected amount spent, the proportion of personalized orders or the net profitability per sale.

In previous chapters, we have learned and explored various types of hypothesis tests and how to carry out these tests. In each of these tests, we

---

[100] In this reality show, entrepreneurs attempt to persuade the show's panelists into investing in their start-up business ventures: http://abc.go.com/shows/shark-tank/episode-guide/season-04/402-episode-402

formulated a hypothesis about the value of a parameter of interest (e.g., $\mu=5$ or $p=0.76$). We then collected data from a single sample and compared the value of our sample statistic to the hypothesized value of the parameter. In this chapter, we introduce and describe hypothesis tests of a population parameter that involve two populations and that require two separately collected samples - a sample from each of the two populations. Appendix A of this book describes how some of these tests can be carried out using the Microsoft Excel *Analysis Toolpak* add-in program, rather than by performing the calculations by hand.

We will begin with the test contrasting two populations' proportions; comparing, for example, the proportion (percent) of male versus female purchasers of a specific product. Next, we describe the test contrasting two populations' variances; comparing, for example, the variance in defect rates for two manufacturing processes or the variance in returns from two investments. We conclude this chapter with the test contrasting two populations' means; comparing, for example, the value of an average order for male versus female purchasers. We are discussing the hypothesis test contrasting two populations' means at the end of this chapter because this test requires the use of the test contrasting two populations' variances.

In each of the hypothesis tests to be introduced in this chapter, the five-step procedure applied earlier in hypothesis testing is followed: (1) formulate the hypotheses, (2) compute the test statistic and p-value, (3) formulate a decision rule, (4) apply the decision rule, and (5) draw your conclusion. In covering each hypothesis test, we also introduce the formula for computing associated confidence intervals and describe how the p-value for the test may be computed using Excel.

Finally, we will also be describing the F-distribution, as this distribution is used in conducting the test comparing the variances of two populations. This chapter covers the following topics:

- Test of Two Populations' Proportions
- Confidence Interval for the Difference between Two Populations' Proportions
- Test of Two Populations' Variances
- The F-Distribution
- Continuing with the Test of Two Populations' Variances
- Confidence Interval for the Ratio of Two Populations' Variances
- Test of Two Populations' Means
- Test of Two Populations' Means: Equal Variances and Independent Samples
- Test of Two Populations' Means: Unequal Variances and Independent Samples
- Test of Two Populations' Means: Paired Samples
- Confidence Interval for the Difference between Two Populations' Means
- Bringing It All Together

UNIT 1

Test of Two Populations' Proportions

We begin our foray into the two-population world with the test of two populations' proportions. According to PEW Research Center: "...half of men and a comparable number of women say they ever play video games on a computer, TV, game console, or portable device like a cellphone. However, men are more than twice as likely as women to identify as 'gamers'. Some 50% of men and 48% of women play video games, while 15% of men and 6% of women say the term 'gamer' describes them well."[101] This is an interesting claim that can be easily tested using a two-populations' proportions test. What we need are two samples and accompanying sample statistics about the proportion of men and women who identify themselves as *gamers*. Let us review in depth the process of this test.

The specific question we seek to answer is: Are men more likely than women to identify themselves as *gamers*? Assume that, in a survey of 250 people who have played video games, 150 respondents were men and 100 respondents were women. Further, assume that 36 of the men identified themselves as *gamers* and that 19 of the women identified themselves as *gamers*. Converting these numbers into proportions, our data tell us that 24% of men (36 out of 150) and 19% of women (19 out of 100) identified themselves as *gamers*. We are interested in finding out whether this difference in proportions between men and women is significant, and we will use $\alpha=0.05$ as our significance level in conducting the test.

Step 1: Formulate Hypotheses

The parameter of interest is the population proportion, but this time we have two groups: men and women. Our research hypothesis (what we want to prove) says that men are more likely to identify themselves as *gamers* than women, meaning that the proportion of men should be higher than the proportion of women, or:

$$H_1: p_m > p_w$$

Note that we use the subscripts *m* and *w* to represent, respectively, *men* and *women*. It is always good practice to use meaningful subscripts as opposed to numerical indices (e.g., $p_1 > p_2$) to avoid confusion.

---

[101] Source: http://www.pewinternet.org/2015/12/15/who-plays-video-games-and-identifies-as-a-gamer/

We rewrite the above hypothesis in a slightly different format:

$$H_0: p_m - p_w \leq 0$$

$$H_1: p_m - p_w > 0$$

to more accurately reflect the alternative hypothesis, which states that there is a difference between the proportion of men who identify themselves as *gamers* and the proportion of women who identify as such, and that this difference is greater than zero. There also is a theoretical justification for formulating the hypotheses this way. In hypothesis testing, we use what we know about the distribution of the parameter of interest to conduct the test. In the case of two populations' proportions, statisticians have determined that the *difference* between the two proportions follows the normal distribution; we thus formulate the hypotheses accordingly.

Step 2: Compute the Test Statistic

As this is a test of proportions, we will be able to use the standard normal distribution, as long as certain criteria are met. Recall from Chapter 9 that the sample size is required to be sufficiently large, specifically that $n*p$ and $n*(1-p)$ are both greater than five. In Chapter 9, we used the hypothesized value of p, reflecting a belief about the true population proportion, to determine whether or not these criteria are met. In the two-population case, we must determine if these criteria hold for *each* of the populations being examined. Checking our sample of men first, with $n_m = 150$ and $\hat{p}_m = 36/150 = 0.24$, both $n_m * \hat{p}_m$ and $n_m * (1 - \hat{p}_m)$ are greater than five. Repeating these calculations for the women's group, with $n_w = 100$ and $\hat{p}_w = 19/100 = 0.19$, both $n_w * \hat{p}_w$ and $n_w * (1 - \hat{p}_w)$ are, again, greater than five.

To compute the test statistic for this test, let us first review how the test statistic for the single population test was computed (from Chapter 9):

$$Z = \frac{(\hat{p} - p)}{\sqrt{\frac{p(1-p)}{n}}}$$

In the two-population case, we use the same formula but replace $\hat{p}$ with $(\hat{p}_1 - \hat{p}_2)$. The general formula to compute the test statistic thus becomes:

$$Z = \frac{(\hat{p}_1 - \hat{p}_2) - (p_1 - p_2)}{\sqrt{\frac{\hat{p}_1(1-\hat{p}_1)}{n_1} + \frac{\hat{p}_2(1-\hat{p}_2)}{n_2}}}$$

where $(\hat{p}_1 - \hat{p}_2)$ is the sample proportions difference, (p_1-p_2) is the hypothesized proportions difference, and $\sqrt{\frac{\hat{p}_1(1-\hat{p}_1)}{n_1} + \frac{\hat{p}_2(1-\hat{p}_2)}{n_2}}$ is the shared standard error. With our *gamers* example:

$$(\hat{p}_1 - \hat{p}_2) = (\hat{p}_m - \hat{p}_w) = 0.24 - 0.19 = 0.05$$

and

$$(p_1 - p_2) = (p_m - p_w) = 0$$

because the null hypothesis states there is no difference between men and women $(p_m - p_w \leq 0)$.

We can simplify the standard error calculation by computing what is termed a **pooled proportion**. Because the null hypothesis assumes the two proportions are equal[102], we can combine the data from the two samples to compute a single proportion of respondents who identify themselves as gamers:

---

[102] If our null hypothesis states that two proportions are equal, then we are essentially saying "there is a *single* true population proportion". What we have in our data are two sample-based estimates of this single true proportion of gamers: one estimate based on a group of men and one based on a group of women. Therefore, we can *pool together* these two estimates to obtain a single average estimate of the hypothesized proportion.

$$\hat{p} = \frac{x_m + x_w}{n_m + n_w} = \frac{36 + 19}{150 + 100} = \frac{55}{250} = 0.22$$

where $\hat{p}$ now represents this *pooled proportion*. In the next example, we solve a problem where the hypothesized difference is not zero and, hence, we do not compute a pooled proportion.

Putting all of the above together, we compute the test statistic:

$$Z = \frac{(\hat{p}_m - \hat{p}_w) - (p_m - p_w)}{\sqrt{\hat{p}(1-\hat{p})\left(\frac{1}{n_m} + \frac{1}{n_w}\right)}} = \frac{0.05 - 0}{\sqrt{0.22 * (1 - 0.22) * \left(\frac{1}{150} + \frac{1}{100}\right)}} = 0.935$$

Note, in particular, that we substituted the pooled proportion, $\hat{p}$, in the denominator of the test statistic formula.

Step 3: Formulate a Decision Rule

As this is an upper-tail test, the decision rule is: reject the null hypothesis if the test statistic falls within the upper-tail rejection region. We can find the critical value defining this rejection region by using Excel or the standard normal distribution table. As we are conducting the test at the α=0.05 level of significance, the critical value is 1.645, and the decision rule becomes: reject the null hypothesis if the value of the test statistic is greater than 1.645. Alternatively, since we are again working with the standard normal distribution, the test's p-value can easily be computed and used for the hypothesis test. Remember that, using the data in this example, the p-value is the probability of finding a test statistic as high as the test statistic (0.935). Using Excel, we can type =1-NORM.S.DIST(0.935,1) to find P(Z>0.935)=0.1749. If we were to follow the p-value approach to hypothesis testing, the decision rule would be: reject the null hypothesis if the p-value (0.1749) is less than the significance level (α=0.05).

Step 4: Apply the Decision Rule

Comparing the test statistic to our critical value, we find 0.935<1.645; therefore, at the 5% level of significance, we fail to reject the null hypothesis, as illustrated in Figure 10-1. We reach this same outcome (i.e., not rejecting the null hypothesis) by comparing the p-value (0.1749) to the value of α (0.05).

**Figure 10-1
Comparing the Test Statistic and Critical Value
for the *Gamers* Example**

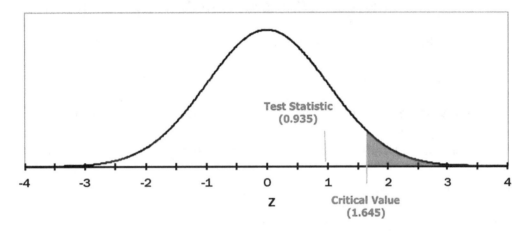

Step 5: Draw Your Conclusion

By failing to reject the null hypothesis, we cannot conclude that there is a significant difference in the proportions of men and women who identify themselves as gamers.

In the hypothesis test just covered, we were interested in learning whether one population's proportion was larger than the other population's proportion. Sometimes, however, we are interested in a more specific question related to the magnitude (or size) of the hypothesized difference. To understand this, consider the following example.

The *New York City Health Department* conducts unannounced inspections of restaurants at least once a year, checking for compliance in food handling, food temperature, employees' hygiene and vermin control. Restaurant inspection results are summarized by grades of A, B or C. If a restaurant does not earn an A on its first inspection (called an *initial inspection*), the *Health Department* does not issue a grade, but instead conducts a second inspection approximately a month later and issues a grade based on this re-inspection.[103] Assume that of 1,000 restaurants inspected in January of one year, 270 received an A grade upon initial inspection. In comparison, in July of the same year, 400 of the 1,000 restaurants inspected received an A grade upon initial inspection. Comparing the proportions of restaurants receiving an A grade on their initial inspections, can you say this number has gone up by more than 5% between January and July (use $\alpha=0.01$)? Again, this question asks not just whether the proportion of A grades has gone up from January to July, but, specifically, whether this proportion has gone up by *more than 5%*.

Step 1: Formulate Hypotheses

We are asked to compare the proportion of restaurants that received an A grade on their initial inspection in January (we'll call it $p_{January}$) to the proportion of restaurants that received an A grade on their initial inspection in July (p_{July}). We are testing to see if the difference between these two proportions is greater than 5%. Hence, our hypotheses are:

$$H_0: p_{July} - p_{January} \leq 0.05$$

$$H_1: p_{July} - p_{January} > 0.05$$

---

[103] Source: http://www.nyc.gov/html/doh/html/services/restaurant-inspection.shtml

This example illustrates the importance, when formulating problems, of using meaningful notation (e.g., January and July) for parameters, as doing so can avoid possible confusion in stating the direction of hypotheses.

Step 2: Compute the Test Statistic

We first check the normal distribution approximation criteria. With $n_{January}=1{,}000$ and $\hat{p}_{January}=270/1000=0.27$, n*p and n*(1-p) are both greater than five. We repeat the above with the July data: $n_{July}=1{,}000$ and $\hat{p}_{July}=400/1000=0.4$; again, n*p and n*(1-p) are both greater than five. As the criteria are met, we can work with the standard normal distribution. The test statistic formula used is:

$$Z = \frac{(\hat{p}_1 - \hat{p}_2) - (p_1 - p_2)}{\sqrt{\frac{\hat{p}_1(1-\hat{p}_1)}{n_1} + \frac{\hat{p}_2(1-\hat{p}_2)}{n_2}}}$$

where $(\hat{p}_1 - \hat{p}_2)$ is the sample proportions difference, (p_1-p_2) is the hypothesized proportions difference, and $\sqrt{\frac{\hat{p}_1(1-\hat{p}_1)}{n_1} + \frac{\hat{p}_2(1-\hat{p}_2)}{n_2}}$ is the shared standard error. In our specific example:

$$(\hat{p}_1 - \hat{p}_2) = (\hat{p}_{July} - \hat{p}_{January}) = 0.4 - 0.27 = 0.13$$

and

$$(p_1 - p_2) = (p_{July} - p_{January}) = 0.05 \text{ (the hypothesized difference)}$$

Putting all of the above together, we obtain the following test statistic:

$$Z = \frac{(\hat{p}_{July} - \hat{p}_{January}) - (p_{July} - p_{January})}{\sqrt{\frac{\hat{p}_{July}(1-\hat{p}_{July})}{n_{July}} + \frac{\hat{p}_{January}(1-\hat{p}_{January})}{n_{January}}}} = \frac{(0.4 - 0.27) - (0.05)}{\sqrt{\frac{0.4 * 0.6}{1000} + \frac{0.27 * 0.73}{1000}}} = 3.826$$

As we are *not* hypothesizing that the populations' proportions are equal, we cannot assume there is a single true population proportion. As a result, we cannot simplify the formula for the test statistic by using the pooled proportion in the denominator. Instead, we work with both of the sample proportions.

Step 3: Formulate a Decision Rule

Again, as this is an upper-tail test, the decision rule is: reject the null hypothesis if the test statistic falls within the upper-tail rejection region. Using the standard normal distribution and α of 0.01, the upper-tail critical value is 2.326. We rephrase the decision rule as: reject the null hypothesis if the test statistic is greater than 2.326.

Step 4: Apply the Decision Rule

Comparing the test statistic to our critical value, we see that 3.826 is greater than 2.326 and we reject the null hypothesis.

Step 5: Draw Your Conclusion

By rejecting the null hypothesis, we conclude that the improvement observed from January to July in the number of restaurants that received an A grade upon initial inspection is likely to be more than 5%, as stated in the alternative hypothesis.

Confidence Interval for the Difference between Two Populations' Proportions

Since we just rejected the above null hypothesis, suppose that we wish to estimate the difference in proportions between restaurants receiving A grades in January and those receiving A grades in July. We can compute the confidence interval using the following formula:

$$(\hat{p}_1 - \hat{p}_2) \pm z_{\alpha/2} \sqrt{\frac{\hat{p}_1(1-\hat{p}_1)}{n_1} + \frac{\hat{p}_2(1-\hat{p}_2)}{n_2}}$$

Using a 98%[104] confidence level, this expression becomes:

$$(\hat{p}_{July} - \hat{p}_{January}) \pm 2.326 \sqrt{\frac{\hat{p}_{July}(1-\hat{p}_{July})}{n_{July}} + \frac{\hat{p}_{January}(1-\hat{p}_{January})}{n_{January}}}$$

or

$$(0.4 - 0.27) \pm 2.326 \sqrt{\frac{0.4 * 0.6}{1000} + \frac{0.27 * 0.73}{1000}}$$

Completing this computation, we obtain a lower confidence limit of 0.081 (or 8.1%) and an upper confidence limit of 0.178 (or 17.8%). As this confidence interval lays beyond (i.e., is higher than) the 5% increase used in the health inspection example, we again conclude, with 98% confidence, that the true difference between the proportion of restaurants receiving an initial grade of A in January and those receiving an initial grade of A in July is greater than 5%.

Unit 1 Summary

- The test of two populations' proportions resembles the test of the single population proportions, but our hypotheses focus on the difference between the two populations, and we rely on data collected from two different samples.

- The critical value is a Z value obtained (via Excel or the standard normal distribution table) using the specified α, as long as both $n*\hat{p}$ and $n*(1-\hat{p})$ are greater than five for both samples.

- Use the following formula to compute the test statistic (also a Z value):

$$Z = \frac{(\hat{p}_1 - \hat{p}_2) - (p_1 - p_2)}{\sqrt{\frac{\hat{p}_1(1-\hat{p}_1)}{n_1} + \frac{\hat{p}_2(1-\hat{p}_2)}{n_2}}}$$

If the difference between the two proportions is hypothesized to be zero, use the **pooled proportion**:

---

[104] The hypothesis test just described was an upper-tail test at $\alpha=0.01$. A 98% confidence interval will generate a similar 1% tail at each end of the distribution and is thus comparable to the test we conducted.

$$\hat{p} = \frac{x_1 + x_x}{n_1 + n_2}$$

In computing the test statistic:

$$Z = \frac{(\hat{p}_1 - \hat{p}_2) - (p_1 - p_2)}{\sqrt{\hat{p}(1-\hat{p})(\frac{1}{n_1} + \frac{1}{n_2})}}$$

- Use the decision rules stated below to help you decide whether or not to reject the null hypothesis.

Table 10-1
Decision Rules for a Hypothesis Test of Two Population Proportions

Type of Test	Rejection Rule
Lower-tail test H_0: $p_1-p_2 \geq$ value H_1: $p_1-p_2 <$ value	Reject the null hypothesis if: Test statistic < lower critical value
Upper-tail test H_0: $p_1-p_2 \leq$ value H_1: $p_1-p_2 >$ value	Reject the null hypothesis if: Test statistic > upper critical value
Two-tail test H_0: $p_1-p_2 =$ value H_1: $p_1-p_2 \neq$ value	Reject the null hypothesis if: Test statistic < lower critical value Or if: Test statistic > upper critical value

- Use the following formula to compute a confidence interval of the two proportions:

$$(\hat{p}_1 - \hat{p}_2) \pm z_{\alpha/2} \sqrt{\frac{\hat{p}_1(1-\hat{p}_1)}{n_1} + \frac{\hat{p}_2(1-\hat{p}_2)}{n_2}}$$

Unit 1 Exercises

1. Compute the test statistic for each case, deciding first whether or not to compute a pooled proportion:

   a. H_0: $p_1-p_2 = 0$; $x_1 = 88$, $x_2 = 93$; $n_1 = 100$, $n_2 = 100$

   b. H_0: $p_1-p_2 \leq 0.35$; $x_1 = 100$, $x_2 = 50$; $n_1 = 250$, $n_2 = 300$

   c. H_0: $p_1-p_2 = 0.5$; $x_1 = 35$, $x_2 = 8$; $n_1 = 60$, $n_2 = 80$

   d. H_0: $p_1-p_2 \geq 0$; $x_1 = 10$, $x_2 = 120$; $n_1 = 8$, $n_2 = 100$

2. Compute the confidence interval of the difference between the two proportions for each case:

   a. $(1-\alpha) = 0.90$; $\hat{p}_1 = 0.7, \hat{p}_2 = 0.5$; $n_1 = 200, n_2 = 200$

   b. $(1-\alpha) = 0.95$; $\hat{p}_1 = 0.043, \hat{p}_2 = 0.02$; $n_1 = 670, n_2 = 500$

   c. $(1-\alpha) = 0.98$; $\hat{p}_1 = 0.86, \hat{p}_2 = 0.73$; $n_1 = 100, n_2 = 100$

3. Customer satisfaction is important for the success of products. Suppose that Table 10-e1 provides the results of a survey collecting data on customer satisfaction of individuals having *iPhones* and on individuals having *Android* phones (the question asked was "are you happy with your phone?" and answers were provided as either "yes" or "no"). Given the statistics in Table 10-e1, is customer satisfaction higher with *Android* users? (Use $\alpha = 0.05$.)

Table 10-e1

	iPhone	Android
Sample Size	500	500
Satisfied with phone	260	310

4. Additional data was gathered to examine two specific phones: the *iPhone 6S* and the new *Samsung Galaxy S7*. Table 10-e2 shows the number of satisfied customers out of a sample of 500 phone owners. Are *iPhone 6S* owners more satisfied than *Galaxy S7* owners? (Use $\alpha = 0.05$.)

Table 10-e2

	iPhone 6S	Galaxy S7
Sample Size	500	500
Satisfied with phone	240	190

5. Companies are always looking for the best way to increase revenue. Advertising on the Internet can be an effective tactic for increasing revenue, but also a big expense. Assume your firm has to choose between advertising on *Google* or on *Facebook* and that the costs of doing so are comparable. Given the data in Table 10-e3, estimate the difference in the proportion of clicks between *Google* and *Facebook*. Is *Google* definitely the better way to go? (Use $\alpha = 0.01$.)

Table 10-e3

	Facebook	Google
Ads	1,000	10,000
Clicked	1	40

6. In an attempt to increase tips, a restaurant manager suggested to his waiters that they may receive more tips if they squatted next to the table when talking to customers instead of towering over customers. At the 5% significance level, and using the data in Table 10-e4, did this suggestion increase the proportion of customers that tipped? Use the standard normal distribution to compute the p-value for this test and compare this p-value to α in order to reach your conclusion.

Table 10-e4

	Did Not Squat	Squatted
Sample Size	200	200
Customers Who Tipped	160	180

7. A further experiment had the waiters briefly touch the shoulder of customers during a meal to see if this increased the proportion of customers who tipped. Test, again using the p-value approach, if touching a customer's shoulder increased the proportion of tipping customers by 10%, compared to not touching a customer's shoulder, using $\alpha=0.05$ and the data in Table 10-e5.

Table 10-e5

	Did Not Touch Shoulder	Touched Shoulder
Sample Size	200	200
Customers Who Tipped	160	187

8. A medical study is aimed at proving that teenagers are more susceptible to placebos than are adults. The proportion of patients who reported feeling better was recorded for both groups, as shown in Table 10-e6. At the 5% significance level, do placebos work better for teenagers?

Table 10-e6

	Teens	Adults
Sample Size	100	100
Reported Feeling Better	32	25

UNIT 2

Test of Two Populations' Variances

The test of two populations' variances is not only useful on its own, but also because it is required in testing two populations' means, which will be covered in Unit 3 of this chapter. In illustrating the procedure for the test of two populations' variances, consider the example of 3D printing. Suppose you would like to purchase a new 3D printer and are in the market for a printer with high accuracy and high consistency. You narrow down the options to two printers, which you would like to test. You have each of the printers create 52 spheres and you measure the diameter of each sphere. You then calculate the standard deviation in diameters for each printer, as a measure of the printer's consistency. Analyzing your data, you find the standard deviation of Printer One to be 0.214 mm and the standard deviation of Printer Two to be 0.296 mm. Can we conclude, at the 5% level of significance, that Printer Two is less consistent than Printer One?

Step 1: Formulate Hypotheses

The parameters of interest are the two printers' printing variances. You would like to find out if Printer Two is less consistent than Printer One by examining the variance in printing accuracy. The alternative hypothesis should, therefore, convey that Printer Two is less consistent than Printer One by stating that σ^2_{II} is greater than σ^2_{I}. Hence, we write the hypotheses as a ratio of the two variances:

H_0: $\sigma^2_{II}/\sigma^2_{I} \leq 1$ (which is the same mathematically as: $\sigma^2_{II} \leq \sigma^2_{I}$)

H_1: $\sigma^2_{II}/\sigma^2_{I} > 1$ (which is the same mathematically as: $\sigma^2_{II} > \sigma^2_{I}$)

The null hypothesis, thus, states that Printer Two's variance is the same as, or less than, Printer One's variance.

Step 2: Compute the Test Statistic

The test statistic for this test is computed simply as the ratio of the two samples' variances:

$$F = \frac{s^2_{II}}{s^2_{I}} = \frac{0.296^2}{0.214^2} = 1.91$$

You need to make sure that the numerator and denominator correspond to those stated in the null hypothesis. Prior to our calculation of the test statistic, we confirmed that the sample variance of Printer Two was the numerator, and the sample variance of Printer One was the denominator.

The next step in our testing process involves formulating a decision rule. As is always the case, we need to know the distribution we are working with in order to formulate our decision rule. For the test of two populations' variances, statisticians have shown that the ratio of two variances follows a distribution known as the *F-distribution* (which was briefly introduced in Chapter 6). We now take a short detour to describe how to work with this distribution, after which we will return to Step 3 of the hypothesis testing procedure.

The F-Distribution

Like the χ^2 distribution, the **F-distribution** is not symmetrical and its shape depends on the number of degrees of freedom of *both* the numerator and the denominator (remember that we are looking at a ratio), as shown in Figure 10-2.

Figure 10-2
The F-Distribution ($\alpha=0.05$, $df_1=10$, $df_2=10$)

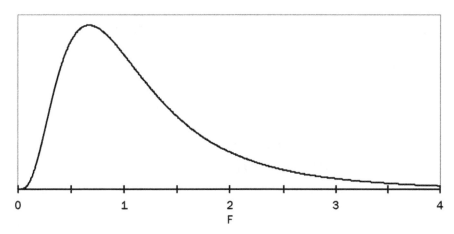

We can use either Excel or the F-distribution table (or, F-table for short) to find values in this distribution. Due to the nature of the F-distribution, the F-table is typically presented as a separate table for each value of α being listed. Most textbooks provide tables for $\alpha=0.05$, $\alpha=0.025$, $\alpha=0.01$ and $\alpha=0.005$. This chapter's Excel file provides an F-table in the worksheet titled 'F-Distribution Table'. In this worksheet, by changing the value of α in cell B1, you are provided with the F-distribution table for this specific α value.

When obtaining values from the F-distribution table, we need to consider degrees of freedom along with a specific α value. As explained earlier, we work with two distinct degrees of freedom (as we are working with samples from two different populations): one for the numerator (calculated as n_1-1) and one for the denominator (calculated as n_2-1).

Table 10-2 provides a partial view of the F-table. This table was created for $\alpha=0.05$ (recall that we use a different table for each value of α). The top row of the table shows the *numerator's* degrees of freedom. The first column provides the *denominator's* degrees of freedom. Assume, for example, that we obtained two samples of sizes $n_1=20$ and $n_2=25$ from our two populations of interest. We can use the table to find, for example, $F_{df1,df2}$ with 19 and 24 degrees of freedom (denoted as $F_{19,24}$) which is the value 2.040 at the intersection of the column titled 19df_1 and the row titled 24df_2, highlighted in Table 10-2.

Table 10-2
Finding Values in the F-Table

	df_1	18	19	20	21	22	23	24	25	26
df_2	...	...	...	...	...	...	...	...	...	...
18	...	2.217	2.203	2.191	2.179	2.168	2.159	2.150	2.141	2.134
19	...	2.182	2.168	2.155	2.144	2.133	2.123	2.114	2.106	2.098
20	...	2.151	2.137	2.124	2.112	2.102	2.092	2.082	2.074	2.066
21	...	2.123	2.109	2.096	2.084	2.073	2.063	2.054	2.045	2.037
22	...	2.098	2.084	2.071	2.059	2.048	2.038	2.028	2.020	2.012
23	...	2.075	2.061	2.048	2.036	2.025	2.014	2.005	1.996	1.988
24	...	2.054	2.040	2.027	2.015	2.003	1.993	1.984	1.975	1.967
25	...	2.035	2.021	2.007	1.995	1.984	1.974	1.964	1.955	1.947
26	...	2.018	2.003	1.990	1.978	1.966	1.956	1.946	1.938	1.929

Note that the table only provides a single value. Specifically, the table provides the upper-tail F value corresponding to a tail of α. That is, the probability of falling to the right of this F value is α. If we wish to find the lower-tail F value for the above example, with $\alpha=0.05$ and with $df_1=19$ and $df_2=24$, we use the formula:

$$F_{(1-\alpha,df1,df2)} = \frac{1}{F_{(\alpha,df2,df1)}}$$

where $F_{(1-\alpha,\,df1,\,df2)}$ is the desired lower-tail F value; that is, an F value that defines a tail of 5% (α) at the left-side of the distribution, or a tail of 95% (1-α) at the right-side. For example, to find the lower-tail F value with $df_1=19$ and $df_2=24$, we use:

$$\frac{1}{F_{(0.05,24,19)}} = \frac{1}{2.114} = 0.473$$

As with the previous distributions we have worked with, you can also use Excel to find values from the F-distribution by using the function =F.INV(α,df$_1$,df$_2$) to obtain the left-tail values, and =F.INV.RT(α,df$_1$,df$_2$) to obtain right-tail values. For example, typing =F.INV(0.05,19,24) returns a value of 0.473, and typing =F.INV.RT(0.05,19,24) returns a value of 2.04, the same values we obtained above using the F-table.

Continuing with the Test of Two Populations' Variances

Step 3: Formulate a Decision Rule

Since this is an upper-tail test, the decision rule is: reject the null hypothesis if the test statistic falls within the upper-tail rejection region. To find the upper-tail critical value, we need to consider our degrees of freedom. We have 52 data points for each of the two printers being compared, or 51 degrees of freedom for each printer. Using Excel, we type =F.INV.RT(0.05,51,51) to obtain the upper-tail critical value of 1.5919. If you wish to use the F-table, use the table for α=0.05 and approximate using 50 degrees of freedom for both samples. You will obtain a critical value of 1.60. Hence, the decision rule becomes: reject the null hypothesis if we obtain a test statistic greater than 1.60.

Step 4: Apply the Decision Rule

Comparing the test statistic to our critical value (see Figure 10-3), the test statistic falls into the upper-tail rejection region (i.e., 1.91 is greater than 1.5919). We thus reject the null hypothesis.

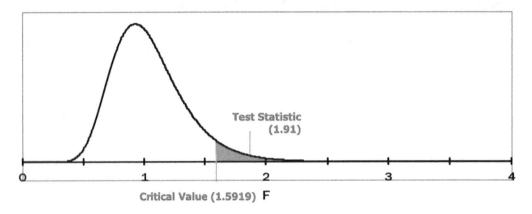

**Figure 10-3
Comparing the Test Statistic with the Critical Value
for the 3D Printers Example**

As has been the case with some other distributions, a specific p-value often can only be computed with the help of statistical software. Applying the p-value approach to hypothesis testing, the decision rule for this upper-tail test would be: reject the null hypothesis if the derived p-value is less than α (here, 0.05). The test statistic computed in Step 2, above, was 1.91. Using Excel, we can type =F.DIST.RT(1.91,51,51) to obtain a derived p-value of 0.0113. As 0.0113 is less than 0.05, we reject the null hypothesis.

Step 5: Draw Your Conclusion

By rejecting the null hypothesis, we conclude that Printer Two is indeed less consistent than is Printer One.

Now, let us go over a two-tail example. The parents of a child with type I diabetes are trying to decide on the most suitable approach for administering insulin to minimize the variance, over time, of their child's glucose levels. The two options they are considering are manually giving shots and using an insulin pump. The pump has the advantage of administering smaller and more accurate doses throughout the day, but it may be less comfortable for a young child. To make this decision, the parents run a two-week test with each of the two options and measure the standard deviation of glucose levels at the end of each two-week period. In analyzing the data, the parents find the standard deviation with giving shots to be 110 mg/dl and the standard deviation with the insulin pump to be 100 mg/dl. At the 1% level of significance, can we conclude that there is a significant difference between these two variances?

Step 1: Formulate Hypotheses

The parameters of interest in this example are the two populations' variances. Specifically, the parents wish to find out if there is significant difference between the two approaches for administering insulin. The alternative hypothesis should convey that the two variances are different, whereas the null hypothesis will assume they are not different. As we explained in the previous example, we write the hypotheses as a ratio of the two variances:

H_0: $\sigma^2_{shots}/\sigma^2_{pump} = 1$ (which is the same as: $\sigma^2_{shots} = \sigma^2_{pump}$)

H_1: $\sigma^2_{shots}/\sigma^2_{pump} \neq 1$ (which is the same as: $\sigma^2_{shots} \neq \sigma^2_{pump}$)

Step 2: Compute the Test Statistic

The test statistic for this test is computed simply as the ratio of the two samples' variances:

$$F = \frac{s^2_{shots}}{s^2_{pump}} = \frac{110^2}{100^2} = 1.21$$

Step 3: Formulate a Decision Rule

Following the critical value approach for a two-tail test, the decision rule is: reject the null hypothesis if the test statistic falls in either the upper-tail rejection region or the lower-tail rejection region. Thus, we need to find the upper- and lower-tail critical values and corresponding rejection regions. Using Excel, we type =F.INV(0.005,13,13) and =F.INV.RT(0.005,13,13) to find 0.2186 and 4.5733, respectively, as the lower and upper critical values. Note that we use a value of 0.005 ($\alpha/2$) for the significance level because this is a two-tail test and that we use thirteen degrees of freedom for each sample because the test period for each sample was two weeks (fourteen days). We restate the decision rule as: reject the null hypothesis if the test statistic is less than 0.219 or greater than 4.57.

Step 4: Apply the Decision Rule

As the test statistic of 1.21 does not fall in either rejection region, we do not reject the null hypothesis.

Step 5: Draw Your Conclusion

By not rejecting the null hypothesis, we can conclude, at the 1% level of significance, that there is no difference in the variances of glucose levels of the two insulin administration approaches.

Confidence Interval for the Ratio of Two Populations' Variances

We can use the above example to compute the confidence interval for the difference in variances of the two insulin delivery approaches. The general confidence interval formula is:

$$LCL = \left(\frac{s_1^2}{s_2^2}\right) \frac{1}{F_{\alpha/2, df_1, df_2}}$$

$$UCL = \left(\frac{s_1^2}{s_2^2}\right) F_{\alpha/2, df_2, df_1}$$

where s_1^2 and s_2^2 are any two sample variances. Using the insulin administration example, a 99% confidence interval for the difference in glucose level variances would be defined by lower and upper confidence limits computed as:

$$LCL = \left(\frac{110^2}{100^2}\right)\frac{1}{F_{0.005,13,13}} = 1.21 * \frac{1}{4.57} = 0.265$$

$$UCL = \left(\frac{110^2}{100^2}\right)F_{0.005,13,13} = 1.21 * 4.57 = 5.53$$

Hence, we can estimate with 99% confidence that the ratio $\sigma^2_{shots}/\sigma^2_{pump}$ is captured by the interval: 0.265 and 5.53.

Unit 2 Summary

- The test of two populations' variances focuses on the *ratio* of the two population variances.

- The following formula is used to compute the test statistic for the test of two populations' variances:

$$F = \frac{s_1^2}{s_2^2}$$

- The critical value is obtained using a specified α from the **F-distribution** and the degrees of freedom for both the numerator and denominator of the test statistic. For two-tail tests, two critical values are obtained using $\alpha/2$.

- Use the decision rules given in Table 10-3 in deciding whether or not to reject the null hypothesis for a test of two populations' variances.

Table 10-3
Decision Rules for the Test of Two Populations' Variances

Type of Test	Rejection Rule
Lower-tail test $H_0: \dfrac{\sigma_1^2}{\sigma_2^2} \geq 1$ $H_1: \dfrac{\sigma_1^2}{\sigma_2^2} < 1$	Reject the null hypothesis if: Test statistic < lower critical value
Upper-tail test $H_0: \dfrac{\sigma_1^2}{\sigma_2^2} \leq 1$ $H_1: \dfrac{\sigma_1^2}{\sigma_2^2} > 1$	Reject the null hypothesis if: Test statistic > upper critical value
Two-tail test $H_0: \dfrac{\sigma_1^2}{\sigma_2^2} = 1$ $H_1: \dfrac{\sigma_1^2}{\sigma_2^2} \neq 1$	Reject the null hypothesis if: Test statistic < lower critical value Or if: Test statistic > upper critical value

- Use the following formulas when estimating a confidence interval for the ratio of two populations' variance at an α level of confidence:

$$LCL = \left(\frac{s_1^2}{s_2^2}\right) \frac{1}{F_{\alpha/2, df_1, df_2}}$$

$$UCL = \left(\frac{s_1^2}{s_2^2}\right) F_{\alpha/2, df_2, df_1}$$

Unit 2 Exercises

1. Test whether the two population variances are equal for each of the cases that follow:

   a. $s^2_1 = 30$, $n_1 = 100$; $s^2_2 = 24$, $n_2 = 200$; $\alpha = 0.05$

   b. $s^2_1 = 120$, $n_1 = 200$; $s^2_2 = 143$, $n_2 = 100$; $\alpha = 0.1$

   c. $s^2_1 = 10$, $n_1 = 50$; $s^2_2 = 12$, $n_2 = 100$; $\alpha = 0.01$

2. Using the two cases given below, explain how an increase in the sample size affects the test statistic:

   a. $s^2_1 = 5$, $n_1 = 50$; $s^2_2 = 15$, $n_2 = 50$; $\alpha = 0.05$

   b. $s^2_1 = 5$, $n_1 = 200$; $s^2_2 = 15$, $n_2 = 200$; $\alpha = 0.05$

3. Compute the confidence interval for the ratio of two populations' variances for each of the cases that follow:

   a. $s^2_1 = 5$, $n_1 = 50$; $s^2_2 = 15$, $n_2 = 50$; $\alpha = 0.05$

   b. $s_1 = 50$, $n_1 = 15$; $s_2 = 70$, $n_2 = 20$; $\alpha = 0.1$

   c. $s_1 = 0.03$, $n_1 = 100$; $s_2 = 0.15$, $n_2 = 100$; $\alpha = 0.01$

4. A factory is testing a new, more energy-efficient machine to fill juice bottles. The new machine will not be selected to replace the current machine if it has a higher variance in fill level than that of the current machine. Given the data for two filling samples provided in Table 10-e7, and applying a 1% significance level, should the new machine replace the current machine?

Table 10-e7

	Current Machine	New Machine
Sample Size	15	15
Variance	0.000592	0.001198

5. Two friends are testing to find out which of them is more consistent at bowling. They record a random sample of their scores, shown in Table 10-

e8. At the 5% significance level, is the veteran more consistent than the rookie?

Table 10-e8

	Rookie	Veteran
Sample Size	15	15
Variance	351.1238	230.6952

UNIT 3

Test of Two Populations' Means

There are two important questions that need to be answered when carrying out a hypothesis test regarding two populations' means. First, are the population variances equal? Second, are the two collected samples (one from each population) independent of each other? In the previous unit, we learned about a hypothesis test that can be used to answer the first of these questions: Are the population variances equal? Regarding the second question about independent samples, two samples are said to be *independent* when data values in one sample have no influence on the probability of the occurrence of data values in the second sample. How these two questions are answered determines how the test statistic and critical value to be used in a test of two populations' means are calculated, as described below.

In describing how to conduct a test of two populations' means, an example is used that includes samples that are independent of one another and samples that are not independent of one another. A professor of business statistics feels that a recent change in public schools' math curriculum has weakened the mathematical foundation of students, and as a result they are less able to handle the material covered in her course. To test this assertion, the professor uses data from three random samples of 33 students who took her course in each of the past three years. She plans to compare students' performances on the final exam in each of these three years. Summary statistics for these three samples are shown in Table 10-4.

Table 10-4
Summary Statistics of Final Exam Scores over Three Years

	Year I	Year II	Year III
Mean	81.24	64.97	62.42
Standard Deviation	11.18	25.90	13.65

Being more specific, the professor wishes to obtain information enabling her to answer three questions. First, at the 5% level of significance, is there evidence of a difference in the average final exam score between Year I and Year III? Second, at the 1% level of significance, is there evidence that the average final exam score for Year II is higher than the average final exam score for Year III? The third question is more complicated. The professor feels that students may have an easier time handling the descriptive statistics part of the course (covered prior to the midterm) than the inferential statistics part (covered after the midterm), and thus she plans to compare the mean grades of a sample of the Year III students on their midterm and final exams. At the 5% level of significance, she will test whether there is evidence of a difference in *each individual student's performance* on the midterm exam and on the final exam.[105]

---

[105] The data needed to answer this third question are introduced later in this unit.

Test of Two Populations' Means: Equal Variances and Independent Samples

Recall from this unit's introduction that, with two *independent* samples, the occurrence of data values in one sample has no influence on the probability of the occurrence of data values in the second sample. In the context of comparing final exam grades for two of the years, this means that we can compare any sample of grades taken from one year with any sample of grades taken from another year, regardless of the students who obtained these grades (that is, assuming no student is repeating the course). *Non-independent* samples, as we will see later in this unit, occur when the groups being compared are matched on some criterion. For example, we can compare, for a sample of students (hence, matching by student), each student's grades on their midterm and final exams.

We answer the professor's first question by testing (at a 1% significance level) for a difference in the mean final exam grades for Year I and Year III. How the test statistic is calculated and the degrees of freedom are determined depends on whether or not we believe that the two population variances are equal. As a consequence, when conducting the test of two populations' means, we add an additional step to our five-step testing procedure - a test of the equality of the two population variances. Step 2a, below, illustrates this test of the equality of variances, which is the F-test covered in Unit 2 of this chapter.

Step 1: Formulate Hypotheses

The parameter of interest is the difference between the two populations' means: $\mu_1 - \mu_2$. The professor was interested in learning whether there is evidence of a *difference* in the average final exam scores for Year I and Year III. Hence, this is a two-tail test with the following hypotheses:

$$H_0: \mu_I - \mu_{III} = 0$$

$$H_1: \mu_I - \mu_{III} \neq 0$$

where μ_I is the mean final exam grade for Year I and μ_{III} is the mean final exam grade for Year III.

Step 2a: Test of Equality of Variances

We test for equality of variances to determine if we can use a *pooled variance*, similar to the notion of a pooled proportion that was used in Unit 1 of this chapter. If we believe that the variances are equal, we can pool them together. If we do not believe the variances are equal, then we cannot pool them together and we compute the test statistic and critical value differently.

We could just eyeball the sample variances to see how close they are, but this leaves far too much room for error. Instead, we use the F-test for testing two populations' variances described in Unit 2.

<u>Step 1: Formulate Hypotheses</u>. The null hypothesis states that the variances are equal. Note that, in our hypotheses statements below, we use the larger of the two sample variances (in this case, the sample variance for Year III) as the numerator. This is a short-cut to conducting the test, as it ensures that the value of the test statistic will be greater than 1.0 and, consequently, fall in the upper-tail of the F-distribution. This allows us to only have to look up the upper-tail critical value, which is more readily obtained from the F-table than is the lower-tail critical value. Thus, the hypotheses being tested are:

$$H_0: \sigma^2_{III}/\sigma^2_I = 1$$

$$H_1: \sigma^2_{III}/\sigma^2_I \neq 1$$

Step 2: Compute the Test Statistic. As we are given the two years' standard deviations, we use them to compute the test statistic:

$$F = \frac{s_{III}^2}{s_I^2} = \frac{13.65^2}{11.18^2} = 1.49$$

Step 3: Formulate a Decision Rule. Although this is two-tail test, we will only be looking at the upper-tail rejection region because of the way we have formulated our hypotheses. The decision rule is: reject the null hypothesis if the test statistic falls into the upper-tail rejection region.

Because this is a two-tail test, we use α/2 in determining this upper-tail critical value. We have 33 exam scores for each exam, or 32 degrees of freedom. We now look up the critical value in Excel by typing =F.INV.RT(0.025,32,32), obtaining a value of 2.0247. Applying the critical value we obtained, the decision rule becomes: reject the null hypothesis if the test statistic is greater than 2.0247.

Step 4: Apply the Decision Rule. Because the test statistic (1.49) is less than the upper-tail critical value (2.0247), we do not reject the null hypothesis.

Step 5: Draw Your Conclusion. By not rejecting the null hypothesis, we assume that the two population variances are equal.

Step 2b: Compute the Test Statistic (Equal Population Variances)

Below is the formula used to compute the test statistic for the test of two populations' means, for the case where the populations' variances are equal:

$$t = \frac{(\bar{X}_1 - \bar{X}_2) - (\mu_1 - \mu_2)}{\sqrt{s_p^2\left(\frac{1}{n_1} + \frac{1}{n_2}\right)}}$$

where $(\bar{X}_1 - \bar{X}_2)$ is the observed difference between the two sample means, $(\mu_1 - \mu_2)$ is the hypothesized difference between the two means, and S_p^2 is the pooled variance:

$$s_p^2 = \frac{(n_1 - 1)s_1^2 + (n_2 - 1)s_2^2}{n_1 + n_2 - 2}$$

In our example comparing the average final exam scores for Year I and Year III, these calculations are:

$$(\bar{x}_1 - \bar{x}_2) = 81.24 - 62.42 = 18.82$$

$$(\mu_1 - \mu_2) = 0$$

$$s_p^2 = \frac{(n_1 - 1)s_1^2 + (n_2 - 1)s_2^2}{n_1 + n_2 - 2} = \frac{(33 - 1) * 11.18^2 + (33 - 1) * 13.65^2}{33 + 33 - 2} = 155.657$$

Since the pooled variance is simply a weighted average of the two sample variances, we can eyeball the above calculation to check that the computed value is indeed between the two samples' variances: Year I's variance is 125 and Year III's variance is approximately 186.

We now compute the test statistic:

$$t = \frac{(\bar{X}_I - \bar{X}_{III}) - (\mu_I - \mu_{III})}{\sqrt{s_p^2(\frac{1}{n_I} + \frac{1}{n_{III}})}} = \frac{18.82 - 0}{\sqrt{155.657(\frac{1}{33} + \frac{1}{33})}} = 6.127$$

Step 3: Formulate a Decision Rule (Equal Population Variances)

For this two-tail test, the decision rule is: reject the null hypothesis if the test statistic falls in either the upper or lower rejection regions. To look up the upper-tail and lower-tail critical values in the t distribution, we use the specified α value (0.05) and determine the degrees of freedom. Since this is a two-tail test, we use α/2=0.025. In the case of equal population variance, the degrees of freedom are computed as: n_1+n_2-2 (or, with our example, 33+33-2=64). Using Excel, we type =T.INV(0.025,64) to obtain the critical values of ±1.9977. The decision rule thus becomes: reject the null hypothesis if the test statistic is greater than 1.9977 or less than -1.9977.

Step 4: Apply the Decision Rule

Comparing the test statistic (6.127) to the upper-tail critical value (1.9977), we see that the test statistic falls into the upper-tail rejection region and we reject the null hypothesis.

Step 5: Draw Your Conclusion

By rejecting the null hypothesis, we conclude that the mean final exam score for Year I is significantly higher than the mean final exam score for Year III.

Test of Two Populations' Means: Unequal Variances and Independent Samples

Still within the situation of having two independent samples, we now address the professor's second question by determining (at a 5% significance level) if the average final exam score for Year II is greater than the average final exam score for Year III.

Step 1: Formulate Hypotheses

The parameter of interest is the difference between the two populations' means. We format the hypotheses as follows:

$$H_0: \mu_{II} - \mu_{III} \leq 0$$

$$H_1: \mu_{II} - \mu_{III} > 0$$

where μ_{II} is the mean final exam grade for Year II and μ_{III} is the mean final exam grade for Year III.

As with the procedure used in answering the professor's first question, we now perform a test of equality of the variances of the two populations, with the outcome determining how the test statistic and the degrees of freedom will be handled.

Step 2a: Test of Equality of Variances

Step 1: Formulate Hypotheses. We specify a null hypothesis stating that the variances are equal. As Year II's variance is larger, we use it as the numerator:

$$H_0: \sigma^2_{II}/\sigma^2_{III} = 1$$

$$H_1: \sigma^2_{II}/\sigma^2_{III} \neq 1$$

Step 2: Compute the Test Statistic. As we are given the standard deviations of both exams, we use these to compute the test statistic:

$$F = \frac{s^2_{II}}{s^2_{III}} = \frac{25.9^2}{13.65^2} = 3.6$$

Step 3: Formulate a Decision Rule. Since we will only be looking at the upper-tail rejection region, the decision rule is: reject the null hypothesis if the test statistic falls into the upper-tail rejection region. Looking up the critical value using Excel, we type =F.INV.RT(0.025,32,32) to obtain a value of 2.0247 (we use 0.025 because this is a two-tail test: we are looking at *equality* of variances). Our decision rule is restated as: reject the null hypothesis if the test statistic is greater than 2.0247.

Step 4: Apply the Decision Rule. Because the test statistic (3.6) is greater than the upper-tail critical value (2.0247), we reject the null hypothesis.

Step 5: Draw Your Conclusion. By rejecting the null hypothesis, we conclude that the variances of the two populations are not equal.

Step 2b: Compute the Test Statistic (Unequal Variances)

Since we have unequal population variances, the sample variances cannot be pooled together in computing a single sample deviation (s^2_p). Instead, the following formula is used to compute the test statistic:

$$t = \frac{(\bar{X}_1 - \bar{X}_2) - (\mu_1 - \mu_2)}{\sqrt{\left(\frac{s_1^2}{n_1} + \frac{s_2^2}{n_2}\right)}}$$

where $(\bar{X}_1 - \bar{X}_2)$ is the observed difference between the two sample means, $(\mu_1 - \mu_2)$ is the hypothesized difference between the two means, S_1^2 is the variance of sample 1, and S_2^2 is the variance of sample 2. Using data from our example:

$$t = \frac{(\bar{X}_{II} - \bar{X}_{III}) - (\mu_{II} - \mu_{III})}{\sqrt{\left(\frac{s_{II}^2}{n_{II}} + \frac{s_{III}^2}{n_{III}}\right)}} = \frac{(64.97 - 62.42) - 0}{\sqrt{\frac{25.9^2}{33} + \frac{13.65^2}{33}}} = 0.5$$

Step 3: Formulate a Decision Rule (Unequal Variances)

As our hypothesis test is an upper-tail test, the decision rule is: reject the null hypothesis if the test statistic falls into the upper-tail rejection region. To find the upper-tail critical value in a t-distribution, we need to have a value for α (specified as 0.01) and a value for the degrees of freedom. When performing a test of two populations' means with unequal variances, the following formula is used in order to account for both samples in computing the degrees of freedom:

$$df = \frac{\left(\frac{s_1^2}{n_1} + \frac{s_2^2}{n_2}\right)^2}{\frac{\left(\frac{s_1^2}{n_1}\right)^2}{n_1 - 1} + \frac{\left(\frac{s_2^2}{n_2}\right)^2}{n_2 - 1}}$$

In our example:

$$df = \frac{\left(\frac{s_{II}^2}{n_{II}} + \frac{s_{III}^2}{n_{III}}\right)^2}{\frac{\left(\frac{s_{II}^2}{n_{II}}\right)^2}{n_{II}-1} + \frac{\left(\frac{s_{III}^2}{n_{III}}\right)^2}{n_{III}-1}} = \frac{\left(\frac{25.9^2}{33} + \frac{13.65^2}{33}\right)^2}{\frac{\left(\frac{25.9^2}{33}\right)^2}{33-1} + \frac{\left(\frac{13.65^2}{33}\right)^2}{33-1}} = 48.5$$

We round down to 48 degrees of freedom and obtain the upper-tail critical value using Excel by typing =T.INV(0.01,48), which returns a value of 2.4066. Now that we know the upper-tail critical value, we restate the decision rule: reject the null hypothesis if the test statistic is greater than 2.4066.

Step 4: Apply the Decision Rule

Comparing the test statistic to the upper-tail critical value, we find the test statistic (0.5) to be less than the critical value (2.4066) and we do not reject the null hypothesis.

Step 5: Draw Your Conclusion

By not rejecting the null hypotheses, we conclude that the mean final exam score for Year II is not significantly higher than the mean final exam score for Year III.

Test of Two Populations' Means: Paired Samples

Finally, we consider the professor's third question of whether there was a significant difference in Year III students' *individual* performances on their midterm and final exams. The key to understanding this question is in recognizing how this question differs from the first two questions.

Remember, with the first two questions, we were able to assume that the two collected samples were independent of each other. This final question, however, asks us to consider how the grades of *individual* students have changed between exams. In other words, we no longer assume that our two samples are independent. Specifically, we expect that students who obtain a relatively high grade on one exam would also obtain a relatively high grade on another exam.

We cannot assume independence between two populations being examined when data values from the two samples are linked in a way likely to associate together data values across the two samples (e.g., two exam scores by the same student; pre- and post-testing results; measures taken from two related people, such as a parent and child, a consultant and client, a manager and employee, and so on; etc.). In such situations, we are dealing with a test of **paired samples**, also referred to as **matched samples**.

Table 10-5 displays the matched samples to be used in examining the professor's third question. In Table 10-5, the first column contains student IDs (numbered from 1 to 33), the second column contains each student's grade on the midterm exam, the third column contains each student's grade on the final exam, and the fourth column contains the calculated difference between these two grades for each student. Our test is conducted on the data provided in the fourth column.

Table 10-5
The Mean Difference between Midterm and Final Exam Scores

Student	Midterm Exam	Final Exam	Difference
1	66	59	7
2	99	90	9
3	97	67	30
4	72	40	32
5	89	74	15
6	94	77	17
7	89	73	16
8	79	46	33
9	87	60	27
10	87	75	12
...	...	...	...
...	...	...	...
...	...	...	...
25	90	61	29
26	84	70	14
27	88	71	17
28	92	71	21
29	82	60	22
30	85	55	30
31	97	85	12
32	73	44	29
33	65	55	10
Mean	81.24	62.42	18.82
Standard Deviation	11.18	13.65	8.74

The parameter of interest for this test is thus the **mean difference**, calculated as the average of the individual difference for each student's scores on the midterm and final exams. Because this is a test of a single mean (that is, the mean difference between two exam scores) with σ (the population's standard deviation) being unknown, we know from Chapter 9 that the t-distribution is used in computing the test statistic and the p-value is used in determining critical values.

Step 1: Formulate Hypotheses

Since the professor suspects that students will perform better on the midterm exam than on the final exam, this hypothesis test is formulated as an upper-tail test. If the final exam was more challenging to students, we would expect to see a decrease in student performance on the final exam relative to the midterm exam. For example, a person scoring a 90 on the midterm would be expected to score lower than 90 on the final, and a person scoring a 50 on the midterm would be expected to score lower than 50 on the final. Calculating the difference between individual exam scores by subtracting the expected lower-score exam from the expected higher-score exam, our hypotheses are:

$$H_0: \mu_d \leq 0$$

$$H_1: \mu_d > 0$$

where μ_d represents the *mean* difference, across all students, between a student's scores on the midterm and final exams. To calculate μ_d, we first compute the difference in grade between the midterm and final for each student and then average these differences across all students, obtaining the result shown in the second-to-last row of the fourth column of Table 10-5.

Step 2: Compute the Test Statistic

The formula used to compute the test statistic for this hypothesis test is:

$$t = \frac{\bar{d} - \mu_d}{s_d/\sqrt{n}}$$

where d is the difference $(x_M - x_F)$ between the two exams for each student, $\bar{d}$ is the mean difference across all 33 students ($\frac{\sum_{i=1}^{n} d_i}{n}$), μ_d is the hypothesized mean difference, s_d is the sample standard deviation of the differences, and n is the number of pairs in the sampled data. Using the data from our example:

$$t = \frac{\bar{d} - \mu_d}{s_d/\sqrt{n}} = \frac{18.82 - 0}{8.74/\sqrt{33}} = 12.37$$

Step 3: Formulate a Decision Rule

As this is an upper-tail test, the decision rule is: reject the null hypothesis if the test statistic falls within the upper-tail rejection region. To obtain the upper-tail critical value from a t-distribution, we need to know α (specified as 0.05) and the degrees of freedom. For this hypothesis test, the number of degrees of freedom is the *number of pairs minus one*. As there are 33 sets of paired student grades, the number of degrees of freedom becomes 33-1=32. Looking up the critical value in Excel by typing =T.INV(0.05,32) returns a value of 1.6939. We thus restate the decision rule: reject the null hypothesis if the test statistic is greater than 1.6939.

Step 4: Apply the Decision Rule

Comparing the test statistic to the upper-tail critical value, the test statistic (12.377) is greater than the critical value (1.6939) and we reject the null hypothesis.

Step 5: Draw Your Conclusion

By rejecting the null hypothesis, we conclude that the mean difference in individual student grades between the midterm and final exams is significantly greater than zero, meaning that the students indeed had an easier time handling the material covered by the midterm exam than that covered by the final exam.

Confidence Interval for the Difference between Two Populations' Means

We calculate confidence intervals for the difference in two populations' means in essentially the same way that we have calculated confidence intervals for other population parameters. Column 2 of Table 10-6 presents the three formulas used in computing confidence intervals for each of the three situations described in this unit, and Columns 3 and 4 of Table 10-6 provide the results of using the formulas with data from this unit's example.

Table 10-6
Confidence Intervals for Two Populations' Means

Type of Test	Confidence Interval Formula	Example (Using $(1-\alpha)=0.95$)	Confidence Interval Limits
Equal Variances, Independent Samples	$(\bar{X}_1 - \bar{X}_2) \pm t_{\alpha/2,df} \sqrt{s_p^2 \left(\frac{1}{n_1} + \frac{1}{n_2}\right)}$ $df = n_1 + n_2 - 2$	$18.82 \pm 1.997 \sqrt{155.657 \left(\frac{1}{33} + \frac{1}{33}\right)}$ $df = 33 + 33 - 2 = 64$	[12.69 ... 24.95]
Unequal Variances, Independent Samples	$(\bar{X}_1 - \bar{X}_2) \pm t_{\alpha/2,df} \sqrt{\left(\frac{s_1^2}{n_1} + \frac{s_2^2}{n_2}\right)}$ $df = \dfrac{\left(\frac{s_1^2}{n_1} + \frac{s_2^2}{n_2}\right)^2}{\frac{\left(\frac{s_1^2}{n_1}\right)^2}{n_1 - 1} + \frac{\left(\frac{s_2^2}{n_2}\right)^2}{n_2 - 1}}$	$2.55 \pm 2.01 \sqrt{\left(\frac{25.9^2}{33} + \frac{13.65^2}{33}\right)}$ $df = \dfrac{\left(\frac{25.9^2}{33} + \frac{13.65^2}{33}\right)^2}{\frac{\left(\frac{25.9^2}{33}\right)^2}{32} + \frac{\left(\frac{13.65^2}{33}\right)^2}{32}} = 48$	[-7.69 ... 12.79]
Paired Samples	$\bar{d} \pm t_{\alpha/2,df} \dfrac{s_d}{\sqrt{n}}$ $df = n-1$	$18.82 \pm 2.04 \dfrac{8.74}{\sqrt{33}}$ $df = 33 - 1 = 32$	[15.72 ... 21.92]

Recall that the hypothesized difference, in the null hypothesis, was zero for each of the tests carried out in answering the three questions raised by the professor in this unit's example. Our conclusion regarding the professor's first question, after rejecting the null hypothesis, was that the average score for Year I was higher than the average score for Year III. Indeed, you can see from Table 10-6 that the hypothesized difference of zero is not included in the computed confidence interval. With the professor's second question, after not rejecting the null hypotheses, we concluded that the average score for Year II was not higher than the average score for Year III. Although this second hypothesis test was conducted at a different level of significance ($\alpha=0.01$ rather than $\alpha=0.05$), you can see from Table 10-6 that the computed confidence interval did include the hypothesized difference of zero. Finally, for the professor's third question, after rejecting the null hypothesis, we concluded that each student's individual performance was greater for the midterm exam than for the final exam. Looking at Table 10-6, you can again observe that the hypothesized value of zero is not included within the computed confidence interval.

Unit 3 Summary

- When conducting tests of two populations' means, we need to first determine whether the samples are independent of each other or not. ***Independent samples*** are selected from two (or more) populations such that the occurrence of values in one sample has no influence on the probability of the occurrence of values in the other sample. ***Paired samples*** are not independent, but instead are ***matched*** on some common attribute, with common examples including collecting data from the same person before and after some treatment or when respondents are paired together based on some common attribute.

- If the two samples are independent, we proceed to test whether the variances are assumed to be equal by adding a new step to the five-step testing process - a test of the equality of variances - as shown in Figure 10-4.

Figure 10-4
Testing Process: Two Populations' Means with Independent Samples

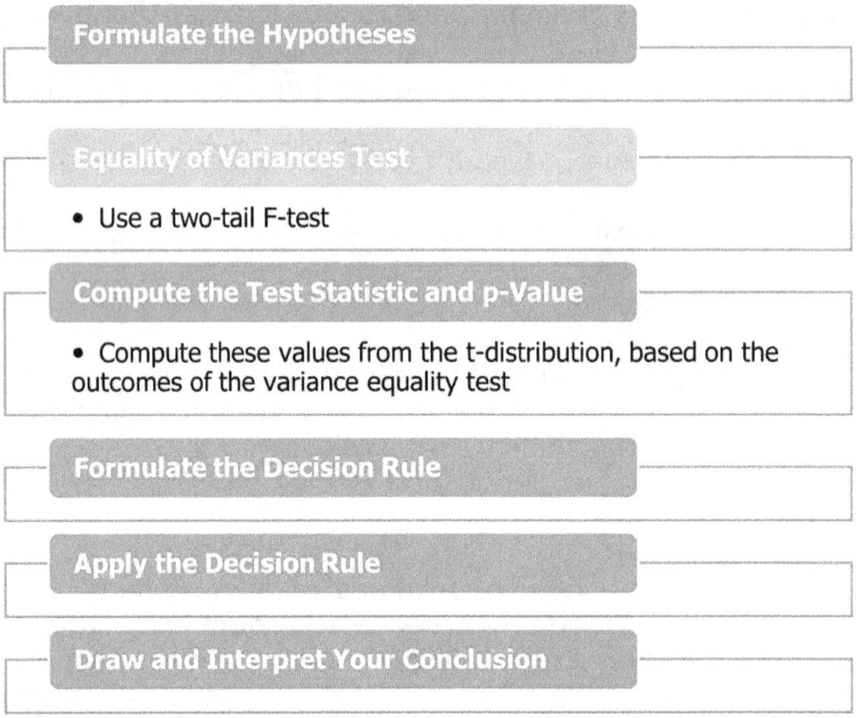

- **Formulate the Hypotheses**

- **Equality of Variances Test**
  - Use a two-tail F-test

- **Compute the Test Statistic and p-Value**
  - Compute these values from the t-distribution, based on the outcomes of the variance equality test

- **Formulate the Decision Rule**

- **Apply the Decision Rule**

- **Draw and Interpret Your Conclusion**

- With *paired samples*, we create a new variable representing the difference for each pair across the two samples, compute the **mean difference** based on this new variable, and proceed with the test.

- Working with the t-distribution, we use α and the number of degrees of freedom to obtain critical values from the t-table. Use the appropriate formula in Table 10-7 to compute the degrees of freedom.

Table 10-7
Degrees of Freedom Computation

Type of Test	Degrees of Freedom Formula
Equal Variances, Independent Samples	$df = n_1 + n_2 - 2$
Unequal Variances, Independent Samples	$df = \dfrac{\left(\dfrac{s_1^2}{n_1} + \dfrac{s_2^2}{n_2}\right)^2}{\dfrac{\left(\dfrac{s_1^2}{n_1}\right)^2}{n_1 - 1} + \dfrac{\left(\dfrac{s_2^2}{n_2}\right)^2}{n_2 - 1}}$
Paired Samples	$df = n - 1$ (where n is the number of pairs)

- Use the formulae in Table 10-8 to compute the test statistic:

Table 10-8
Test Statistics Computation

Type of Test	Test Statistic Formula
Equal Variances, Independent Samples	where $t = \dfrac{(\bar{X}_1 - \bar{X}_2) - (\mu_1 - \mu_2)}{\sqrt{s_p^2 \left(\dfrac{1}{n_1} + \dfrac{1}{n_2}\right)}}$ $s_p^2 = \dfrac{(n_1 - 1)s_1^2 + (n_2 - 1)s_2^2}{n_1 + n_2 - 2}$
Unequal Variances, Independent Samples	$t = \dfrac{(\bar{X}_1 - \bar{X}_2) - (\mu_1 - \mu_2)}{\sqrt{\dfrac{s_1^2}{n_1} + \dfrac{s_2^2}{n_2}}}$
Paired Samples	$t = \dfrac{\bar{d} - \mu_d}{s_d / \sqrt{n}}$

- Use the rules in Table 10-9 to help you decide whether or not to reject the null hypothesis.

Table 10-9
Decision Rules for a Test of Two Populations' Means

Type of Test	Rejection Rule
Lower-Tail Test $H_0: \mu_1 - \mu_2 \geq$ value $H_1: \mu_1 - \mu_2 <$ value For paired samples: $H_0: \mu_d \geq$ value $H_1: \mu_d <$ value	Reject the null hypothesis if: Test statistic < critical value
Upper-Tail Test $H_0: \mu_1 - \mu_2 \leq$ value $H_1: \mu_1 - \mu_2 >$ value For paired samples: $H_0: \mu_d \leq$ value $H_1: \mu_d >$ value	Reject the null hypothesis if: Test statistic > critical value

Two-Tail Test
$H_0: \mu_1 - \mu_2 =$ value
$H_1: \mu_1 - \mu_2 \neq$ value
 	
For paired samples:
$H_0: \mu_d =$ value
$H_1: \mu_d \neq$ value
	Reject the null hypothesis if:
	  Test statistic < lower critical value
	Or if:
	  Test statistic > upper critical value

- Use the formulae in Table 10-10 to compute confidence intervals.

Table 10-10
Confidence Interval Formulae

Type of Test	Confidence Interval Formula
Equal Variances, Independent Samples	$(\bar{X}_1 - \bar{X}_2) \pm t_{\alpha/2, df} \sqrt{s_p^2 \left(\frac{1}{n_1} + \frac{1}{n_2}\right)}$
Unequal Variances, Independent Samples	$(\bar{X}_1 - \bar{X}_2) \pm t_{\alpha/2, df} \sqrt{\left(\frac{s_1^2}{n_1} + \frac{s_2^2}{n_2}\right)}$
Paired Samples	$\bar{d} \pm t_{\alpha/2, df} \frac{s_d}{\sqrt{n}}$

Unit 3 Exercises

1. Find the test statistic, the critical value, and state the decision rule for Questions 1.a., 1.b. and 1.c (assume these are all independent samples):

   a. $\bar{x}_1 = 53$, $s^2_1 = 130$, $n_1 = 100$; $\bar{x}_2 = 47$, $s^2_2 = 144$, $n_2 = 200$
   upper-tail test; $\alpha = 0.05$

   b. $\bar{x}_1 = 53$, $s^2_1 = 200$, $n_1 = 100$; $\bar{x}_2 = 47$, $s^2_2 = 250$, $n_2 = 200$
   upper-tail test; $\alpha = 0.05$

   c. $\bar{x}_1 = 53$, $s^2_1 = 130$, $n_1 = 200$; $\bar{x}_2 = 47$, $s^2_2 = 144$, $n_2 = 400$
   upper-tail test; $\alpha = 0.05$

   d. Compare your answers to Question 1.a. and Question 1.b. What is the effect of an increase in sample variances?

   e. Compare your answers to Question 1.a. and Question 1.c. What is the effect of an increase in sample size?

   f. Conduct the three tests again at the 1% level of significance. What is the effect of the change in α?

   g. Conduct the three tests again as a two-tail test with $\alpha = 0.05$. What is the difference between the one-tail and the two-tail tests?

2. Determine the degrees of freedom for each case (assume independent samples):

   a. $s^2_1 = 3$, $n_1 = 50$; $s^2_2 = 5$, $n_2 = 75$

   b. $s^2_1 = 14$, $n_1 = 100$; $s^2_2 = 11$, $n_2 = 150$

   c. $s^2_1 = 8$, $n_1 = 50$; $s^2_2 = 5$, $n_2 = 50$

3. Test the following hypotheses using: (1) critical values, and (2) confidence intervals:

   a. $H_0: \mu_1 - \mu_2 \leq 0$; $H_1: \mu_1 - \mu_2 > 0$; $\alpha = 0.05$
      $\bar{x}_1 = 44$, $s^2_1 = 22.5$, $n_1 = 60$; $\bar{x}_2 = 42$, $s^2_2 = 25$, $n_2 = 45$

   b. $H_0: \mu_1 - \mu_2 \geq 0$; $H_1: \mu_1 - \mu_2 < 0$; $\alpha = 0.01$
      $\bar{x}_1 = 6$, $s^2_1 = 16$, $n_1 = 140$; $\bar{x}_2 = 7$, $s^2_2 = 10$, $n_2 = 80$

   c. $H_0: \mu_1 - \mu_2 \geq 0$; $H_1: \mu_1 - \mu_2 < 0$; $\alpha = 0.01$
      $\bar{x}_1 = 30$, $s^2_1 = 56$, $n_1 = 100$; $\bar{x}_2 = 32$, $s^2_2 = 68$, $n_2 = 80$

4. A *New York Yankees* fan is arguing with a *Boston Red Sox* fan, claiming the Yankees score, on average, more runs per game. Construct a 99% confidence interval for the difference in the average runs scored per game for the two teams. Data collected from eight games played by each team is provided in Table 10-e9. Can you say that the *Yankees* score more runs per game than the *Red Sox*? (To answer this question, you need to check whether the value of zero falls within the computed confidence interval.)

Table 10-e9

	Yankees	Red Sox
	4	4
	8	7
	2	2
	5	0
	3	9
	6	4
	10	7
	1	2
Mean	4.88	4.38
Variance	9.27	9.41

5. A student desiring to build some muscle over the summer has to choose between two programs. The first program involves a daily training regime, while the second program involves training only twice a week - but with much greater intensity. After looking through some Internet forums on muscle-building, he randomly records the muscle weight, in pounds, that others have gained in one month through each of the two programs. The data are shown in Table 10-e10. Should he choose the twice-a-week (but more intensive) training program? (Use a 5% significance level.)

Table 10-e10

	Gain with the Once-a-Week Program	Gain with the Twice-a-Week Program
	3	4
	4	6
	2	3
	5	1
	1	7
	3	3
	4	5
	2	1
	5	6
	7	2
	2	4
	1	7
Mean	3.25	4.08
Variance	3.30	4.63

6. Households are becoming more and more tech savvy and are believed to own more computers today than five years ago. Using the data in Table 10-e11, can you conclude at the 1% significance level that households own more computers today?

Table 10-e11

	Household Computer Ownership Five Years Ago	Household Computer Ownership Today
Sample Size	200	200
Mean	1.8	2.2
Variance	2	2.9

7. A sample of newly-released movies and their rating scores (out of 30), as voted by audiences in the US and in England, is shown in Table 10-e12. Conduct a test at the 5% significance level to see if ratings are lower in England.

Table 10-e12

	The Boss	Zootopia	Batman: The Killing Joke	Deadpool	Batman v Superman	The Lobster	Keanu
US	30	25	23	26	25	25	24
England	29	22	26	23	22	23	21

8. Bill and Kristen are both avid readers and are curious to see which of the two has checked out (and read) more books from their local public library. They are able to gain access to the library's records and, using the most-recent ten-month time period, they record the number of books each checked out, as shown in Table 10-e13. At the 99% confidence level, estimate the mean difference between the number of books that Bill has checked out and the number of books that Kristen has checked out.

Table 10-e13

	Jan	Jun	May	Nov	Apr	Dec	Mar	Jul	Feb	Sep
Bill	5	7	6	5	8	5	4	2	4	6
Kristen	4	4	3	4	5	3	6	5	3	7

UNIT 4

Bringing It All Together

We have covered a variety of hypothesis tests in Chapters 7, 8 and 9 and again, here, in Chapter 10. In describing these tests, we have used a common five-step procedure as a guide for understanding the similarities and differences across the different tests. In concluding this four-chapter coverage of hypothesis testing, we now provide a decision aid to help you in determining the specific hypothesis test to use in a given situation. The End-of-Chapter Practice exercises that follow this final unit of the chapter require the use of the hypothesis tests that have been covered; however, you need to figure out by yourself which specific test should be carried out. Appendix A describes how to use the Microsoft Excel *Analysis ToolPak* in performing some of these tests.

The first question to ask, when choosing among the types of hypothesis tests, is whether you are dealing with one or with two populations. The simplest way to answer this question is to see how many samples are described in the question. If you are provided with data taken with a single sample, then you are probably being asked to compare a sample statistic (i.e., mean, variance or proportion) to a specified value and you would conduct a one-population test. If you have data representing two different samples, then you are probably being asked to compare a sample statistic across the two populations and you would conduct a two-population test.

Figure 10-5 provides a decision aid (here, in the form of a decision tree) to help you navigate among the different tests of single population parameters. Going from left-to-right with this decision tree, first identify the parameter of interest (i.e., mean, proportion or variance). Next, ensure that all necessary conditions are met and identify the proper distribution. Finally, the right-hand side of the tree provides the formulae to use for each case.

Figure 10-5
Decision Tree for One-Population Tests

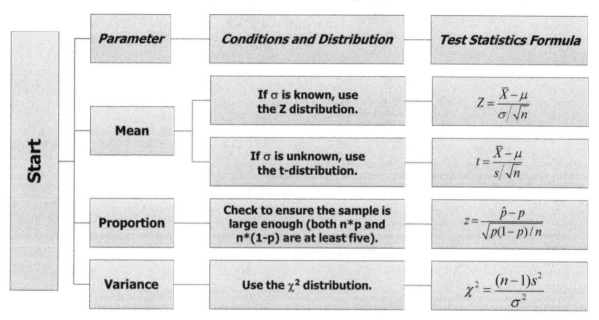

Figure 10-6 provides a similar decision tree for the two-populations tests. This tree is a little more complex, as certain situations require that decisions be made about the use of pooled parameters and about the appropriate formula to use in computing the degrees of freedom associated with a test.

Figure 10-6
Decision Tree for Two-Populations Tests

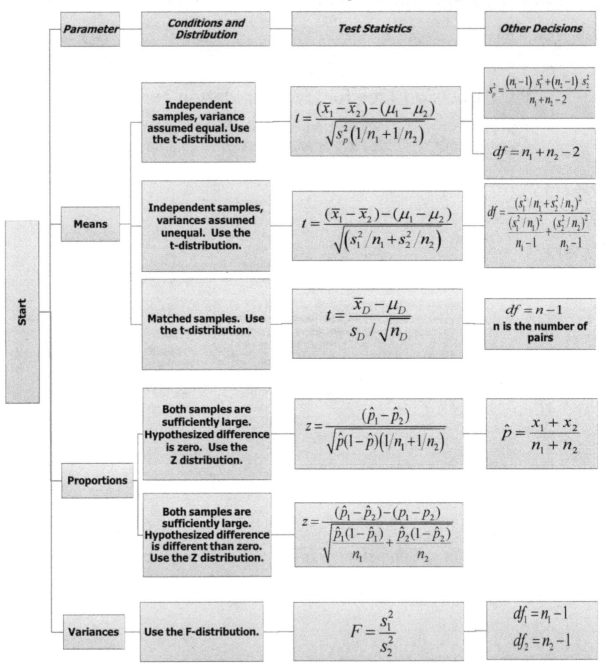

END-OF-CHAPTER PRACTICE

1. A low-fat yogurt processing plant requires the variance in the amount of fat in their yogurt products to be no more than 0.45. A manager wants to test that the process is working as required and takes a random sample of 50 yogurt cups. The variance is calculated to be 0.51. At a significance of 5%, does the manager need to have a technician investigate the manufacturing process?

2. A new printer ink cartridge is being advertised as having the ability to print 700 pages before a replacement cartridge must be bought, with a standard deviation of twenty pages. To test this, eighteen ink cartridges were bought and the standard deviation was determined to be 29. Is there anything unusual about the claim?

3. ABC Company is planning to open a new location of its retail outlet in City Z. However, they are also considering City Y as an alternative. If City Y has significantly different sales volume than City Z, they will need to reconsider their plans of expansion. To analyze the two cities, the project manager gathers information on the sales levels of 50 stores in each city. City Z averages sales of $250,000 each month with a variance of $8,500. City Y averages sales of $275,000 each month with a variance of $9,500. Is there a significant difference between City Y and City Z? Use a 5% significance level (conduct your calculations using thousands of dollars as your unit).

4. A steel rod production company is in the process of buying a new machine. To test out the abilities of the machines, the company tested the production of 25 steel rods. They found that Machine A produced rods with variance in the diameter of 6.2 mm^2, while Machine B's rods had diameter variance of 8.4 mm^2. Does this suggest that Machine A has lower variability in its production? (Use $\alpha=0.01$.)

5. A bank manager claims that the standard deviation in the waiting time to be served by a teller is less than 4.6 minutes. A random sample of ten visitors to the branch was taken and the standard deviation of the waiting time was found to be four minutes. At 5% significance, can the manager's claim be supported?

6. A marketing agency is helping to design a new advertising campaign for a retail store which claims that, although they have many visitors to their store, the number of people who actually buy their products is very small. Prior to the advertising campaign, out of a random sample of 178 individuals who visited the store, only 25 actually purchased an item. After the campaign, 47 out of a random sample of 229 people entering the store had purchased an item. At a 5% significance level, has the campaign helped the store?

7. ABC Company has just released their new line of super camera batteries which are designed to last for longer periods of time - perfect for vacations, as a battery-charger will not be needed. The company promises that these batteries last fifteen hours with normal usage. To test this out, a photography enthusiast took a random sample of twelve of the batteries and tried them out. She found that the average time the batteries in the sample group lasted was twelve hours. Assuming that the time the batteries last is normally distributed with a known standard deviation of 3.4 hours, do you believe the claim made by the company? Use a 5% significance level.

8. An economist believes that the unemployment percentage in New Zealand is higher than the unemployment percentage in Australia. The Australian unemployment rate stands at 10%. In a sample of 250,000 people in New Zealand, 35,000 were unemployed. Based on this information, can the economist conclude at a 1% significance level that the unemployment rate in New Zealand is higher than that in Australia?

9. A small business has recently started a website, a *Twitter* page, a *Facebook* page, and advertises itself on *Google*. Comparing revenue (in thousands of dollars) from before and after the change, shown in Table 10-e14, determine at a 5% significance level whether these Internet marketing actions have increased sales.

Table 10-e14

Month	Before	After
January	45	48
February	43	43
March	37	40
April	41	42
May	39	43
June	36	36
July	35	38
August	33	30
September	34	32
October	37	41
November	40	40
December	49	55

10. Sue has been using one email address for both business and personal contacts, but is wondering if she should get a second email address in order to separate her mail. She decides to get the second email address if more than 20% of her inbox turns out to be personal emails. A random sample of

30 emails in her inbox finds that nine email messages are personal. Using a 5% significance level, does Sue get a second email address?

11. *Facebook* has become a very popular website among senior citizens. A random sample of 65 senior Americans showed a sample average of 189 friends with a sample standard deviation of 25 friends. Estimate the number of *Facebook* friends senior Americans have (use a 99% confidence level). Can it be concluded at the 1% significance level that senior Americans have at least 200 *Facebook* friends?

12. A professor teaches two sections of a statistics course: one section on Monday morning and another on Wednesday afternoon. The students in both sections have just taken their midterm exams, with the Monday morning class of 34 students having an average score of 70 and a standard deviation of fourteen. The Wednesday class has 50 students and scored a mean of 75 with a standard deviation of twelve. Can the professor say that there is a difference with the two sections' performances on the midterm exam? (Use $\alpha=0.05$.)

13. It has been claimed by educational experts that the number of hours spent doing homework is associated with students' performances on exams. To test this, two groups of students in a class were questioned about the amount of time they spent doing homework. Group A, of 25 students, had marks in the top half of the class while the Group B, similarly of 25 students, was comprised of students with marks in the lower half of the class. Group A students averaged five hours doing homework, with a variance of 45 minutes^2; while Group B students averaged 2.5 hours doing homework, with a variance of one hour^2. At the 1% level of significance, is there sufficient difference between the two groups to support the claim about a relationship between time spent on homework and exam performance?

14. Company ABC has undertaken mentorship and support programs to increase the number of women in leadership positions within the firm. Five years ago, of the 100 senior management positions, 33 of them were filled by women. Today, as the company expands, they now have 175 senior management positions and 68 of them are occupied by women. Have the company's new gender-equality programs had an effect? (Use $\alpha=0.05$.)

15. As technology improves, more and more children are receiving smartphones from their parents. In 2014, a random sample of 50 elementary school students was taken and 25 stated that they had a cell phone, with ten of these cell phone users having a smartphone. In 2016, a similar survey of a random sample of 50 elementary school students was undertaken. 35 students had a cell phone, of which 21 were smartphones. Answer the following two questions: First, has cell phone usage increased among elementary school students? Second, has smartphone usage increased among the students using cell phones by more than 10%? (Use $\alpha=0.05$.)

16. A local farmer claims that her fresh apple juice contains fewer calories than the juice from a juice company that produces its juice from concentrate. To test this claim, a dietician took 25 juice bottles (all sixteen ounces) from both the farmer and the juice company and measured their caloric content. The farmer's apple juice had an average of 116 calories per bottle, with a standard deviation of 45. The company's juice had an average calorie content of 124 per bottle, with a standard deviation of 36. Can the dietician support the farmer's claims using a 5% significance level?

17. Was the 2016 Euro soccer championship in France exciting? Some will say "no" because not that many goals were scored during the games. Given the sample of goals scored by eight teams in both the 2016 and 2012 Euro championships in Table 10-e15, determine at the 1% significance level if the mean goal scores per team decreased from 2012 to 2016.[106,107]

Table 10-e15

Year	Spain	Germany	Italy	Portugal	England	Russia	Sweden	France
2012	12	10	6	6	5	5	5	3
2016	5	7	6	9	4	2	1	13

18. In finance, variance is used as a measurement of the risk associated with an investment. A risk-averse investor wants to determine if there is a difference in the variability, or risk, of two investment portfolios. The investor looked at the monthly returns, over fifteen months, of the two portfolios and calculated that Portfolio A has a variance of 24, while Portfolio B has a variance of 32. Is Portfolio B riskier than Portfolio A, using a 5% significance level?

19. A dairy processing plant samples twenty half-gallon cartons of milk and finds that the variance in the amount of fat in their milk is 0.32. A competitor also samples twenty half-gallon cartons of milk and finds that the variance in the amount of fat in their milk is only 0.25. Is there evidence of a significant difference in the variability of the fat content in the milk of the two competitors? (Use $\alpha=0.01$.)

20. A new restaurant aims to provide faster, more efficient service and states that the variance in the time taken to serve customers is only 3.7 minutes^2. They determined this figure by measuring the variance in the time taken to serve ten customers. An older restaurant in the area did the same thing and

---

[106] Source: http://www.uefa.com/uefaeuro/season=2012/statistics/round=15172/teams/type=goalsscored/index.html
[107] Source: http://www.uefa.com/uefaeuro/season=2016/statistics/round=2000448/teams/index.html

found the variance in the time to serve ten customers to be 4.5 minutes². At a 5% significance level, is there a difference in the variance of the time to serve customers between the two restaurants?

21. Soccer is a sport that is increasingly becoming popular in North America. During the 2006 *World Cup*, a random sample of 450 individuals was taken and 275 said they were watching this sporting event. In 2010, another random sample of 500 individuals was taken and 350 stated that they were watching the *World Cup*. Has soccer viewership increased by more than 5% over this four-year period? (Use $\alpha=0.1$.)

22. Has the number of people going to movie theaters increased over the past five years? To test this, customer information from 25 random theaters was studied. The figures showed that in 2006 the average number of moviegoers each week was 374, with a variance of 87. In 2011, each week averaged 475 moviegoers, with a variance of 105. Has there been a significant (at a 5% level) change in moviegoers from 2006 to 2011?

23. Major car accidents often occur on highways due to cars travelling at differing speeds. A state highway engineer sampled the speeds of 55 cars on two major highway routes. The standard deviation of the speeds was 13.1 mph with Highway A and 9.5 mph with Highway B. The engineer wants to know if there is evidence that the variability of the speed is higher on Highway A. (Use $\alpha=0.05$.)

CHAPTER 11: Chi-Square Tests

Thus far, we have dealt mostly with quantitative data in this book. However, qualitative, or categorical, data can be equally valuable to researchers and practitioners alike. For example, we may wish to study which of three web browsers do people prefer to use or whether their choice of a browser is related to the brand of computer that they have. With categorical data (such as proportions, which we have previously encountered in this book), a very useful measurement involves the percentage of observations falling into each category.

As an example, consider a startup company creating financial apps for smartphones. The owners of the company need to select the smartphone operating system for the first versions of their financial apps. Ideally, the selected operating system would hold the largest market share (i.e., the largest number of potential sales opportunities) among smartphone operating systems. A national market research study[108] observed the market shares held by the leading smartphone operating systems in February 2016 to be 53.1% for *Android*, 43.1% for *Apple*, 2.8% for *Microsoft*, and 1% for *BlackBerry*. This, then, is what we might *expect* to be the market shares for smartphone operating systems. However, after taking a new sample of 100 smartphone users within the target demographic market for the startup's financial apps, it was observed that there were 60 *Android* users, 30 *Apple* users, five *Blackberry* users, and five *Microsoft* users.

How can the startup company test to determine if the market shares observed in this just-taken sample significantly differ from the expected market

---

[108] Source: https://www.comscore.com/Insights/Rankings/comScore-Reports-February-2016-US-Smartphone-Subscriber-Market-Share

shares, i.e., those observed in the February 2016 national market research study? The test we apply to make such a determination examines the extent to which category percentages with the observed data (i.e., the just-taken survey of smartphone users) *fit* the category percentages of the expected data. This test, called the *goodness of fit* test, makes use of the chi-square distribution and applies a test statistic reflecting the difference between an observed distribution (the observed category percentages) and an expected distribution (the expected category percentages). Essentially, the expected distribution represents the test's null hypothesis.

Another test we can conduct that makes use of categorical data is the *test of independence*, also known as the *test of contingency tables*. This test examines whether two qualitative variables are related to each other. For example, we may wish to find out whether the market shares held by smartphone operating systems vary for users living in three different countries. Here, one of the variables of interest is Smartphone Operating System and the other is Country of Residence. Both of these are qualitative variables and we could use the test for independence to determine whether a relationship exists between the two variables. Again, we would compare an *expected* distribution with the distribution *observed* from sample data taken in each of the three countries.

Both the goodness of fit test and the test of independence are based on the same premise: to see if an observed distribution conforms to an expected distribution. We first cover the goodness of fit test and then the test of independence. This chapter covers the following topics:

- Goodness of Fit Test

- Goodness of Fit Test for Normality
- Sample Size Considerations
- Test of Independence

UNIT 1

Goodness of Fit Test

Suppose a market analyst in early 2014 was investigating penetration rates of Internet browsers in the United States. As of September 2013, *Internet Explorer* had 12% of the market, *Firefox* had 28%, *Chrome* had 53%, *Safari* had 4%, and *Other* browsers had a 3% penetration rate.[109,110] The market analyst believed that *Safari* should have a higher share on college campuses, observing that Macs and iPads are quite popular with college students. The analyst surveyed 200 local university students about the primary Internet browser each preferred to use. The results indicated that fifteen students preferred to use *Internet Explorer*, 60 preferred to use *Firefox*, 110 preferred to use *Chrome*, and fifteen preferred to use *Safari*. None of the students preferred to use any other browser. Can the researcher conclude, at a 5% significance level, that the 2013 national shares of Internet browsers reflect those preferred in early 2014 on the university campus?

The above question is essentially a hypothesis test employing the following hypotheses:

H_0: The campus distribution of browsers is the same as the national one.

H_1: The campus distribution of browsers is different than the national one.

---

[109] Source: http://www.w3schools.com/browsers/browsers_stats.asp
[110] In case you are wondering, the same source indicates that in May 2016 *Chrome* had 71.4% of the market, followed by *Firefox* (16.9%), *Explorer* (5.7%), *Safari* (3.6%), and *Opera* (1.2%).

If the market analyst is able to reject the null hypothesis, then the analyst will have proven her claim about there being different penetration rates of browsers for college students (at least for this local university and similar campuses).

As discussed earlier, this hypothesis test is called a **goodness of fit** test because it measures *how well* an observed distribution *fits with*, or matches, an expected distribution. We follow the usual five-step hypothesis-testing procedure with Step 1, the formulation of hypotheses, already covered.

Step 2: Compute the Test Statistic

Before computing a test statistic, it is helpful to review the expected and observed distributions associated with this example. In Table 11-1, Column 1 contains the five categories of browsers, Column 2 contains the market share of each browser obtained through a national survey (the notation p_i represents the market share percentages obtained from the national survey), and Column 3 contains the observed number of browsers obtained from surveying 200 students at the local university (the notation o_i represents the observed data). Finally, Column 4 shows the number of browsers we might expect, based on a sample size of n=200, if the observed results from the national survey market shares were to hold for the local university students (the notation e_i represents the expected data). For example, if the national market shares reflect the browsers preferred on this campus, then we would expect to see 12% of the 200 students, or 24 students, preferring to use *Internet Explorer* and 4% of the 200 students, or eight students, preferring to use *Safari*.

Table 11-1
Browser Example: Expected Distribution vs. Observed Distribution

Browser	National Market Share p_i	Number of Browsers Observed o_i	Number of Browsers Expected e_i
Internet Explorer	0.12	15	0.12*200 = 24
Firefox	0.28	60	0.28*200 = 56
Chrome	0.53	110	0.53*200 = 106
Safari	0.04	15	0.04*200 = 8
Other	0.03	0	0.03*200 = 6
Total		200	200

As you know by now, it is not surprising to find survey data differing somewhat from what has been hypothesized. In other words, we should not expect to find *exactly* 24 *Internet Explorer* users in our sample of 200 students. The question we always ask is: "How significant is the difference between our sampled data values and our hypothesized data values?" In order to answer this question, as with the hypothesis tests covered earlier, we need to compute a test statistic and a critical value.

The test statistic we use is a χ^2 (chi-square) value computed as follows:

$$\chi^2 = \sum_{i=1}^{k} \frac{(o_i - e_i)^2}{e_i}$$

where o_i is the observed cell frequency for category i, e_i is the expected cell frequency for category i, and k is the number of categories (in our browser example, the categories include four identified browsers and an Other category, thus k=5).

Table 11-2, below, shows the step-by-step computation of the test statistic in our example. The first four columns were previously shown in Table 11-1. Column 5 computes each of the numerators of the test statistic; and, Column 6 computes the components of the test statistic, which are then summed in the bottom row of Column 6 to compute the test statistics (the χ^2 value) of 15.937.

Table 11-2
Browser Example: Computing the Test Statistic (χ^2 Value)

Browser	National Market Share	Number of Browsers Observed	Number of Browsers Expected	(Observed − Expected)²	(Observed − Expected)² / Expected
	p_i	o_i	e_i	$(o_i - e_i)^2$	$(o_i - e_i)^2 / e_i$
Internet Explorer	0.12	15	0.12*200 = 24	(15-24)² = 81	81/24 = 3.375
Firefox	0.28	60	0.28*200 = 56	(60-56)² = 16	16/56 = 0.286
Chrome	0.53	110	0.53*200 = 106	(110-106)² = 16	16/106 = 0.151
Safari	0.04	15	0.04*200 = 8	(15-8)² = 49	49/8 = 6.125
Other	0.03	0	0.03*200 = 6	(0-6)² = 36	36/6 = 6
Total		200	200		Σ = 15.937

Step 3: Formulate a Decision Rule

Next, we compare this test statistic to a critical value from the χ^2 distribution. We also need to identify the rejection region (Is this an upper-tail or lower-tail test?). For this test, if the null hypothesis is true, then the test statistic would be low in value (as the observed and expected values would not differ greatly from each other, the numerators in Column 5 of Table 11-2 would each be low in value). Therefore, in order to conclude that the observed and expected distributions are significantly different, we would need to see a high chi-square value. This means that all goodness-of-fit tests are upper-tail tests. The upper-tail critical value (obtained from the χ^2 distribution) is $\chi^2_{\alpha,k-1}$, where α is the specified level of significance, k is the number of categories, and k-1 is the number of degrees of freedom associated with the test. The decision rule is: reject the null hypothesis if the test statistic is greater than the critical value. For our Internet browser example, the value of this upper-tail critical value ($\chi^2_{0.05,4}$) is 9.49. We restate the decision rule as: reject the null hypothesis if the test statistic is greater than 9.49.

Step 4: Apply the Decision Rule

Since the value of the test statistic (15.937) is greater than the value of the critical value (9.49), we reject the null hypothesis.

Step 5: Draw Your Conclusion

Since we rejected the null hypothesis, we conclude that, at the 5% level of significance, the local university students' market shares of Internet browsers are different from those of the national distribution.

Goodness of Fit Test for Normality

We can also use the goodness of fit test to find out if a data sample follows a normal distribution. The **goodness of fit test for normality** is useful as many statistical tests are based on the assumption that the underlying population's data are normally distributed.

When testing for normality, we use the standard normal distribution table to derive the expected probabilities of specific data value intervals and then apply these probabilities in constructing a set of expected data values that are compared with a set of sampled data values. The easiest way to do so is by breaking up the frequency distribution into intervals of *one standard deviation from the mean*, as illustrated in Table 11-3. In order to better understand this approach, recall that by using the empirical rule, we know that for any normal distribution, approximately: 68% of the observations lie within ±1 standard deviation from the mean, 95% are within ±2 standard deviations from the mean, and practically all the data (99.7%) lie within ±3 standard deviations from the mean. Therefore, we work with these same intervals of ±1 standard deviations, ±2 standard deviations, and ±3 standard deviations in Table 11-3.

Table 11-3
Expected Number of Data Values in One Standard Deviation Intervals for a Population that Follows the Normal Distribution

Standard Deviations from the Mean	Probability (from the Standard Normal Table)	Expected Frequency (n=100)	Expected Frequency (n=250)
More than 2	P(Z>2) = 0.0228	0.0228*100 = 2.28	0.0228*250 = 5.7
Between 1 and 2	P(1<Z<2) = 0.1359	0.1359*100 = 13.59	0.1359*250 = 33.975
Between 0 and 1	P(0<Z<1) = 0.3413	0.3413*100 = 34.13	0.3413*250 = 85.325
Between -1 and 0	P(-1<Z<0) = 0.3413	0.3413*100 = 34.13	0.3413*250 = 85.325
Between -2 and -1	P(-2<Z<-1) = 0.1359	0.1359*100 = 13.59	0.1359*250 = 33.975
More than -2	P(Z<-2) = 0.0228	0.0228*100 = 2.28	0.0228*250 = 5.7

The first column of Table 11-3 lists the six intervals. The second column provides the probability of an observation falling in each of these intervals (these values are taken from the standard normal distribution table). For example, the probability of a value being more than two standard deviations above the mean (any mean) is: P(Z>2)=1-0.9772=0.0228; the probability of a value being between one and two standard deviations from the mean is: P(1<Z<2)=P(Z<2)−P(Z<1)=0.9772−0.8413=0.1359; and, the probability of a value being between zero and one standard deviations from the mean is: P(0<Z<1)=P(Z<1)−P(Z<0)=0.8413−0.5=0.3413. Since the normal distribution is symmetric, the bottom part of the table is a mirror image of the top part.

The third and fourth columns of Table 11-3 illustrate how to find the expected number of data values within each interval for, respectively, sample sizes of n=100 and n=250. For example, if we take a sample of 250 observations, we would *expect* to find that 5.7 observations are at least two standard above the mean, 33.975 observations are between one and two standard deviations above the mean, and so on.

Let us proceed with an example to fully illustrate the goodness of fit test for normality. A professor wishes to determine, at a 5% level of significance, if the final exam grades for the students in her large section of a multi-section class are normally distributed with a mean of 70 and a standard deviation of 8. The professor constructs a frequency distribution for the grades of 400 students in her section (shown in Table 11-4).

Table 11-4
Professor Example: Final Exam Grade Distribution

Grade Range	Number of Students with an Exam Score in the Grade Range
86 and higher	11
78 to <86	60
70 to <78	141
62 to <70	130
54 to <62	50
Below 54	8
	Σ=400

Step 1: Formulate Hypotheses

To find out whether the professor's grades follow the specified distribution, we hypothesize:

H_0: Final exam grades are normally distributed ($\mu=70$, $\sigma=8$).

H_1: Final exam grades are not normally distributed ($\mu=70$, $\sigma=8$).

Step 2: Compute the Test Statistic

Constructing Table 11-5 helps in computing the test statistic. The category *86 and higher* represents grades that are two or more standard deviations above the expected mean (the expected mean is 70 with a standard deviation of 8;

therefore, 70+2*8=86). Similarly, the category *78 to <86* represents grades between one and two standard deviations above the mean, and so on. Column 3 provides the probabilities explained earlier in Table 11-3. Column 4 is the expected number of students in each grade category based on a class size of 400 and the hypothesized distribution. Column 5 shows the observed number of students in each grade category. Finally, Columns 6 and 7 illustrate the computation of the test statistic. As computed in the last row of Column 7, the test statistic is: 1.918.

Table 11-5
Professor Example: Computing the Test Statistic

Grade Range	Standard Deviations from the Mean	Prob.	Expected (n=400)	Observed	(Observed – Expected)^2	(Observed – Expected)^2 / Expected
86 and higher	More than 2	0.0228	0.0228*400 = 9.12	11	(11-9.12)^2 = 3.53	3.53/9.12 = 0.388
78 to <86	Between 1 and 2	0.1359	0.1359*400 = 54.36	60	(54.36-60)^2 = 31.81	31.81/54.36 = 0.585
70 to <78	Between 0 and 1	0.3413	0.3413*400 = 136.52	141	(136.52-141)^2 = 20.07	20.07/136.52 = 0.147
62 to <70	Between -1 and 0	0.3413	0.3413*400 = 136.52	130	(136.52-130)^2 = 42.51	42.51/136.52 = 0.311
54 to <62	Between -2 and -1	0.1359	0.1359*400 = 54.36	50	(54.36-50)^2 = 19	19/54.36 = 0.350
Below 54	More than -2	0.0228	0.0228*400 = 9.12	8	(9.12-8)^2 = 1.25	1.25/9.12 = 0.138
						Σ = 1.918

Step 3: Formulate a Decision Rule

As all goodness-of-fit tests are upper-tail tests, the decision rule is: reject the null hypothesis if the test statistic is greater than the upper-tail critical value. By typing =CHISQ.INV.RT(0.05,5) (because α=0.05 and there are six categories), we obtain the critical value of 11.07. Hence, the decision rule becomes: reject the null hypothesis if the test statistic is greater than 11.07.

Step 4: Apply the Decision Rule

Since the test statistic (1.918) is less than the critical value (11.07), we do not reject the null hypothesis.

Step 5: Draw Your Conclusion

By not rejecting the null hypothesis, we cannot conclude, at the 5% level of significance, that the final exam grades follow a different distribution than the hypothesized normal distribution.

Sample Size Considerations

Before we move to practice what we have learned about the goodness of fit test, it is important to raise an important issue about sample sizes with goodness of fit tests. The **rule of five** states that sample sizes need to be large enough so that *each cell* in the *expected* column has a value of five or more. Going back to the example just covered, but now using a sample size of 100 rather than a sample size of 400, the expected values for the first and last categories would be quite small (0.0228*100=2.28). To overcome this problem, either a larger sample must be obtained or categories need to be combined. In Table 11-6, the top and bottom categories have each been combined with an adjacent category. After combining these categories, we can proceed with the testing procedure, but now using just the remaining four categories.

Table 11-6
The Rule of Five

Standard Deviations from the Mean	Probability	Expected Frequency (n=100)
~~More than 2~~	~~P(Z>2) = 0.0228~~	~~0.0228*100 = 2.28~~
More than 1	P(Z>1) = ~~0.1359~~ 0.1587	0.1587*100 = 15.87
Between 0 and 1	P(0<Z<1) = 0.3413	0.3413*100 = 34.13
Between -1 and 0	P(-1<Z<0) = 0.3413	0.3413*100 = 34.13
More than -1	P(Z<-1) = ~~0.1359~~ 0.1587	0.1587*100 = 15.87
~~More than -2~~	~~P(Z<-2) = 0.0228~~	~~0.0228*100 = 2.28~~

Unit 1 Summary

- The **goodness of fit** test examines the fit between observed and expected frequencies for categorical data. It is used to test whether the data follows some hypothesized distribution.

- The **goodness of fit test for normality** is a special application of the goodness of fit test that is used to assess whether or not a data sample follows a specific normal distribution.

- Use the following formula to compute the χ^2 (chi-square) test statistic:

$$\chi^2 = \sum_{i=1}^{k} \frac{(o_i - e_i)^2}{e_i}$$

- Use α and the number of degrees of freedom (number of categories in the frequency table minus one) to look up the critical value in the χ^2 table.

- Goodness of fit tests are always upper-tail tests, with an upper-tail critical value of $\chi^2_{\alpha,k-1}$, where k is the number of categories in the frequency table. If the observed frequencies are very different from the expected frequencies, we will obtain a very large test statistic. The decision rule, therefore, is: reject the null hypothesis if the test statistic is greater than the critical value.

- The **rule of five** states that the expected frequency for each cell must be at least five. If this rule is not met, either obtain a larger sample or combine the low-value categories with adjacent categories.

Unit 1 Exercises

The first five questions ask you to test a hypothesized distribution, given the observed sample provided in each corresponding table. For each question, the null hypothesis provides the *expected distribution* proportions. As an aid to explaining how to proceed, the steps taken to answer the first question are provided. You are on your own for Questions 2, 3, 4 and 5.

1. H_0: $p_1=0.05$, $p_2=0.25$, $p_3=0.1$, $p_4=0.3$, $p_5=0.3$ (Use $\alpha=0.05$.)

Table 11-e1

	P1	P2	P3	P4	P5
Observed Frequency	1	26	7	34	32

Solution:

a. Sum all observed frequencies to obtain: n=100

b. Find the expected frequencies in each cell, given the null hypothesis:

Table 11-e1a

	P1	P2	P3	P4	P5
Expected Frequency	0.05*100 = 5	0.25*100 = 25	0.1*100 = 10	0.3*100 = 30	0.3*100 = 30

c. Compute the test statistic:

$$\chi^2 = \sum_{i=1}^{k} \frac{(o_i - e_i)^2}{e_i} = \frac{(1-5)^2}{5} + \frac{(26-25)^2}{25} + \frac{(7-10)^2}{10} + \frac{(34-30)^2}{30} + \frac{(32-30)^2}{30} = 4.81$$

d. Look up the critical value: $\chi^2_{0.05,4} = 9.49$

e. Compare the test statistic (4.81) to the critical value (9.49).

f. As the test statistic is less than the critical value, we do not reject the null hypothesis. Insufficient evidence exists to conclude that the distribution is different than the one hypothesized.

g. Alternately, compute the p-value: =CHISQ.DIST.RT(4,81,4). The p-value is 0.3704, which is greater than α (0.05). Consequently, we do not reject the null hypothesis.

2. H_0: $p_1=0.25$, $p_2=0.25$, $p_3=0.45$, $p_4=0.05$ (Use $\alpha=0.05$.)

Table 11-e2

	P1	P2	P3	P4
Observed Frequency	46	57	79	18

3. H_0: $p_1=0.3$, $p_2=0.5$, $p_3=0.2$ (Use $\alpha=0.05$.)

Table 11-e3

	P1	P2	P3
Observed Frequency	35	54	11

4. H_0: $p_1=0.1$, $p_2=0.2$, $p_3=0.15$, $p_4=0.4$, $p_5=0.15$ (Use $\alpha=0.05$.)

Table 11-e4

	P1	P2	P3	P4	P5
Observed Frequency	16	56	72	180	76

5. Change the sample size in Question 4 to 200 by dividing each observed frequency by two. Then, repeat the test. Have your results changed? Why?

6. Are all engineering majors equally popular? After surveying 1,000 engineering students, the following majors were observed (see Table 11-e5). What can you conclude about the popularity of engineering majors? (Use $\alpha=0.05$.)

Table 11-e5

Engineering Major	Frequency
Computer	270
Electrical	240
Mechanical	310
Civil	180
Total	1000

7. Students working in part-time jobs reported an average income of $2,000 per month with a standard deviation of $500. A sample of 500 students revealed the distribution of income shown in Table 11-e6. At the 0.01 level

of significance, can you conclude that the population is normally distributed, with a mean of $2,000 and a standard deviation of $500?

Table 11-e6

Income ($)	Frequency
Less than 1,000	9
1,000 to less than 1,500	63
1,500 to less than 2,000	165
2,000 to less than 2,500	180
2,500 to less than 3,000	71
3,000 or more	12
Total	500

UNIT 2

Test of Independence

A second type of chi-square test that is frequently used is the **test of independence** to determine if a relationship exists between two categorical variables.[111] In order to illustrate this test, an example will be used that seeks to determine if there is a relationship between children's digital screen time and obesity. Consider the following two statements from recent news:

> "Children aged five to 16 spend an average of six and a half hours a day in front of a screen compared with around three hours in 1995."[112]

> "Childhood obesity has more than doubled in children and quadrupled in adolescents in the past 30 years."[113]

Table 11-7 provides (hypothetical) data for 100 ten-year-old children regarding their weight category and screen time.

---

[111] You may recall the similar concept of *correlation* reviewed in Chapter 5. Correlation measures the existence and strength of the relationship between two quantitative variables, whereas the chi-square statistic is used to study the existence of a relationship between two categorical variables.
[112] Source: http://www.bbc.com/news/technology-32067158
[113] Source: https://www.cdc.gov/healthyschools/obesity/facts.htm

Table 11-7
Obesity Example: Observed Weight and Screen Time

Weight	Screen Time (hours per day)			Total
	Less than Three	**Three to Six**	**More than Six**	
Obese	1	9	20	**30**
Overweight	20	15	15	**50**
Normal Weight	10	5	5	**20**
Total	**31**	**29**	**40**	**100**

Table 11-7 is an illustration of a **contingency table** used to describe the relationship between two variables of interest. With this obesity example, the two variables of interest are: Children's Weight and Screen Time. For example, the table indicates that, of the children who are considered obese (30 in total), the majority (twenty in total) spent more than six hours in front of a screen each day. In contrast, half of the normal weight kids (ten out of the twenty children) spent fewer than three hours per day in front of screens. Our objective with this example is to test, at the 1% level of significance, whether these two variables (i.e., the weight of children and the average screen time) are *related* or are *independent* of each other.

We begin by assuming that the variables are **independent** - in other words, that no relationship exists between the two variables. Based on this assumption, we construct a table of *expected frequencies*, as follows. First, look at the bottom row of Table 11-7. It shows that 31 out of the 100 children surveyed (or 31%) had fewer than three hours of screen time per day, 29% had three to six hours of screen time per day, and 40% had more than six hours of screen time per day. Now take these percentages (31%, 29%, and 40%) and apply them to each row in the table in order to compute expected cell values if the two variables are, in fact, independent. For example, we would expect 31% of the 30 obese children to have fewer than three hours of screen time per day: 0.31*30=9.3. Similarly, we would expect 31% of the 50 overweight children to have fewer than three hours of screen time per day: 0.31*50=15.5. And, we would expect 31% of the 20 normal weight kids to have fewer than three hours of screen time per day: 0.31*20=6.2. The remainder of the table (shown as Table 11-8) is constructed in a similar manner.

Table 11-8
Obesity Example: Expected Weight and Screen Time

Weight	Screen Time (hours per day)			Total
	Less than Three	Three to Six	More than Six	
Obese	0.31*30 = 9.3	0.29*30 = 8.7	0.4*30 = 12	30
Overweight	0.31*50 = 15.5	0.29*50 = 14.5	0.4*50 = 20	50
Normal Weight	0.31*20 = 6.2	0.29*20 = 5.8	0.4*20 = 8	20
Total	31	29	40	100

The general formula used in computing expected cell frequencies in contingency tables is (division by the sample size simply turns each cell's content into a percentage):

$$e_{ij} = \frac{(i^{th} \text{ row total}) * (j^{th} \text{ column total})}{\text{Total sample size}}$$

Now that we know how to compute expected cell frequencies in a contingency table, we are able to conduct the test of independence following the familiar five-step procedure.

Step 1: Formulate Hypotheses

We are interested in finding whether children's weight and screen time are related. Thus, our hypothesis test should determine if a relationship exists between these two variables. We begin by assuming that there is no such relationship, with the alternative being that there is a relationship:

H₀: Children's weight is independent of screen time.

H₁: Children's weight is not independent of screen time.

If we are able to reject the null hypothesis, then we can conclude that a relationship does exist between children's weight and screen time.

Step 2: Compute the Test Statistic

We use the following formula to compute the test statistic:

$$\chi^2 = \sum_{i=1}^{r}\sum_{j=1}^{c} \frac{(o_{ij} - e_{ij})^2}{e_{ij}}$$

where r is the number of rows, c is the number of columns, o_{ij} is the observed cell frequency, e_{ij} is the expected cell frequency, and we use (r-1)*(c-1) degrees of freedom. This is basically the same formula in computing the test statistic for the goodness of fit test, except that we now have a contingency table and take into account the expected number of cells across multiple rows and columns. Tables 11-7 and 11-8, respectively, provide the observed and expected cell frequencies for our example. Table 11-9 demonstrates how the components of the test statistic are computed for the nine cells comprising the contingency table. Summing these nine components yields the test statistic:

$$\chi^2 = 7.41+1.31+2.33+0.01+0.02+0.11+5.33+1.25+1.13 = 18.89$$

Table 11-9
Obesity Example: Computing the Components of the Test Statistic

Weight	Screen Time (hours per day)		
	Less than Three	Three to Six	More than Six
Obese	(1-9.3)²/9.3 = 7.41	(9-8.7)²/8.7 = 0.01	(20-12)²/12 = 5.33
Overweight	(20-15.5)²/15.5 = 1.31	(15-14.5)²/14.5 = 0.02	(15-20)²/20 = 1.25
Normal Weight	(10-6.2)²/6.2 = 2.33	(5-5.8)²/5.8 = 0.11	(5-8)²/8 = 1.13

Step 3: Formulate a Decision Rule

As with the goodness of fit test, the test of independence is an upper-tail test; thus, we reject the null hypothesis when the test statistic is sufficiently large (i.e., when the differences between observed and expected values are high in value). Hence, our decision rule is: reject the null hypothesis if the test statistic is greater than the critical value. For this example, the number of degrees of freedom is computed as (since the contingency table has three rows and three columns):

$$df = (r-1)*(c-1) = (3-1)*(3-1) = 4$$

Applying the specified significance level of 1% and looking up the critical value in Excel, we obtain: $\chi^2_{0.01,4} = 13.27$.

Step 4: Apply the Decision Rule

As the test statistic (18.89) is greater than the critical value (13.27), we reject the null hypothesis.

Step 5: Draw Your Conclusion

By rejecting the null hypothesis, we conclude, at the 1% level of significance, that children's weight is related to the number of hours of screen time per day.

Unit 2 Summary

- A **contingency table** can be used in assessing the relationship between two variables as part of the **test of independence** for two categorical variables. Specifically, we are testing to see whether a relationship exists between the two variables. If no relationship is found to exist, we say that the two variables are **independent** of each other.

- The null and alternative hypotheses for the test of independence are:
    - H_0: The variables are independent of each other.
    - H_1: The variables are not independent of each other.

- Use the following formula to compute the test statistic:

$$\chi^2 = \sum_{i=1}^{r}\sum_{j=1}^{c} \frac{(o_{ij} - e_{ij})^2}{e_{ij}}$$

- Use α and the number of degrees of freedom for the test of independence to look up the critical value in the χ^2 table. For this test, the degrees of freedom are computed as (r-1)*(c-1), where r is the number of rows in the contingency table and c is the number of columns.

- The test of independence is an upper-tail test. If the observed frequencies are significantly different from the expected frequencies, we will obtain a very large test statistic. The upper-tail critical value is obtained from the χ^2 table: $\chi^2_{\alpha,(r-1)*(c-1)}$. The decision rule is: reject the null hypothesis if the test statistic is greater than the critical value.

Unit 2 Exercises

1. Conduct a chi-square test at α=0.05 to determine whether variables A (three categories) and B (two categories), as shown in Table 11-e7, are independent.

Table 11-e7

	A1	A2	A3
B1	3	5	4
B2	7	6	5

2. Increase the sample size in Question 1 by doubling every cell value and repeat the test. What is the effect of an increase in sample size?

3. Complete the unfinished contingency table (see Table 11-e8) and test whether the variables are independent. (Use α=0.05.)

Table 11-e8

	A1	A2	Total
B1	9		
B2		4	
B3	7		11
Total		11	33

4. The candidates for the 2016 U.S. election used social media to interact with supporters. Each candidate has a unique style, reflected in part by differences in their use of different social media channels. Given data on *Facebook* and *Twitter* posts from January of 2016, provided in Table 11-e9[114], can you infer at the 1% significance level that a relationship exists between the candidates and the social media channels used?

---

[114] Source: https://www.socialbakers.com/elections-2016/us

Table 11-e9

	Trump	Cruise	Clinton	Sanders
Facebook	236	275	323	149
Twitter	525	1100	687	874

5. A survey of residents from three states asked for the make of the car each respondent drove. Using the data in Table 11-e10, is there enough evidence at the 5% significance level to conclude that there is a relationship between state of residence and type of car?

Table 11-e10

	California	Vermont	Kentucky
Jeep	10	21	13
BMW	34	7	2
Chevy	14	25	31

END-OF-CHAPTER PRACTICE

1. A high school has been making the transition to become an Advanced Placement (AP) school, offering courses to students that can eventually be transferred for university credits after the students pass the standardized AP exams. Now offering three AP courses, the school notes the class (Grades 9, 10, 11 or 12) of each student currently taking each course. Using the data shown in Table 11-e11, is there a relationship between class level and students taking AP courses, at a 10% significance level?

Table 11-e11

	Grade 9	Grade 10	Grade 11	Grade 12
English	34	30	14	10
Mathematics	30	29	28	10
History	30	16	15	15

2. LEGO has been making construction sets targeted at film franchises (i.e., films that have multiple installments). In order to assess the success of this business strategy, LEGO surveys past purchasers of franchise LEGO construction sets and tests for a sales relationship between movie franchises and each subsequent film in a franchise. Using the data shown in Table 11-e12, what can LEGO conclude at $\alpha=0.05$?

Table 11-e12

	Pirates of the Caribbean	Harry Potter	Batman
1st Movie	29	30	21
2nd Movie	20	26	15
3rd Movie	10	30	5

3. A study of sports-related injuries comparing males and females surveyed athletes and asked which type of injury they sustained. Is the data shown in Table 11-e13 sufficient to show, at a 5% significance level, a relationship between gender and type of sports injury?

Table 11-e13

	Knee	Shoulder	Head	Back
Male	18	13	10	9
Female	15	18	2	15

4. The researcher in the previous question has received new information on how long each of the injuries lasted, as shown in Table 11-e14. Is this data sufficient to show, at a 5% significance level, a relationship between recovery time and type of injury?

Table 11-e14

	Knee	Shoulder	Head	Back
One to Two Weeks	14	16	1	5
Two to Four Weeks	11	10	3	12
More than Four Weeks	8	5	8	7

5. Political polls are important, but often misleading.[115] In the 2011 federal election in Canada, the *Liberals* conducted research to predict the support for the five major contending parties. The results of this prediction are shown in the first row of Table 11-e15. After the election, the actual voting data were recorded (shown in the second row of Table 11-e15).[116] Test, at the 1% significance level, whether the *Liberals*' statisticians had produced a sufficiently good prediction of voter support prior to the election.

---

[115] Source: http://www.nytimes.com/2015/06/21/opinion/sunday/whats-the-matter-with-polling.html?_r=0
[116] Source: http://en.wikipedia.org/wiki/Canadian_federal_election,_2011

Table 11-e15

	Liberals	Conservatives	NDP	BLOC	Green
Expected Votes	25%	36%	27%	9%	3%
Actual Votes	2,783,175	5,832,401	4,508,474	889,788	576,221

6. The manager of a baseball team assigns the lineup in a particular order, expecting a certain proportion of hits to come out of each spot in the lineup. Before the season begins, he summarizes his expectations; after one month of play, he records the actual hits received from each spot in the batting lineup. Using the data provided in Table 11-e16, have the players' batting performances mirrored the manager's expectations? (Use $\alpha=0.05$.)

Table 11-e16

	Leadoff	Second	Third	Fourth	Fifth	Sixth	Seventh	Eighth	Ninth
Expected Proportion of Hits	0.12	0.19	0.2	0.12	0.09	0.08	0.07	0.07	0.06
Actual Number of Hits	25	24	15	25	20	10	15	9	7

7. A medical researcher has concluded a study examining relief of headaches with a soon-to-be-released medication. With a sample of 400 subjects, she records how long it took for each subject to obtain relief from a headache, as shown in Table 11-e17. At a 1% significance level, do these data support the researcher's expectations?

Table 11-e17

	Less than Ten Minutes	Ten to 30 Minutes	30 to 45 Minutes	45 to 60 Minutes	More than 60 Minutes
Expected Proportion	0.13	0.46	0.32	0.08	0.01
Study Results	30	201	125	40	4

8. A high school math teacher has carefully crafted his final exam in order to get the grade distribution shown in the first row of Table 11-e18. Do the actual exam grades, at a 5% significance level, align with the teacher's expectation?

Table 11-e18

	Final Exam Grades		
	A or B	*C or D*	*F*
Expectation	13	12	5
Actual Grades	18	10	2

9. Four big television networks (N1, N2, N3 and N4) fight for viewers every night, but there is some overall consistency in viewer distribution, as shown in the expected market share (see Table 11-e19). With a new show, one network hopes to increase its market share relative to the other networks. Given the data provided by 1,000 viewers, shown in Table 11-e19, test whether the networks' market shares have indeed changed since the new show aired. Use the 5% level of significance.

Table 11-e19

	N1	*N2*	*N3*	*N4*
Expected Market Shares	28%	31%	23%	18%
Observed Market Shares after the Airing of the New Show	22%	24%	29%	25%

10. One of the greatest novelists of the Victorian period from 1837 to 1901 was Charles Dickens. He wrote fifteen novels and hundreds of short stories, all while editing an English journal. His most famous work, *A Tale of Two Cities*, sold over 200 million copies worldwide and is still celebrated as one of the best literary works of all time. With his books still selling today, an English professor surveyed 50 students about their favorite Dickens novel amongst a choice of four novels. The results of this survey are shown as the observed percentages in Table 11-e20. The professor has read all four novels and creates her own prediction (also shown in Table 11-e20). Test to see if her prediction holds at the 5% significance level.

Table 11-e20

	A Tale of Two Cities	*Great Expectations*	*Bleak House*	*Hard Times*
Prediction	40%	24%	19%	17%
Observed	37%	14%	21%	28%

11. Interested to further pursue the topic examined in Question 10, the professor decides to also ask her students to indicate their college major.

The course she teaches consists of majors from only four fields of study (Law, History, English and Political Science). The four novels have distinct plots exposing very different issues and she is interested in determining whether a novel's theme is related to the (student) reader's major. Given the data provided in Table 11-e21, test whether these two variables (novel theme and reader's college major) are independent. (Use $\alpha=0.05$.)

Table 11-e21

	A Tale of Two Cities (War/Revolution)	Great Expectations (Personal Growth)	Bleak House (Justice)	Hard Times (Class Struggle)
Law	7	3	8	4
History	15	4	6	3
English	5	6	2	9
Political Science	10	1	5	12

CHAPTER 12: Analysis of Variance

Charity runs are events organized by service organizations to fundraise for various causes including health-related maladies such as diabetes and breast cancer. Participants in charity runs often reach out to their friends and colleagues for donations on a participant's behalf. Suppose that we are interested in determining which types of charitable causes produce higher fundraising amounts. To do so, we examine data on the average amounts fundraised by participants in three different runs: a cancer charity run, a diabetes charity run, and a stroke charity run. In other words, we wish to determine whether there is a significant difference in the average amounts of money raised by runners in these three types of charity runs.

In Chapter 10, we learned how to compare means of two populations using the t-test. In this case, however, we are comparing the means of more than two populations. Of course, rather than learning a new statistical technique, we could instead conduct a series of two-populations t-tests (i.e., test for a difference between the cancer run and the diabetes run, a difference between the cancer run and the stroke run, and a difference between the diabetes run and the stroke run). One problem with this approach is that it is labor intensive, especially as the number of groups we wish to compare grows. For example, if we were interested in comparing five types of charity runs, we would have to make ten two-populations comparisons. A second problem is that undertaking multiple two-populations t-tests increases the probability of a type I error (i.e., rejecting a true null hypothesis). For these reasons, it is preferable to test for differences in three (or more) populations' means using *a single test*. The technique for doing so is called *analysis of variance* and is the focus of this chapter. If you had only two groups to compare, you could either use the analysis of variance test or a two-populations t-test (assuming equal population variances).

An *ANalysis Of VAriance* (*ANOVA*) only tells us whether all means are the same or whether at least one of the means is significantly different from the other means. If we conclude that the means are not the same, we need to follow up with additional testing to explore which of the means differ from the rest. This follow-up testing is called a *post hoc* analysis.[117]

In addition to testing for differences in the means of multiple populations (or, groups), ANOVA can simultaneously test for multiple effects (or, the effects of more than one variable of interest). For example, we can use ANOVA to determine if either or both the type of charitable cause and the city in which a run took place make a difference on the amount of funds raised. This chapter covers the following topics:

- Terminology and Assumptions
- One-Way ANOVA
- Two-Way ANOVA
- Randomized Block ANOVA Design
- Two-Factors ANOVA
- Interpreting Interaction Effects

---

[117] There are several statistical procedures for *post hoc* analysis, which is the testing of all possible pair-wise combinations to determine *which* population means differ. These procedures are included in most standard statistical software packages. Detailed discussion of *post hoc* analysis procedures is beyond the scope of this book.

UNIT 1

Terminology and Assumptions

Analysis of variance (or **ANOVA**) is one of the most widely used statistical techniques. Using ANOVA, we test the impact of a certain independent variable, called a **factor**, on some **dependent variable**. For example, we might examine the effect of a new teaching method (factor) on students' performance (dependent variable) or the effect of the type of charitable cause (factor) on the amount of money being fundraised (dependent variable). Often, especially when the research approach involves an experimental design, the factor is referred to as a *treatment*. A factor in ANOVA is typically set up so that it consists of multiple **levels**, with each level representing a different factor category. For example, we can study the effect of three teaching methods: online teaching, in-class lecturing, and a seminar format. Each of these categories thus represents a level of the factor Teaching Method. Similarly, the three types of charity runs (i.e., a cancer run, a diabetes run and a stroke run) represent three levels of a Cause factor.

For the charity runs example mentioned above, we can take three random samples of data (one sample for each charity, i.e., each level of the Cause factor), collecting the amount fundraised (the dependent variable) by each runner. These data are presented in Table 12-1. We then compare the means of the three types of charity runs, or groups, to see if there are significant differences in the dependent variable that can be attributed to the different levels of the Cause factor.

Table 12-1
Fundraising Amounts: Three Types of Charity Runs

Cancer Charity Run	Diabetes Charity Run	Stroke Charity Run
$669	$153	$354
$270	$432	$179
$425	$754	$251
$232	$140	$146
$262	$452	$418
$687	$319	$295
$633	$807	$215
$331	$987	$431
$745	$132	$477
$704	$958	$202
$491	$863	$335
$733	$643	$298
$205	$566	$498
$360	$502	$308
$706	$802	$324
$650	$782	$186
$696	$504	$410
$325	$705	$170
$583	$990	$218
$733	$625	$148
Mean: $522.00	**Mean: $605.80**	**Mean: $293.15**

The question to be answered in this charity runs example is: are there significant differences in the average amounts fundraised for each of the three charity runs? To answer this question, we conduct a hypothesis test. Our hypotheses are:

H_0: $\mu_{cancer} = \mu_{diabetes} = \mu_{stroke}$

H_1: At least one of the means differs from the others.

The null hypothesis states that the three group means are the same. There are always as many means to be compared as there are levels of the factor being assessed. For example, if our factor has five levels, we would compare the means of five groups. The alternative hypothesis does not require that *all* means differ from each other, but simply that *at least one* of the means differs. If we find that at least one of the means differ, then we can conclude that at least some of the levels of our factor have an effect on the mean value of the dependent variable. Using the charity runs example, we may eventually determine that only the cancer run has a higher mean than the other two types of charity runs. This finding still implies that the factor Cause had an effect on the dependent variable Fundraised Amount.

It is important to emphasize that the ANOVA test covered in this chapter only tells us whether or not one or more of the group means differ from the other group means. In other words, it tells us whether the factor has a significant effect on the dependent variable. It does not tell us which of the factor level means differ. Instead, as was noted in the introduction, follow-up (also referred to as *post hoc*) procedures can be employed to further explore the nature of the differences among the factor level means.

Before presenting the steps involved in the ANOVA test, we note three assumptions that are required to hold in order to conduct an ANOVA test:

1. All populations[118] are normally distributed.
2. The population variances are equal.
3. The observations are independent of each other.

In our example, these assumptions mean that the distribution of fundraised amounts for each of the types of charity runs follows the normal distribution, that the variances for these three population distributions can be assumed equal, and that the observations comprising each of the three samples are independent of each other. Having independent observations with these three samples means that

---

[118] Populations, here, refers to the full set of data in a given level of the factor. Recall that for our charity runs example, we collected data from three *samples* of runners. The assumption about population here refers to the fundraised amounts by *all* participants in the given run.

the runners in the samples were randomly selected and that none of these runners were matched on any criteria (e.g., the same runner does not appear in more than one sample, none of the runners were paired with others of similar running ability or experience, etc.). Note that while these assumptions are important, ANOVA can still be carried out if they are violated. Specifically, ANOVA is not very sensitive for normality or variance equality violations, especially if the sample sizes are relatively similar. One can also employ data transformation to improve these assumptions (as explained in Chapter 6, we can use transformations such as taking the square root of each data point or the log of each data point to normalize data).

Finally, the dependent variable in ANOVA must be *quantitative*, but the factors must be treated as being qualitative (while the data collected to represent factors can be either *quantitative* or *categorical*). Specifically, because our hypotheses are about the *mean* of the dependent variable, we need to be able to compute a mean, which requires the dependent variable to be quantitative. Factors, on the other hand, are treated as qualitative because we always split the population into distinct groups (levels). We can, however, work with quantitative factors, if needed, by creating artificial groupings. For example, if we would like to include age as a factor, we can create distinct groups categorized as differing age ranges. (Note that if we want to use quantitative factors, then ANOVA may not be our analysis method of choice; instead, we would most likely use regression analysis, which is covered in Chapter 13.) To summarize, ANOVA is most frequently used with a quantitative dependent variable and one or two categorical factors with two or more levels.

Unit 1 Summary

- **ANOVA** stands for **Analysis of Variance** and it is a statistical technique used to test differences between means of two or more populations.

- The **dependent variable** is the variable whose mean is thought to vary systematically across two or more populations. This dependent variable should be quantitative.

- ANOVA includes independent variable(s), termed **factor(s)**, which are thought to significantly affect the dependent variable. A factor has two or more categories, or **levels**. The levels of the factor vary categorically.

- To conduct an ANOVA test, three assumptions must hold:
  - All populations are normally distributed.
  - The population variances are equal.
  - The observations are independent of each other.

Unit 1 Exercises

For Questions 1-5, a research project is described. Identify the *dependent variable*, the *factor*, and the *levels* of the factor.

1. A study of the impact of age on income. The researchers collected data on adults using three age categories: 20 to 30, 30 to 50, and 50 to 65.

2. A study of the impact of graphical quality on time spent playing a computer game. Researchers record the time, in minutes, that children spent playing three computer games: one game with minimal 2D graphics, one with high 2D graphic quality, and one with 3D graphic quality.

3. A study of the impact of fund managers' years of experience on fund performance. Researchers collected data on fund managers' years of experience (measured as: less than five years, five to ten years, and over ten years) on mutual fund performance.

4. A study of the number of hits a video gets on *YouTube* based on the type of video (humorous, inspirational, musical, and informational).

5. How much time should you invest in a sport given a desire to participate in the sport at a collegiate level? Researchers studied the age at which athletes on college teams began to play five different sports: football, baseball, track and field, swimming, and hockey.

UNIT 2

One-Way ANOVA

Recall that our objective in the charity runs example, shown again in Table 12-2, is to test (using a 5% significance level) whether the charitable cause behind each of the three runs affects the amount fundraised by runners. This test is termed a **one-way ANOVA** (or a *single-factor ANOVA*) because we only have a single factor (Cause) and we are studying the impact of this factor on the dependent variable (Fundraised Amount). Toward the bottom of Table 12-2, the mean and standard deviation for each type of charity run are provided. The **grand mean**, calculated as the average of all 60 observations in the dataset, is provided at the very bottom of the table. This grand mean has a value of $473.65 and represents the average of fundraised amounts across all sixty runners. Of course, each individual observation differs from this grand mean – some are above it and others are below it. We already know that these deviations represent the variance in the data, but precisely what are the sources of this variance?

Table 12-2
One-Way ANOVA: Charity Runs Data Set

Cancer Charity Run	Diabetes Charity Run	Stroke Charity Run
$669	$153	$354
$270	$432	$179
$425	$754	$251
$232	$140	$146
$262	$452	$418
$687	$319	$295
$633	$807	$215
$331	$987	$431
$745	$132	$477
$704	$958	$202
$491	$863	$335
$733	$643	$298
$205	$566	$498
$360	$502	$308
$706	$802	$324
$650	$782	$186
$696	$504	$410
$325	$705	$170
$583	$990	$218
$733	$625	$148
Mean: $522.00	**Mean: $605.80**	**Mean: $293.15**
Standard Deviation = $198.00	**Standard Deviation = $274.53**	**Standard Deviation = $111.19**
Grand Mean: $473.65		

With this example, the variance in the data comes from two possible sources: the differences *between* the three causes (the three *levels* of the factor, reflecting a belief that certain charitable causes may be more important to people); and, differences *within* each cause (each *level* of the factor), what we term *error* or *residual* variance. In this charity runs example, the residual error simply reflects an expectation that the individual runners will raise differing amounts of money.

The ANOVA test looks at the ratio of the *between-level* (or, *factor*) and *within-level* (or, *residual*) variances. If the between-level variance is significantly larger than the within-level variance, then it can be concluded that significant differences exist between the three group means (the means of the three charitable causes); as a result, the null hypothesis (i.e., equality of the group means) is rejected. Let us begin, as always, with properly setting up the hypotheses to be tested.

Hypotheses

As explained earlier, the hypotheses in ANOVA compare the means of several groups. Here, each *group* represents a level of the factor whose effect on a dependent variable is being examined. With a one-way ANOVA, the null hypothesis states that all the group means are the same:

$$H_0: \mu_1 = \mu_2 = \mu_3 = \ldots = \mu_n$$

The alternative hypothesis does not require that *all* of the group means differ from each other, but rather that *at least one* of the group means will differ. Hence, we use a statement in our alternative hypothesis such as:

H_1: At least one of the group means differs from the others.

In the charity runs example, our goal was to find out whether differences exist in the amounts of funds raised between the three charity runs, with each run representing a different cause. We thus have three groups, and the hypotheses are:

$$H_0: \mu_{cancer} = \mu_{diabetes} = \mu_{stroke}$$

H_1: At least one of the group means differs from the others.

As with other hypothesis testing procedures, we need to calculate a test statistic.

The Test Statistic

As explained earlier, our test focuses on comparing two sources of variation regarding the grand mean. Hence, in ANOVA tests we analyze what is called the **sum of squares**, or a representation of the squared deviations from the mean.

The **sum of squares between groups (SSB)** examines the variation due to the different groups (or, levels of the factor). It is computed based on deviations between each of the three group means and the grand mean. Specifically, for each group we compute the squared distance between that group's mean and the grand mean, and then we sum these squared distances across all groups:

$$SSB = \sum_{j=1}^{k} n_j (\bar{X}_j - \bar{\bar{X}})^2$$

where k is the number of groups, $\bar{X}_j$ is the mean of group j, $\bar{\bar{X}}$ is the grand mean, and n_j is the size of group j (in our example: k=3, $\bar{X}_1 = \$522.00$, $\bar{X}_2 = \$605.80$, $\bar{X}_3 = \$293.15$, $\bar{\bar{X}} = \$473.65$, and $n_1=n_2=n_3=20$).

The **sum of squares within groups (SSW)** accounts for the variation due to error (or residual). That is, it accounts for the fact that the observations are not identical. It is calculated as the weighted sum of the variances computed for each of the factor levels:

$$SSW = \sum_{j=1}^{k} \sum_{i=1}^{n_j} (x_{ij} - \bar{X}_j)^2 = \sum_{j=1}^{k} (n_j - 1) s_j^2$$

In order to understand the second part of the above equation, recall that the variance (s^2) is computed as:

$$s^2 = \frac{\sum_{i=1}^{n}(x_i - \bar{X})^2}{n - 1}$$

We can reorganize the above equation to be:

$$(n - 1)s^2 = \sum_{i=1}^{n}(x_i - \bar{X})^2$$

This is the term that we have substituted above in the equation for SSW. For our example: $n_1=n_2=n_3=20$, $s_1=\$198$, $s_2=\$274.53$ and $s_3=\$111.19$.

Finally, the *sum of squares total (SST)* is simply computed as the sum of the *within* sums of squares (SSW) and the *between* sums of squares (SSB):

$$SST = SSB + SSW$$

The test statistic is the ratio of the mean sum of squares between and within groups. The mean sum of squares for the effect of the factor (the *Mean Squares Between*, or *MSB*) is computed by dividing SSB by the number of groups (k) minus one:

$$MSB = \frac{SSB}{k-1}$$

Next, the *Mean Square Within* (or *MSW*) is computed by dividing SSW by the number of observations (n) minus the number of the groups (k):[119]

$$MSW = \frac{SSW}{n-k}$$

Now, we can compute our test statistic, referred to in ANOVA as the **F-statistic**, as:

$$F = \frac{MSB}{MSW}$$

A visually-easier way to go through the above computation of the test statistic is to use the ANOVA table shown in Table 12-3. In Table 12-3, the Total Row is the sum of the two rows above it: SSB+SSW=SST and (k-1)+(n-k)=n-1. The Mean Squares Column is computed by dividing the Sum of Squares Column by the Degrees of Freedom Column. Finally, the F-statistic is computed by dividing MSB by MSW.

---

[119] If we have three groups, then: $(n_1-1)+(n_2-1)+(n_3-1)$ adds up to n-3, or more generally: n-k.

Table 12-3
The ANOVA Table

Source of Variation	Sum of Squares	Degrees of Freedom	Mean Squares	F-statistic
Factor (between)	SSB	k-1	MSB=SSB/(k-1)	F=MSB/MSW
Error (within)	SSW	n-k	MSW=SSW/(n-k)	
Total	SST	n-1		

Let us bring together all of the above formulas to compute the test statistic in the charity runs example. From the data provided in Table 12-2, we know that there are a total of 60 runners, hence n=60. We also know that the Cause factor has three levels (i.e., three different charity runs); hence, k=3. Based on this information, we can partially fill the ANOVA table, now shown in Table 12-4.

Table 12-4
Partial ANOVA Table: Charity Runs Example

Source of Variation	Sum of Squares	Degrees of Freedom	Mean Squares	F-statistic
Factor (between)	SSB	3-1=2	MSB=SSB/2	F=MSB/MSW
Error (within)	SSW	60-3=57	MSW=SSW/57	
Total	SST	60-1=59		

Now, we need to compute SSB, SSW, and SST. Let us begin with SSB. We have three groups with twenty runners within each group ($n_1=n_2=n_3=20$). The means for the three groups and the grand mean have already been computed and are provided in Table 12-2. Using the equation to compute SSB we obtain:

$$SSB = \sum_{j=1}^{k} n_j (\bar{X}_j - \bar{\bar{X}})^2$$

$$= 20 * (522 - 473.65)^2 + 20 * (605.8 - 473.65)^2 + 20 * (293.15 - 473.65)^2$$

$$= 1{,}047{,}631.9$$

Next, we compute SSW. Again, from Table 12-2, we already have a value for the standard deviation of each of the charitable runs. SSW is computed as:

$$SSW = \sum_{j=1}^{k}(n_j - 1)s_j^2 = (n_1 - 1)s_1^2 + (n_2 - 1)s_2^2 + (n_3 - 1)s_3^2$$

$$= (20 - 1) * 198^2 + (20 - 1) * 274.45^2 + (20 - 1) * 111.19^2$$

$$= 2{,}411{,}803.75$$

Returning to the ANOVA table, we can insert the computed values for SSB and SSW (as shown in Table 12-5) and compute the remaining values in the table. As shown in Table 12-5, the value of the test statistic is 12.38.

Table 12-5
Completed ANOVA Table: Charity Runs Example

Source of Variation	Sum of Squares	df	Mean Squares	F-statistic
Factor (between)	1,047,631.9	3-1=2	MSB=1,047,631.9/2= =523,815.95	F=523,815.95/ 42,312.35=12.38
Error (within)	2,411,803.75	60-3=57	MSW=2,411,803.75/57= =42,312.35	
Total	3,459,435.65	60-1=59		

Next, we need to find the critical value and formulate the decision rule about whether or not to reject the null hypothesis.

Critical Value and Decision Rule

As the ANOVA test statistic is a ratio of two variances, it is distributed according to an F-distribution, with $df_1=k-1$ and $df_2=n-k$ degrees of freedom. Although the hypotheses focus on the equality of the population means, remember that the test itself is a test of two variances. Specifically, the numerator represents the variance that is explained by our *factor* (i.e., *between* levels), whereas the denominator represents the *error* variance (i.e., *within* levels). Therefore, we will reject the null hypothesis and conclude that there are significant differences between the population means if this ratio is significantly high (if more variance is explained by the factor than is attributed to error). For this reason, the ANOVA test is always an *upper-tail test*, and the critical value is obtained from the F-distribution[120] with α as specified and with k-1 and n-k degrees of freedom: $F_{(\alpha,k-1,n-k)}$.

In the charity runs example, the critical value obtained for the ANOVA test is computed as: $F_{0.05,2,57}$=3.1588 (we use the Excel function =F.INV.RT(0.05,2,57) to obtain this value.)

Our decision rule becomes: reject the null hypothesis if the test statistic is greater than the critical value of 3.1588. Because the test statistic (12.38) is much greater that the critical value (3.1588), we reject the null hypothesis and conclude that there is a significant difference in the mean amount fundraised for at least one of the three charity runs.

Using Excel

Of course, the above computations can become cumbersome when working with large datasets. Therefore, we generally use Excel's *Analysis ToolPak* (or other statistical software) to conduct ANOVA tests.

In this chapter's Excel file, open the worksheet titled 'Charity Runs'. From the Data Analysis menu[121], choose 'ANOVA: Single Factor'. Figure 12-1 displays the menu that opens after you choose the ANOVA test and click 'OK'. Here, you are asked to specify several components needed in conducting the test:

- Input:
    - Input Range: Select the cells A1:C21, which hold the data for the three charity runs.

---

[120] Chapter 10 covers how to obtain critical values in the F-distribution.
[121] Remember, you need to have installed the *Analysis ToolPak* onto your version of Excel in order to be able to see and use the Data Analysis menu. Appendix A explains how to install and use the *Analysis ToolPak*.

- o Grouped By: The two options are 'Columns' and 'Rows' to indicate how your data are laid out. In our example, the data for each run is in a separate column and, thus, we select: 'Columns'.
- o Check the box titled 'Labels in first row' to indicate that you have column titles in Row 1.
- o Alpha: 0.05 is the default level. Since this is the significance level at which we wish to conduct our test, we can leave this value as is.
- Output options:
- o Pick the options you prefer. Here we selected to place the output table in cell E1 on the same worksheet.
- Click 'OK'.

Figure 12-1
Excel Screen Shot: Setting Up the Charity Runs ANOVA Test

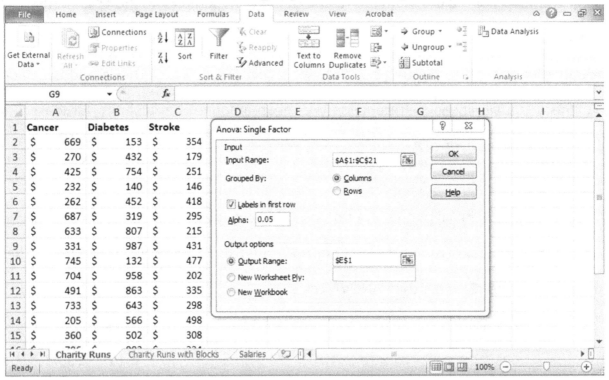

Figure 12-2 provides the screen capture of the output obtained from Excel. There are two parts to this output. The columns of the top table provide summary statistics for the data. For each of the three charity runs, you are given:

- Count: The sample size for each group.

- Sum: The sum of the fundraised amount in each group.

- Average: The average fundraised amount in each group.
- Variance: The variance in fundraised amount for each group.

The bottom table is the ANOVA table, which is similar to Table 12-5 covered above. In this ANOVA table, Excel provides both the p-value and the critical value (F crit) associated with the test, so there is no need to look these values up in the F-table. Looking at this bottom output table, we can immediately compare the test statistic (F) of 12.38 to the critical value (F crit) of 3.1588 and conclude that we can reject the null hypothesis. Alternatively, as the p-value is very low (five decimal places from zero), we likewise conclude that the null hypothesis should be rejected.

Figure 12-2
Excel Screen Shot: Charity Runs ANOVA Output

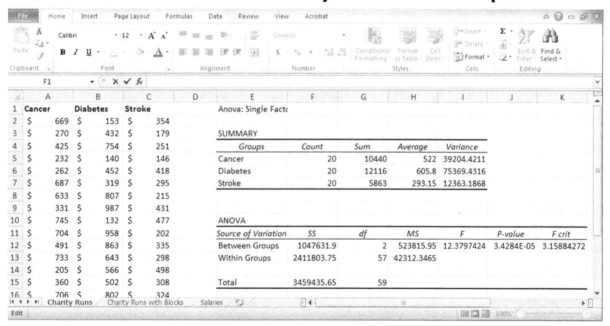

Unit 2 Summary

- **One-way ANOVA** is used when there is a single independent variable (*factor*) having two or more levels (or groups).

- To understand the effect of the factor on the dependent variable, we study the sources of variation from the **grand mean** (the mean of all observations in our data set) by looking at the **sum of squares**. The **sum of squares** within groups (SSW)* represents variation within groups whereas the *sum of squares between groups (SSB)* represents variation between groups.

- Hypotheses:
  - $H_0: \mu_1 = \mu_2 = \mu_3 = \ldots = \mu_k$
  - H_1: At least one of the group means differs from the others.

- The test statistic is referred to as the **F-statistic** and is computed as:

$$F = \frac{MSB}{MSW}$$

- The critical value comes from the F-distribution with α as specified and with k-1 and n-k degrees of freedom.

- This is an upper-tail test, so we reject the null hypothesis if the test statistic is greater than the critical value. Working with Excel, we are also given the test's p-value; using the p-value approach to hypothesis testing, we reject the null hypothesis when the p-value is less than the specified α.

Unit 2 Exercises

1. Compute the ANOVA table given the data in Table 12-e1.

Table 12-e1

	Factor Levels			
	1	2	3	4
n	10	12	8	11
$\bar{x}$	37	42	40	36
s^2	10	10	10	10

2. Repeat Question 1, changing the four sample sizes, respectively, to $n_1=12$, $n_2=14$, $n_3=10$ and $n_4=13$. What is the effect of an increase in sample sizes on the test statistic?

3. Compute the ANOVA table given the data in Table 12-e2.

Table 12-e2

	Factor Levels		
	1	2	3
n	10	15	10
$\bar{x}$	8	5	6
s^2	5	5	5

4. Repeat Question 3, changing the three sample means, respectively, to 58, 55 and 56. What is the effect of an increase in sample means on the test statistic?

5. Compute the ANOVA table given the data in Table 12-e3.

Table 12-e3

	Factor Levels		
	1	2	3
n	6	8	10
$\bar{x}$	13	12	16
s^2	8	8	8

6. Repeat Question 5, changing the three sample variances to 10 each. What is the effect of an increase in sample variances on the test statistic?

7. Marketing employees in different salary brackets were asked to provide the number of weeks of annual vacation each receives. Does the data in Table 12-e4 provide sufficient evidence to conclude that there are differences in vacation time between the salary-bracket groups? (Use $\alpha=0.05$.)

Table 12-e4

$40,000 - $49,999	$50,000 - $59,999	$60,000 - $69,999	$70,000 - $79,999
4	5	5	8
4	4	5	6
5	6	7	7
3	4	4	6
2	3	3	5
2	3	5	5
3	4	6	6
3	4	3	6
2	2	4	8
4	3	5	7

8. A family with four teenage drivers buys one car for all of them to use. For the first month, the parents record how many miles each teenager drives every time he/she uses the car. (Each teenager used the car eight times this first month.) Given the sample data shown in Table 12-e5, can the parents conclude that the car is not being shared equally? (Use $\alpha=0.01$.)

Table 12-e5

Eric	Alyssa	John	Emily
10	17	13	12
6	10	11	17
8	16	5	10
10	16	8	17
8	17	6	10
11	6	13	9
9	9	8	14
11	11	12	12

9. The PTAs of three schools are weighing (and calculating the percentage of) each school's weekly garbage that is placed in recycling bins. They believe that the higher this percentage, the more environmentally friendly the school. Each school provides a sample of these percentages for seven random weeks, as shown in Table 12-e6. At the 5% significance level, are the recycling percentages unequal?

Table 12-e6

School A	School B	School C
43	40	73
36	55	58
41	49	48
50	35	54
21	53	71
53	38	58
23	45	72

10. Total Harmonic Distortion (THD) is a measured difference in audio levels between the input audio signal's harmonics and the harmonics of the output audio signal of an amplifier or other audio device. For quality speakers, the THD is less than 1%, but some THDs can range up to 5% (or even higher). Given the samples provided for each of four speaker brands in Table 12-e7[122], is there a difference in the brands' mean THD at the 1% significance level?

---

[122] Source: http://www.tech-faq.com/total-harmonic-distortion.html

Table 12-e7

Panasonic	Logitech	SONY	Bose
1	3	2	1
3	4	1	2
5	2	3	1
2	3	1	1

11. Fill in the blanks in the ANOVA table provided in Table 12-e8 (based on the information provided). Would you reject the null hypothesis at the 5% level of significance?

Table 12-e8

Source of Variation	Sum of Squares	df	Mean Squares	F-statistic	P-value	F critical
Factor (Between)	0.04912				0.9308	3.88529
Error (Within)		12				
Total	4.13536	14				

UNIT 3

Two-Way ANOVA

In Unit 2, we described how ANOVA is used to assess the effect of a factor (with two or more levels) on a dependent variable. In addition, ANOVA also enables us to test the effect of more than one factor on a dependent variable. For example, we may wish to study the effect of product price (Factor 1) and advertising strategy (Factor 2) on sales volume (dependent variable), or the effect of a website's design (Factor 1) and the website's content quality (Factor 2) on the number of visitors to the website (dependent variable), or the effect of training method (Factor 1) and supervisor's management style (Factor 2) on employee job performance (dependent variable). All of these examples require a design called a two-factors *ANOVA* or, more commonly, a **two-way ANOVA**.

A two-way ANOVA is used in two distinct ways. First, we may wish to determine the effects of two separate factors on a dependent variable. Consider, for example, that the Dean of a business school is interested in studying the effect of teaching method on students' satisfaction for a large-lecture course. Here, Teaching Method is a factor with three levels: online only, in-class only, and blended (a combination of online and in-class teaching). In studying the effect of teaching method on students' satisfaction, the Dean designs an experiment. In this experiment, the students registering for the course are broken up into three groups, with each group taking the course via a different teaching method. At the end of the course, each student's satisfaction with the course is measured. The Dean, however, has to assign a different instructor to each course, and she knows that the instructor assigned to a course is very likely to influence the students' satisfaction with the course. Thus, the Dean needs to account for the person teaching the course. In other words, Instructor becomes a second factor (in addition to Teaching Method) likely to affect the students' satisfaction. We will describe this type of analysis later in this unit.

A second way in which two-way ANOVA is used is when we are not actually interested in learning about *both* effects, but wish to control for one of the factors. For example, we may already know that one factor will (or is very likely to) affect the dependent variable and are interested in determining if a different factor also affects the dependent variable. In such situations, we use a two-way ANOVA to learn about the effect of the second factor while controlling for the effect of the first factor. To illustrate this case, consider the same example of the three teaching methods. Let us assume that the same instructor teaches all three sections so we no longer have to worry about the instructor effect. Assume, however, that we also believe that students' capability to do well in the course is likely to make a difference in the students' course satisfaction. To overcome this problem, what we can do is *match* students based on their grades in a pre-requisite course. That is, we will not compare three random samples of users, but rather we will compare three *matched* samples. Specifically, we will compare more-capable students to other more-capable students under each of the three teaching methods and less-capable students to other less-capable students. This design is called a *randomized block ANOVA* design.

We begin this unit by describing how the randomized block ANOVA test is conducted. Then, we will describe how to conduct a two-way ANOVA test in which the effects of two factors on a dependent variable are tested.

Randomized Block ANOVA Design

Let us go back to the three charity runs example in order to illustrate this type of two-way ANOVA design. Table 12-6 replicates the data for the three runs, but with an important change. Now, the table does not provide data on 60 different individuals competing in one run of their choice. Rather, as the added first column indicates, these are the fundraising amounts raised by twenty runners, each of whom is competing in the three different types of charity runs. For example, John is runner number five and he raised $262 for the cancer charity, $452 for the diabetes charity, and $418 for the stroke charity; and, Barbara is runner number eighteen and she raised $325 for the cancer charity, $705 for the diabetes charity, and $170 for the stroke charity.

Table 12-6
Randomized Block ANOVA: Charity Runs Data Set

Runner	Cancer Charity Run	Diabetes Charity Run	Stroke Charity Run
1	$669	$153	$354
2	$270	$432	$179
3	$425	$754	$251
4	$232	$140	$146
5	$262	$452	$418
6	$687	$319	$295
7	$633	$807	$215
8	$331	$987	$431
9	$745	$132	$477
10	$704	$958	$202
11	$491	$863	$335
12	$733	$643	$298
13	$205	$566	$498
14	$360	$502	$308
15	$706	$802	$324
16	$650	$782	$186
17	$696	$504	$410
18	$325	$705	$170
19	$583	$990	$218
20	$733	$625	$148

The fact that we are now dealing with the same people competing in all three runs makes an important difference. Specifically, we are no longer looking at the variance across 60 random runners. Instead, we are now looking at the variance across twenty runners, each of whom competes in three charity runs. Thus, some portion of the overall variance in our data is likely to be associated with the individual fundraising abilities of these twenty runners. This is termed a **blocking** effect. Blocking means that we are controlling for (or blocking) the variance in the dependent variable that is due to runners' individual abilities to fundraise. As explained earlier, we use blocking because we expect the runners to vary in their fundraising abilities. However, we are not interested in investigating this effect in this research project; we are only interested in the effect of the charitable cause on fundraised amounts. Such experimental designs, i.e., where the rows represent repeated or matched observations, are termed a *randomized block ANOVA* (or, in Excel, a ***Two-Factor ANOVA without Replication***).

The randomized block ANOVA primarily differs from the one-way ANOVA in the sources of variation that must be taken into consideration when conducting a test. In covering one-way ANOVA in Unit 2, we attributed some of the variation in the grand mean to the factor levels (the variance *between* the three levels of the Cause factor) and the rest was considered error (the *within* variation), because we worked with independent samples. In the randomized block design, we need to consider *three* sources of variation: the variation that can be explained by the Cause factor (the *between* variance), the variation that can be explained by the runners (the *blocking* variance), and the variation that we are not able to explain (the *within* variance). The blocking source of variation did not exist in the one-way ANOVA.

With this difference in mind, the procedure for conducting the test is similar to that followed in Unit 2: We formulate our hypotheses, compute the test statistic, and compare the test statistic to a critical value in order to draw a conclusion.

Because we have added a blocking effect, we now have two sets of hypotheses. The hypotheses for the effect of the Cause factor are:

H_0: $\mu_1 = \mu_2 = \mu_3$

H_1: At least one of the means differs from the others.

The hypotheses for the effect of blocking are:

H_0: $\mu_1 = \mu_2 = \mu_3 = \ldots = \mu_{20}$

H_1: At least one of the means differs from the others.

Note that we compare twenty means in the second set of hypotheses, as there are twenty different runners.

The total sum of squares (SST) now consists of three components:

$$SST = SSB + SSBL + SSW$$

where *SSBL* is the *sum of squares due to blocking* (in this example, accounting for the differences between the twenty different runners). As calculating SSB, SSBL and SSW by hand would be rather cumbersome, it is preferable to use Excel in carrying out this test. Prior to showing how this is done, we first describe the ANOVA table for the randomized block ANOVA.

The ANOVA table (see Table 12-7) shows the calculation of the test statistic for the randomized block ANOVA. In particular, note that there is now a row in the table representing the variation due to blocking. The first column identifies the sources of variation in the data: Blocking, Factor, Error, and Total. The second column contains the sums of squares for each of these sources of variation. As with one-way ANOVA, SST is calculated as the sum of the other three sums of squares. The third column provides the degrees of freedom associated with each sum of squares. Here, b refers to the number of blocks (twenty in our example), so there are b-1 degrees of freedom associated with blocking. Note that a different formula, accounting for the number of blocks, is used in computing the number of degrees of freedom associated with the error sum of squares. The fourth column shows the mean squares for the three sources of variations. Finally, the fifth column computes the test statistic. Because we are now examining two sets of hypotheses, we have two test statistics and will conduct two hypothesis tests.

Table 12-7
The ANOVA Table for a Randomized Block ANOVA

Source of Variation	Sum of Squares	Df	Mean Squares	F-statistic
Blocking	SSBL	b-1	MSBL=SSBL/(b-1)	F=MSBL/MSW
Factor (between)	SSB	k-1	MSB=SSB/(k-1)	F=MSB/MSW
Error (within)	SSW	n-k-b+1	MSW=SSW/(n-k-b+1)	
Total	SST	n-k		

We are now ready to use Excel in carrying out the test. In this chapter's Excel file, open the worksheet titled 'Charity Runs with Blocks'. Select 'Data Analysis' from the 'Data' menu. Now, choose the menu option: 'Anova: Two-Factor Without Replication'. Figure 12-3 displays the menu that opens after you select this test and click 'OK'. Here, you are asked to specify several components needed in conducting the test:

- Input:
  - Input Range: Select the cells A1:D21, which hold the data for the three charity runs, as well as for the blocks (i.e., runners).
  - Check the box titled 'Labels' to indicate that you have column titles in Row 1.
  - Alpha: 0.05 is the default level. Since this is the significance level at which we wish to conduct our test, we leave this value as is.
- Output options:
  - Pick the options you prefer. Here we selected to place the output table in cell F1 on the same worksheet.
- Click 'OK'.

**Figure 12-3
Excel Screen Shot: Setting Up the Charity Runs
Randomized Block ANOVA Test**

Figure 12-4 provides the screen capture of the output obtained from Excel. As with the one-way ANOVA, there are two parts to this output. The columns across the top table provide summary statistics for each of the twenty blocks (i.e., runners in this example) and each of the three charity runs:

- Count: The number of observations.
- Sum: The sum of the fundraised amount.
- Average: The average fundraised amount.
- Variance: The variance in fundraised amount.

The bottom table is the ANOVA table for the randomized block design. The first column identifies the sources of variation. Here, the first row in the first column is labeled Rows and corresponds to the blocks (SSBL). You can see this by noting that it has nineteen degrees of freedom, as there are twenty runners. Also, if you consider how the data were laid out in the Excel worksheet, the runners' data is organized as rows, whereas the charitable cause data is organized as columns. The second row is labeled Columns and corresponds to the Cause factor, or the three charity runs (SSB). The third row in the ANOVA table is labeled Error and corresponds to the error component (SSW). Finally, the last row in this first column is labeled Total and corresponds to the total variation in the data. The second column presents the sums of squares (SS). (You can check to see that: SSBL+SSB+SSW=SST.) The third column provides the degrees of freedom (df) associated with each of the sources of variation. The fourth column presents the mean squares (MS), obtained by dividing the sum of squares by the degrees of freedom. For example:

$$MSBL = SSBL/df = 762{,}156.32/19 = 40{,}113.49$$

The fifth column provides the test statistics (F) to be used in testing both sets of hypotheses:

$$\text{Test Statistic}_{blocking} = MSBL/MSW = 40{,}113.49/43{,}411.77 = 0.92$$

$$\text{Test Statistic}_{factor} = MSB/MSW = 523{,}815.95/43{,}411.77 = 12.07$$

Finally, the fifth and sixth columns, respectively, provide the p-values (P value) and critical values (F crit) to be used in drawing conclusions about the two sets of hypotheses. Recall that the ANOVA test is always an upper-tail test. The critical value provided in the Excel output table for blocking comes from the F-distribution with $\alpha=0.05$, a degrees of freedom numerator of b-1, and a degrees of freedom denominator of n-k-b+1. The critical value for the Cause factor comes from the F-distribution with $\alpha=0.05$, a degrees of freedom numerator of k-1, and a degrees of freedom denominator of n-k-b+1.

Figure 12-4
Excel Screen Shot: Charity Runs Randomized Block ANOVA Output

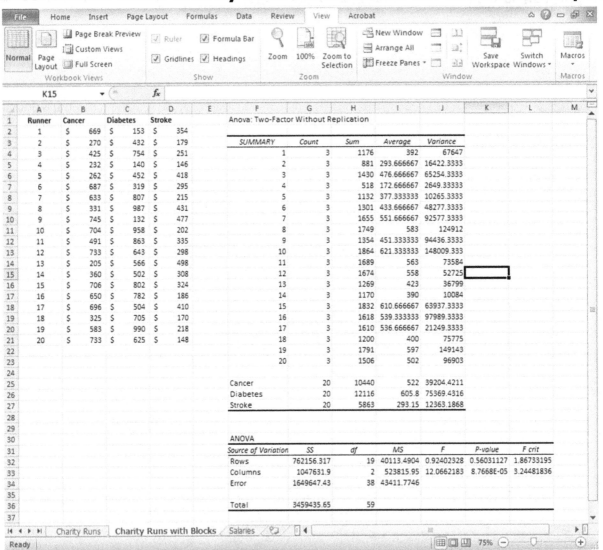

Given the Excel results provided in Figure 12-4, we are now ready to draw our conclusions regarding this blocking version of the charity runs example. The decision rule is: reject the null hypothesis if the test statistic is greater than the critical value. Hence, we will reject the null hypothesis for blocking if the test statistic is greater than 1.87, and we will reject the null hypothesis for the Cause effect if the test statistic is greater than 3.24. From the Excel output results provided in Figure 12-4, we see that the test statistic for the blocking effect (0.92) is less than the test's critical value (1.87); therefore, we do not reject the null hypothesis and are unable to conclude that the runners were a significant factor affecting the mean fundraised amount. Comparing the test statistic for the Cause effect (12.07) to this test's critical value (3.24), we reject the null hypothesis and conclude that there is a significant difference in the fundraised amounts of the three types of charity runs. You could also reach these same two conclusions by comparing each test's p-value to the specified alpha of 0.05.

Summarizing our findings for this randomized block ANOVA test, we conclude that the cause of the charity run (cancer, diabetes, or stroke) makes a difference in the fundraised amount, but that there is no significant difference between the amounts fundraised by the different runners.

Redoing the Analysis as a One-Way ANOVA

In the example just used to illustrate the randomized block ANOVA, we suspected that some of the variation in the grand mean of the fundraised amount was due to the fundraising abilities of individual runners. The analysis did not support this suspicion (i.e., we were unable to reject the null hypothesis with respect to blocking). What this means is that we were incorrect in using the randomized block ANOVA design and that we should have conducted the test as a one-way ANOVA.

Why is this important? When we use the randomized block ANOVA design, we take some of the variation previously attributed to error and assign it to the blocking effect. Mathematically, this affects the calculation of the primary factor's test statistic, which is computed as MSB/MSW. Hence, if a randomized block ANOVA is performed and the blocking effect is found not significant, the test needs to be redone as a one-way ANOVA.

Redoing the analysis as a one-way ANOVA table is straightforward, as shown in Tables 12-8a and 12-8b. Moving from Table 12-8a to Table 12-8b, we simply add SSW+SSBL to form our revised SSW. We also add together the degrees of freedom (df) of the error and the blocks. We then compute MSB, MSW and F. Looking at the revised Table 12-8b, we can see that the F-statistic is high (12.38). Comparing this value to a critical value ($F_{0.05,2,57}$=3.1588), we reject the null hypothesis and conclude that there is a significant effect of Cause on Fundraised Amount.

Table 12-8a
Original Randomized Block ANOVA Table

Source of Variation	Sum of Squares	df	Mean Squares	F-statistic
Blocking	SSBL = 762,156.32	19	MSBL = 40,113.49	MSBL/MSW = 0.92
Factor (Between)	SSB = 1,047,631.90	2	MSB = 523,815.95	MSB/MSW = 12.07
Error (Within)	SSW = 1,649,647.43	38	MSW = 43,411.77	
Total	SST = 3,459,435.65	59		

Table 12-8b
Reconstructed One-Way ANOVA Table

Source of Variation	Sum of Squares	df	Mean Squares	F-statistic
Factor (between)	SSB = 1,047,631.90	2	MSB = SSB/df =1,047,631.90/2 =523,815.95	F = MSB/MSW =523,815.95/ 42,312.35 =12.38
Error (within)	SSW = SSW(old) + SSBL =1,649,647.43+762,156.32 =2,411,803.75	38+19=57	MSW = SSW/df =2,411,803.75/57 =42,312.35	
Total	SST = 3,459,435.65	59		

Making Sure You Understand

Let us review another (hypothetical) example before moving on, in order to ensure that the concepts introduced so far are clear. A business student wishes to compare, at a 5% significance level, expected salary levels in three possible careers: a financial analyst, a marketing consultant, and a business analyst. She collects data on recent salaries as published on a career website, as shown in Table 12-9. She further suspects that these salaries are likely to vary significantly for different cities in the United States; so, she adds this information to the data table. In this example, Career is the primary factor of interest, while City is a blocking effect. Given the existence of an expected blocking effect, she uses a randomized block ANOVA design to test if there are significant differences between mean salary levels for the three career choices.

Table 12-9
Salary Levels for Different Careers, Controlling for Location (City)

City	Financial Analyst	Marketing Consultant	Business Analyst
New York	$84,000	$92,000	$92,000
Chicago	$76,000	$83,000	$83,000
LA	$75,000	$82,000	$82,000
Miami	$70,000	$77,000	$76,000
Dallas	$70,000	$76,000	$76,000
Seattle	$67,000	$74,000	$74,000

As always, we begin by formulating our hypotheses. The first set of hypotheses concerns the Career effect:

H_0: $\mu_1 = \mu_2 = \mu_3$

H_1: At least one of the means differs from the others.

The second set of hypotheses concerns the blocking effect (City):

H_0: $\mu_1 = \mu_2 = \mu_3 = \mu_4 = \mu_5 = \mu_6$

H_1: At least one of the means differs from the others.

We use Excel to run the randomized block ANOVA test (Anova: Two-Factor without Replication), obtaining the results shown in Figure 12-5. You can run the test on your own to make sure you understand how to use Excel for this test. The data for this example are provided in the worksheet titled 'City Salaries' in this chapter's Excel file.

Figure 12-5
Excel Screen Shot: Salary Levels, Randomized Block ANOVA Output

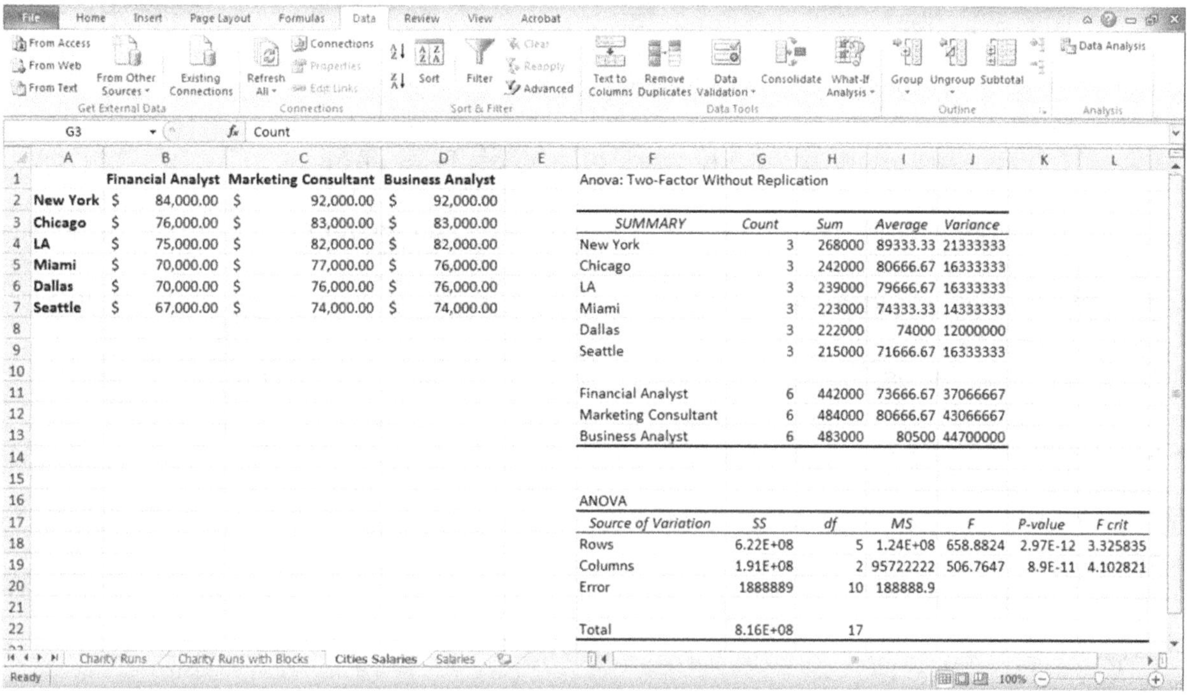

	Financial Analyst	Marketing Consultant	Business Analyst
New York	$84,000.00	$92,000.00	$92,000.00
Chicago	$76,000.00	$83,000.00	$83,000.00
LA	$75,000.00	$82,000.00	$82,000.00
Miami	$70,000.00	$77,000.00	$76,000.00
Dallas	$70,000.00	$76,000.00	$76,000.00
Seattle	$67,000.00	$74,000.00	$74,000.00

Anova: Two-Factor Without Replication

SUMMARY	Count	Sum	Average	Variance
New York	3	268000	89333.33	21333333
Chicago	3	242000	80666.67	16333333
LA	3	239000	79666.67	16333333
Miami	3	223000	74333.33	14333333
Dallas	3	222000	74000	12000000
Seattle	3	215000	71666.67	16333333
Financial Analyst	6	442000	73666.67	37066667
Marketing Consultant	6	484000	80666.67	43066667
Business Analyst	6	483000	80500	44700000

ANOVA

Source of Variation	SS	df	MS	F	P-value	F crit
Rows	6.22E+08	5	1.24E+08	658.8824	2.97E-12	3.325835
Columns	1.91E+08	2	95722222	506.7647	8.9E-11	4.102821
Error	1888889	10	188888.9			
Total	8.16E+08	17				

As you can see from the output in Figure 12-5, the Career factor (Columns) and the City factor (Rows) are both highly significant (both test statistics are much greater than the corresponding critical values, and both p-values are much lower than the specified alpha level of 0.05). In other words, both are influential in determining the salary that a business student majoring in one of these three careers might expect to receive upon graduation. In this example, therefore, the student was justified in blocking for the City effect (i.e., we rejected the null hypothesis for blocking), and we conclude that significant salary level differences do exist for at least one of the three career choices.

Two-Factors ANOVA

In the examples covered so far in this unit, we have illustrated how to examine the significance of a single factor (the three charity runs and the three career choices), controlling for a blocking variable (the twenty runners and the six cities). Often, we wish to study the effect of two separate factors on some dependent variable. At the beginning of this unit, we provided some examples of studies involving two separate factors: a study of the effect of both product price (Factor 1) and advertising strategy (Factor 2) on sales volume (dependent variable), the effect of both a website's design (Factor 1) and the website's content (Factor 2) on the number of visitors to the website (dependent variable), or the effect of both training method (Factor 1) and supervisor's management style (Factor 2) on employee job performance (dependent variable). Such examples require a design called a ***two-factors ANOVA***.

To understand this type of analysis, consider the following hypothetical example. The Human Resources (HR) manager at a large technology company is evaluating the effectiveness of different supervisory management styles and different training methods on employees' performance. Specifically, the HR manager wishes to assess the effectiveness of two different management styles. The first style might be called *laissez-faire* and reflects work area supervisors acting mainly as mentors, but only when asked to do so by work area employees, and the employees self-managing themselves. The second style might be called *autocratic* and reflects the work area supervisors making all of the important work area decisions, without much regard for the employees' inputs. The HR manager believes that the laissez-faire style works better in his industry and should result, on average, in higher work area performance.

The HR manager also desires to compare the use of three different training methods applied in enabling employees to catch up on recent developments in their industry: in-class training with an exam (In-Class), online training with an exam (Online (Exam)), and online training without an exam (Online (No Exam)). In general, the HR manager believes that in-class training is likely to prove most effective. Finally, the HR manager also suspects that some form of *interaction* might arise between these two factors (Management Style and Training Method). An **interaction** effect exists when the effect of one factor on the dependent variable changes for different levels of the other factor. Specifically, the HR manager thinks that online training (with or without an exam) would work better with the laissez-faire management style, as employees working under a laissez-faire supervisor would be more comfortable working on their own.

We have three sets of hypotheses to test with this example. First, we have the hypotheses for the effect of Management Style:

$$H_0: \mu_{\text{laissez-faire}} = \mu_{\text{autocratic}}$$

$$H_1: \mu_{\text{laissez-faire}} \neq \mu_{\text{autocratic}}$$

Second, we have the hypotheses for the effect of Training Method:

$H_0: \mu_{\text{in-class}} = \mu_{\text{online(exam)}} = \mu_{\text{online(noexam)}}$

H_1: At least one of the Training Method means differs from the others.

Finally, we have the hypotheses for the interaction effect between Management Style and Training Method:

H_0: No interaction exists between Management Style and Training Method.

H_1: An interaction exists between Management Style and Training Method.

To test these three sets of hypotheses, the HR manager collected data on the performance scores (out of 100%) of 60 employees. The data are collected within a department having six work areas, with a supervisor with a laissez-faire management style managing three of the work areas and a supervisor with an autocratic management style managing the other three work areas. These data are shown in Table 12-10 and in the worksheet titled 'Performance' in this chapter's Excel file. Note that there are six groups in the table, with ten data points in each group. These ten data points for each group are called **replications**. You can visualize this table as a room with 60 people in it, divided into six groups of ten: the first group attends an in-class training session and has a laissez-faire manager; the second group also attends an in-class training session, but has an autocratic manager; and so on.

Table 12-10
Two-Factors ANOVA: HR Manager Data Set

	In-Class Training	Online (Exam)	Online (No Exam)
Laissez-Faire	71	62	63
	85	73	60
	86	72	73
	67	83	63
	85	90	64
	83	75	69
	66	71	81
	77	83	85
	70	85	72
	67	85	89
Autocratic	84	55	71
	81	80	71
	69	73	66
	71	61	77
	79	71	58
	88	66	73
	62	55	61
	66	67	73
	84	61	71
	85	66	63

The ANOVA table (Table 12-11) shows the calculation of the test statistic for the two-factors ANOVA design. We have five rows in the table to represent the five sources of variation: Factor A, i.e., Management Style; Factor B, i.e., Training Method; Interaction, i.e., the interaction between Factor A and Factor B; Error (Within), i.e., the variation due to error; and, of course, Total, i.e., the total variation in the data. The first column identifies these five sources of variation. The second column contains the sums of squares for each source of variation, with SST calculated as the sum of all other sums of squares. The third column shows the degrees of freedom associated with each sum of squares. In these degrees of freedom formulae: (a-1) represents the number of degrees of freedom of Factor A (number of levels minus one); (b-1) represents the degrees of freedom of Factor B;

(a-1)*(b-1) represents the degrees of freedom for the interaction test; (n-(a*b)) represents the degrees of freedom for the error variance; and, (n-1) represents the total degrees of freedom. The fourth column shows the mean squares for the four sources of variations. Finally, the fifth column computes the three test statistics used, respectively, in assessing our three sets of hypotheses.

Table 12-11
ANOVA Table for a Two-Factors ANOVA design

Source of Variation	Sum of Squares	df	Mean Squares	F-statistic
Factor A	SSA	a-1	MSA=SSA/(a-1)	F=MSA/MSW
Factor B	SSB	b-1	MSB=SSB/(b-1)	F=MSB/MSW
Interaction	SSAB	(a-1)*(b-1)	MSAB=SSAB/(a-1)*(b-1)	F=MSAB/MSW
Error (Within)	SSW	n-(a*b)	MSW=SSW/(n-(a*b))	
Total	SST	n-1		

We use Excel to run the ANOVA test. In Excel, this test is called **ANOVA: Two-Factor with Replication**. Figure 12-6 provides the screen shot for the input window for this test:

- Input:
  - Input Range: Select the cells A1:D21, which hold the data and labels for rows and columns. Note that Excel assumes that these row and column labels are included and, if these labels are not included in the specified range, you may receive either an error message (if you exclude the top row) or inaccurate results (if you exclude the first column).
  - Rows per sample: This requires you to input the number of *replications*, which is 10 in our example. In Excel, you must have the same number of replications for all samples. If this is not the case with the data collected, you will need to use a different statistical software package.
  - Alpha: 0.05 is the default level. Since this is the significance level at which we wish to conduct our test, we can leave this value as is.
- Output options:
  - Pick the options you prefer. Here we selected to place the output table in cell F1 on the same worksheet.
- Click 'OK'.

Figure 12-6
Excel Screen Shot: Setting Up the
HR Manager Two-Factors ANOVA Test

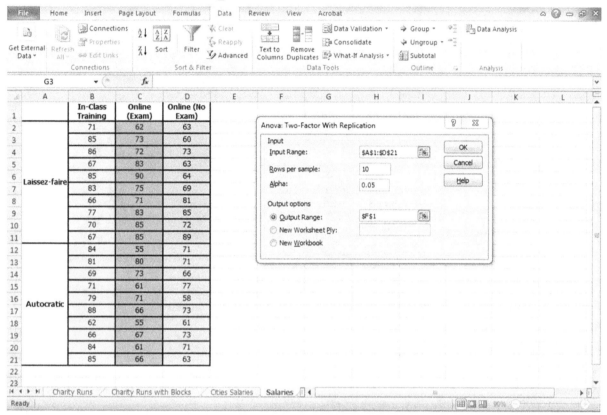

Figure 12-7 provides the screen capture of the output obtained from Excel. The very top table provides summary statistics by training method for the three groups (i.e., work areas) being managed by the laissez-faire supervisor. You can see just by looking at these summary statistics that mean performance is highest for the Online (Exam) group (77.9), followed by the In-Class group (75.7), and finally, the Online (No Exam) group (71.9).

Figure 12-7
Excel Screen Shot: HR Manager Two-Factors ANOVA Output

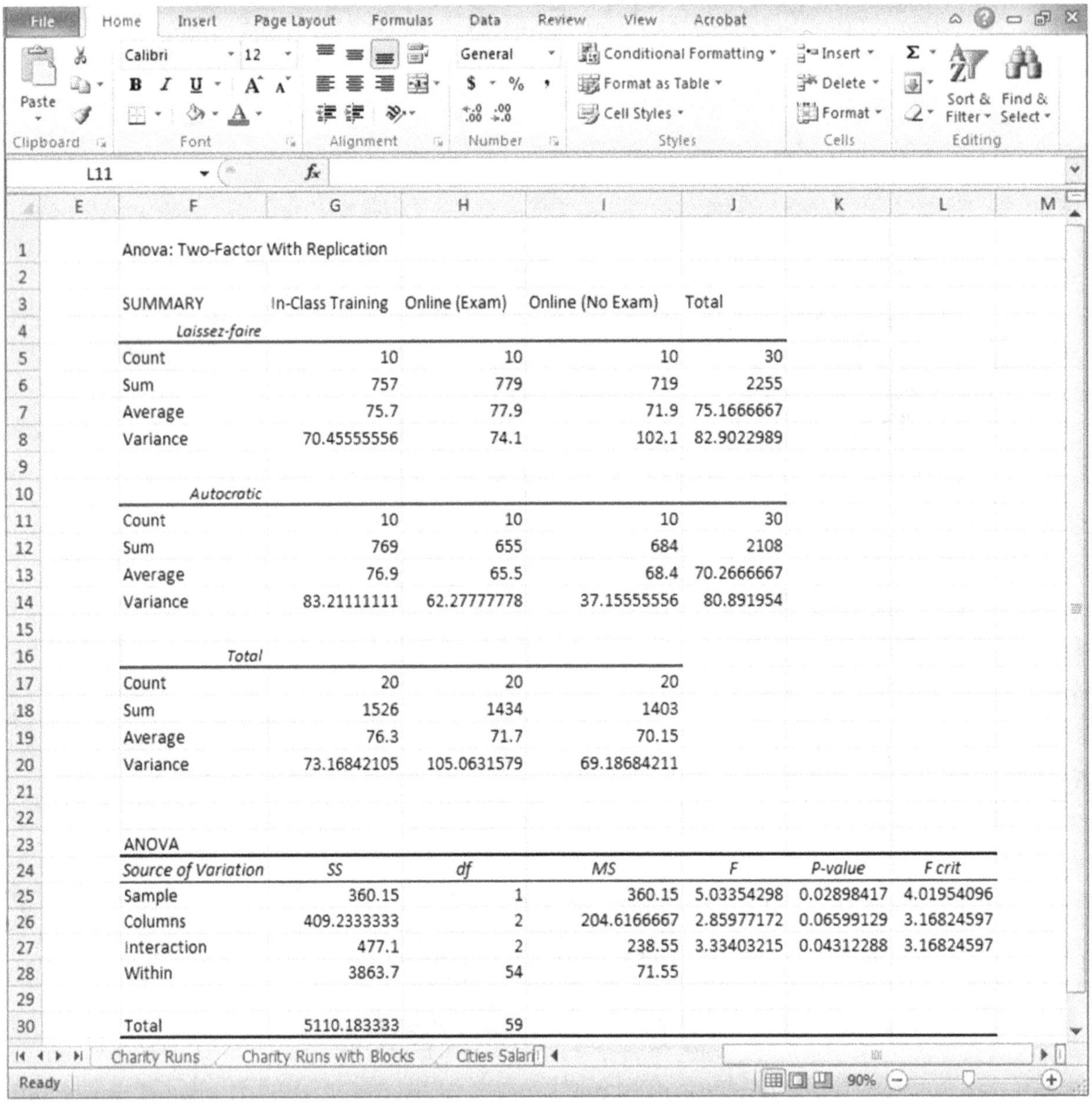

Anova: Two-Factor With Replication				
SUMMARY	In-Class Training	Online (Exam)	Online (No Exam)	Total
Laissez-faire				
Count	10	10	10	30
Sum	757	779	719	2255
Average	75.7	77.9	71.9	75.1666667
Variance	70.45555556	74.1	102.1	82.9022989
Autocratic				
Count	10	10	10	30
Sum	769	655	684	2108
Average	76.9	65.5	68.4	70.2666667
Variance	83.21111111	62.27777778	37.15555556	80.891954
Total				
Count	20	20	20	
Sum	1526	1434	1403	
Average	76.3	71.7	70.15	
Variance	73.16842105	105.0631579	69.18684211	

ANOVA						
Source of Variation	SS	df	MS	F	P-value	F crit
Sample	360.15	1	360.15	5.03354298	0.02898417	4.01954096
Columns	409.2333333	2	204.6166667	2.85977172	0.06599129	3.16824597
Interaction	477.1	2	238.55	3.33403215	0.04312288	3.16824597
Within	3863.7	54	71.55			
Total	5110.183333	59				

The next-lower table in Figure 12-7 provides the same summary statistics for the three groups with the autocratic supervisor. Here, the highest mean performance was obtained by the In-Class group (76.9), followed by the Online (No Exam) and Online (Exam) groups (68.4 and 65.5, respectively).

The third table from the top provides the combined summary statistics for all 60 employees across the three training methods. Overall, we see the highest mean performance was obtained for the In-Class groups (76.3), followed by the Online (Exam) groups (71.7), and then the Online (No Exam) groups (70.15). But, of course, we need to find out if these differences are statistically significant, and we obtain this information from the bottom table in the Excel output.

The bottom table in Figure 12-7 is the ANOVA table, which provides the statistics used to test to see whether the differences between group means are significant. The ANOVA table has five rows. The first row, titled Sample, corresponds to the two management styles (Factor A: Management Style). The second row, titled Columns, corresponds to the three training methods (Factor B: Training Method). The third row, titled Interaction, corresponds to the hypotheses regarding the existence of an interaction between Management Style and Training Method. We will interpret the meaning of this interaction effect shortly. The fourth row, titled Within, is our error, or residual, variance. Finally, the fifth row is the Total row.

There are three test statistics in this output (the values for F provided in the fifth column of the bottom table). First, the test statistic for the difference in mean performance between employees under the two management styles is 5.034. Second, the test statistic for the difference in mean performance between the three training methods is 2.86. Finally, the test statistic for the interaction effect is 3.33.

In order to draw conclusions about these tests, we need to compare these three test statistics to their respective critical values (obtained from the F-distribution). The three critical values are provided in the seventh column (F crit) of this bottom table, with each critical value representing a 5% upper-tail rejection region in the F-distribution. The critical value for the test of management styles is 4.02 and is computed with 1 and 54 degrees of freedom. The critical value for the test of training methods is 3.17 and is computed with 2 and 54 degrees of freedom. The critical value for the interaction test is also 3.17, computed again with 2 and 54 degrees of freedom. Indeed, you can see that these latter two critical values are the same in the output table.

Looking at the test statistic for the Management Style effect, the value of 5.034 is greater than the critical value of 4.02 and, therefore, we reject the null hypothesis and conclude that, at a 5% level of significance, evidence shows that performance of employees differs under the two management styles. The test statistic for the Training Method effect is 2.86, which is lower than the test's critical value of 3.17. Thus, at a 5% level of significance, we cannot reject the null hypothesis. However, if you examine this test's p-value (given in the sixth column of the bottom output table), you see a value of 0.066. While we are unable to reject the null hypothesis at a 5% level of significance, we would reject it at a 10% level of significance (where we would use $\alpha=0.1$), as the p-value of 0.066 is less than 0.1. Therefore, we can conclude, at a 10% level of significance, that employee mean performance differs under the different training methods.

Finally, the test statistic for the interaction effect is 3.33, which is greater than this test's critical value of 3.17. Thus, at a 5% level of significance, we reject the null hypothesis and conclude that an interaction is present between Management Style and Training Method.

Interpreting Interaction Effects

Earlier in this unit, we explained that the existence of an interaction between two factors means that the effect of one factor on a dependent variable changes based on the levels of the second factor. This is illustrated graphically through the interaction plot depicted in Figure 12-8 for the HR manager example. An ***interaction plot*** is a graph depicting how two independent variables interact to affect the dependent variable. This is drawn by plotting the effect of one variable on the dependent variable (placing this variable on the x-axis) and then creating a different graph (or, line) for each level of the second variable (this is shown by the three different lines in Figure 12-8). In this figure, the y-axis represents performance values, and the x-axis has two points indicated: one point for each of the two management styles. The three lines (one with *dots*, one with *dashes*, and one with *both* dots and dashes) connecting the different markers (the *triangles*, the *squares*, and the *diamonds*) represent the three training methods. Recognizing that Management Style is not a continuous variable, we are simply using these three lines to visualize the nature of the relationships between employee performance and each of the two management styles, under each of the three training methods.

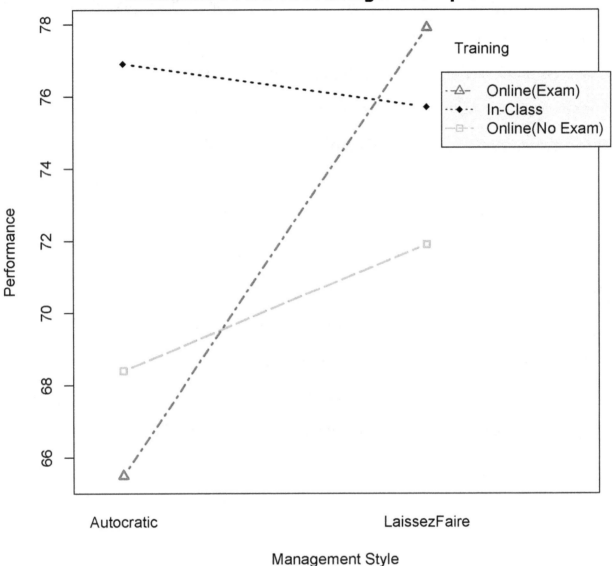

**Figure 12-8
Interaction Plot: HR Manager Example**

The next step in interpreting this interaction effect is to examine the interaction plot to gain insight into how the nature of the relationship between Management Style and employee performance changes under the different types of Training Method. For groups having an autocratic supervisor (i.e., the three data points on the left-side of the interaction plot in Figure 12-8), the lowest performance was obtained by the group given the Online (Exam) training, followed by the group given the Online (No Exam) training, and finally the group given the In-Class training.

In the absence of an interaction effect, we would expect to see this same ordering of outcomes for groups having a laissez-faire supervisor. However, as depicted in this interaction plot by the three data points on the right-side of the plot, the ordering of mean group performance associated with each training method differs for the groups having a laissez-faire supervisor, with the highest performance now observed with the group given the Online (Exam) training. You can also see that the slopes of the dotted line (diamond markers) and the dashed line (square markers) differ, indicating that groups given the In-Class training performed better under an autocratic supervisor, while groups given the Online (No Exam) training performed better under a laissez-faire supervisor.

To summarize, what we learn from the interaction plot in Figure 12-8 is that the effect of training method on performance depends on whether employees have an autocratic or a laissez-faire supervisor. The most effective training method for employees having an autocratic supervisor is In-Class training, followed by Online (No Exam) and, lastly, by Online (Exam). For employees having a laissez-faire supervisor, the most effective training method is Online (Exam) training, followed by In-Class training, and, lastly, by Online (No Exam) training.

The important message to take away from this example is: if we find that an interaction effect is present (that is, if we reject the null hypothesis for interaction when conducting a two-factors ANOVA), we cannot interpret the effects (on a dependent variable) of the individual factors alone. Instead, the influence of each of the two factors on the dependent variable can only be interpreted given a specified level of the other factor.

Consequently, if an interaction effect is found to be significant after carrying out a two-factors ANOVA, separate one-way ANOVA tests (one test for each of the two factors) need to be carried out. With these one-way ANOVA tests, the level of the factor whose effect on the dependent variable is *not* being assessed is held constant. This is illustrated for the HR Manager example in Tables 12-12 and 12-13. Table 12-12 provides the results of a one-way ANOVA for the subset of employees having an autocratic supervisor, while Table 12-13 provides the results of a one-way ANOVA for the subset of employees having a laissez-faire supervisor. By splitting the data set into these two subsamples, we are able to assess the effect of Training Method on employee performance while keeping the level of Management Style constant. Interestingly, while the results provided in Table 12-12 indicate that a significant effect of Training Method on performance is present under the autocratic management style, no such significant effect is present under the laissez-faire management style. Can you explain this lack of significance given the interaction plot shown in Figure 12-8?

Table 12-12
One-Way ANOVA Table for the Autocratic Management Style

Source of Variation	Sum of Squares	df	Mean Squares	F-statistic	P-value	F crit
Factor (between)	702.0667	2	351.0333	5.7658	0.00822	3.354
Error (within)	1643.8	27	60.8815			
Total	2345.867	29				

Table 12-13
One-Way ANOVA Table for the Laissez-Faire Management Style

Source of Variation	Sum of Squares	df	Mean Squares	F-statistic	P-value	F crit
Factor (between)	184.2667	2	92.1333	1.1206	0.3408	3.354
Error (within)	2219.9	27	82.2185			
Total	2404.167	29				

Unit 3 Summary

- In **two-way ANOVA**, the effects of two factors (Factor A and Factor B) on a dependent variable are tested.

- One type of two-way ANOVA is a **randomized block ANOVA** (referred to as a **Two-Factor ANOVA without Replication** in Excel). This test examines for the effect of a primary factor while controlling for the effect of another factor. This use of a control factor is termed **blocking**.

- There are two sets of hypotheses being tested in the randomized block ANOVA design:
    - The hypotheses for the effect of the primary factor with k levels are:
        - $H_0: \mu_1 = \mu_2 = \ldots = \mu_k$
        - H_1: At least one of the means differs from the others.
    - The hypotheses for the effect of a blocking factor with b levels are:
        - $H_0: \mu_1 = \mu_2 = \mu_3 = \ldots = \mu_b$
        - H_1: At least one of the means differs from the others.

- The test statistic for the effect of the primary factor with a randomized block ANOVA is:
$$F_{factor} = \frac{MSB}{MSW}$$

- The test statistic for the effect of the blocking factor with a randomized block ANOVA is:
$$F_{blocks} = \frac{MSBL}{MSW}$$

- The critical values for these two tests come from the F-distribution. The degrees of freedom are k-1 for the factor, b-1 for the blocks, and n-k-b+1 for the error term.

- As these are upper-tail tests, we reject the null hypothesis if the test statistic is greater than the critical value ($F_{statistic} > F_{critical}$). Working with Excel, we are also given a test's p-value and can apply the p-value rejection rule, where the null hypothesis is rejected when the p-value is less than the specified α.

- The ANOVA table with a randomized block ANOVA is:

Source of Variation	Sum of Squares	df	Mean Squares	F-statistic
Blocking	SSBL	b-1	MSBL=SSBL/(b-1)	F=MSBL/MSW
Factor (between)	SSB	k-1	MSB=SSB/(k-1)	F=MSB/MSW
Error (within)	SSW	n-k-b+1	MSW=SSW/(n-k-b+1)	
Total	SST	n-1		

- A second type of two-way ANOVA is a **two-factors ANOVA** (referred to as **ANOVA: Two-Factor with Replication** in Excel) used to test the effect of two different factors (Factor A and Factor B) on a dependent variable, as well as the possibility of interactions between the two factors. **Replications** mean that for each level of the factors, we are taking repeated measures of the dependent variable (so we have more than one observation per cell in the data table).

- **Interaction** means that the effect of one factor on the dependent variable changes for different levels of the other factor. Interaction is visualized through an **interaction plot** that graphs the effect of one independent variable on the dependent variable for each level of the second independent variable. If we find that interaction exists, we need to redo the test as separate one-way ANOVA tests. Here, each one-way ANOVA tests the effect of one factor, holding the levels of the other factor constant.

- There are three sets of hypotheses being tested in a two-factors ANOVA design:
  - The hypotheses for the effect of Factor A with a levels are:
    - H_0: $\mu_1 = \mu_2 = \ldots = \mu_a$
    - H_1: At least one of the means differs from the others.
  - The hypotheses for the effect of Factor B with b levels are:
    - H_0: $\mu_1 = \mu_2 = \ldots = \mu_b$
    - H_1: At least one of the means differs from the others.
  - The hypotheses for the interaction effect are:
    - H_0: There is no interaction between Factor A and Factor B.
    - H_1: There is an interaction between Factor A and Factor B.

- The test statistic for the effect of Factor A with a two-factors ANOVA is:
$$F_{factorA} = \frac{MSB_A}{MSW}$$

- The test statistic for the effect of Factor B with a two-factors ANOVA is:

$$F_{factorB} = \frac{MSB_B}{MSW}$$

- The test statistic for the interaction effect with a two-factors ANOVA is:

$$F_{interaction} = \frac{MSB_{AB}}{MSW}$$

- All critical values come from the F-distribution. The degrees of freedom are a-1 for Factor A, b-1 for Factor B, (a-1)*(b-1) for the interaction effect, and n-(a*b) for the error term.

- As these are all upper-tail tests, we reject the null hypothesis if the test statistic is greater than the critical value ($F_{statistic} > F_{critical}$). Working with Excel, we are also given a test's p-value and can apply the p-value rejection rule, where the null hypothesis is rejected when the p-value is less than the specified α.

- The ANOVA table with a two-factors ANOVA is:

Source of Variation	Sum of Squares	df	Mean Squares	F-statistic
Factor A	SSA	a-1	MSA=SSA/(a-1)	F=MSA/MSW
Factor B	SSB	b-1	MSB=SSB/(b-1)	F=MSB/MSW
Interaction	SSAB	(a-1)*(b-1)	MSAB=SSAB/(a-1)(b-1)	F=MSAB/MSW
Error (within)	SSW	n-(a*b)	MSW=SSW/(n-ab)	
Total	SST	n-1		

Unit 3 Exercises

1. Complete the partially-filled ANOVA table given as Table 12-e9. Then answer the three following questions:

   a. How many levels of Factor A were tested?

   b. How many replicates were there?

   c. What is your conclusion about each of the three tests (Factor A, Factor B, and interaction)?

Table 12-e9

Source of Variation	Sum of Squares	df	Mean Squares	F-statistic	P-value	F critical
Factor A	302083333.2	3			0.879	2.87
Factor B	529166666.7	2			0.561	3.26
Interaction	779166666.7	6			0.939	2.36
Error (within)		36				
Total	17847916667					

2. A store manager at the mall is considering how she can increase sales through product placement (i.e., where on the store's retail shelves products are placed) and the music genre played at the store. She measured average sales under three different product shelving arrangements (Placement) and five different music genres (Genre). Partial ANOVA table results are shown in Table 12-e10, below.

   a. Complete the ANOVA table.

   b. How many replicates were there?

   c. Clearly list each applicable set of hypotheses to be tested and then draw a conclusion regarding each of these sets of hypotheses.

Table 12-e10

Source of Variation	SS	df	MS	F-statistic	P-value	F critical
Placement	985624			6.528505989	0.002720698	
Genre		4	324064.25		0.004034908	
Interaction				1.253279022	0.285008895	
Within		60	75486.18333			
Total						

3. For each of the studies described below, decide whether you should use the randomized block ANOVA design or the two-factors ANOVA design.

   a. What affects the success of *YouTube* videos? To understand why some videos receive a large number of views whereas others do not, researchers set up an experiment measuring the number of views of videos in three categories: humorous, inspirational, and pets. For each of the videos, they measured the number of views. The researchers worried that shorter videos may be viewed more than longer videos and, therefore, they controlled for the effect of video length in their study.

   b. A study on the effects of preparation level and emotional maturity on the success of children transitioning from pre-school to kindergarten.

   c. As a strategy to increase the effectiveness of *Twitter* for business firms, consultants suggested that a firm should directly ask their customers to retweet the firm's tweets, so that these tweets reach a larger audience. To test whether this strategy works, the consultants conducted a study in which one group of customers was asked to retweet a tweet and a second group of customers was not asked to retweet the same tweet.

The number of retweets in each group was measured. In addition, the study also examined the effect of using humor in a tweet on the likelihood of a customer actually retweeting the tweet. To include humor in the study, the consultants sent humorous tweets to half of the customers in each of the two groups and sent non-humorous tweets to the other half of customers in the two groups.

4. The *Harry Potter* film franchise is the second highest grossing film series of all time (after Marvel superhero films), with eight films produced between 2001 and 2011. The first movie alone grossed $974,755,371 worldwide. In the worksheet titled 'Harry Potter' in this chapter's Excel file, box office revenues for all eight movies across four countries are provided.[123] At the 5% significance level, test to see if box office revenues differ across the four countries.

5. The worksheet in this chapter's Excel file titled 'Swimming' provides the race times attained by young swimmers in a swim meet. Times are grouped by gender (female or male swimmers) and stroke (freestyle, backstroke, breaststroke, and butterfly). Are there significant differences in swimmers' racing times that can be attributed to a swimmer's gender and to a race's stroke?

END-OF-CHAPTER PRACTICE

1. There is great competition between today's latest smartphones. One way of measuring the popularity of phones with consumers is by recording the number of times each day that phones are searched for on Google. The worksheet titled 'Phones' provides a daily search index for three different smartphone brands (*iPhone*, *Samsung Galaxy*, and *Blackberry*) for a 90-day period. Choose a random sample of 30 days and conduct a test to see whether there are differences in the popularities of these three phones. Note that daily variations in online searches are common; be sure to take this into consideration as you select the most appropriate testing procedure for examining differences in smartphone brand popularity. (Use $\alpha=0.05$.)

2. Approval ratings are one way to measure public opinion on the current president. An approval rating is determined by a polling which indicates the percentage of respondents to an opinion poll who approve of a particular person or program. The worksheet titled 'Presidents' provides a sample of United States Presidential approval ratings.[124]

---

[123] Source: http://www.boxofficemojo.com/
[124] Source: http://www.gallup.com/poll/116677/Presidential-Approval-Ratings-Gallup-Historical-Statistics-Trends.aspx#2

a. Is there a difference in approval ratings between each of the four years of Presidential first-terms?

b. Coincidentally, these presidents were re-elected for a second term of four years. Redo Question 2.a. for Presidential second-terms.

3. The worksheet titled 'Armed Forces' provides data on the percent of armed forces personnel out of a country's total labor force.[125] The data are provided for various countries for four years, from 2008 until 2011. Perform the most appropriate ANOVA test to see if there are differences in the percentage of armed forces personnel that can be attributed to the year for which these data refer.

4. The worksheet titled 'Obesity' provides data on the number of overweight or obese children in New York State.[126,127] The data are organized by region and grade level.

    a. Perform a one-way ANOVA to test whether there are differences in the number of *overweight or obese* children in elementary schools between the New York State regions.

    b. Perform a two-factors ANOVA on the number of *overweight or obese* children using the New York State region as one factor and grade level as a second factor. Be sure to test to determine if these two factors interact.

5. The worksheet titled 'School Ranking' provides data on various indicators of fifty top-ranked business schools. On which indicators (one or more) does the data provide evidence to conclude that there is a difference between ranked business schools in California, New York, Pennsylvania, and Massachusetts?

6. A not-for-profit organization operates walk-in clinics in three neighborhoods of a city. To better service patients, the organization would like to know more about the patients who use the clinics' services. The worksheet titled 'Age' provides the ages for three random samples of patients, with each sample taken at one of the three clinics. Is there sufficient evidence to conclude that the age of patients differs between the three clinics?

---

[125] Source: http://data.worldbank.org/indicator/MS.MIL.TOTL.TF.ZS
[126] Source: https://health.data.ny.gov/Health/Student-Weight-Status-Category-Reporting-Results-B/es3k-2aus
[127] The worksheet titled 'Obesity_Raw Data' provides the data as it appears on the NY State health website. If you would like to practice your Excel skills, you can start with this data set and organize it first to the table format needed to conduct the ANOVA test. Note that in the 'Obesity' worksheet, we use sample subsets from each region. We selected these samples randomly by using Excel's random numbers generator.

7. The worksheet titled 'Chimpanzee' provides data on the time it takes to teach ten different signs to four chimpanzees.[128] Is there sufficient evidence to conclude that there is a difference in the innate abilities of these four chimpanzees to learn signs? Is blocking necessary?

8. In the 2002 *Winter Olympic Games*, a scandal rocked the Figure Skating community. A tight competition between Russian pair skaters Elena Berezhnaya and Anton Sikharulidze and Canada's Jamie Salé and David Pelletier ended in a major judging controversy that resulted in the Russian skaters being awarded the Gold medal and Canadian skaters the Silver medal. It was later determined that the French judge had been pressured to vote for the Russian pair as part of a deal to obtain votes for the French ice dance couple in a later event. Responding to media and public pressure, Salé and Pelletier's medal was upgraded to a Gold medal, which they shared with the Russian pair skaters. The spreadsheet titled 'Figure Skating' holds data from the 2010 *Winter Olympic Games* in Vancouver, Canada.[129] The data describes judges' scores for the leading pair skaters who competed in the Figure Skating event, for both the short- and long-programs (there were nine judges). Using these data for the 2010 *Winter Olympic Games*:

   a. Is there evidence to conclude that there is a significant difference between the judges' scores on the short-program (suggesting the possibility of a judging controversy similar to that which occurred in the 2002 *Winter Olympic Games*)?

   b. Conduct a two-factors ANOVA on the scores assigned to the pairs of skaters using Judges and Program as factors. What do you conclude?

---

[128] Source: ftp://ftp.nist.gov/pub/dataplot/other/reference/ATKINSON.DAT
[129] Source: http://www.isuresults.com/results/owg2010/

CHAPTER 13: Regression Analysis

Economists look at consumers' spending on goods (e.g., clothes) and services (e.g., financial advice), termed *consumption*, as a function of consumers' incomes, such that a certain percentage of each additional dollar that is earned will be spent on goods and services. More specifically, this percentage is known as the *marginal propensity to consume* and it is determined by collecting data about individuals' incomes and spending on goods and services.

In order to study the relationship between income and consumption, we collect data on income and consumption for a random sample of people and then apply regression analysis - the focus of this final chapter - to these data. We use **regression analysis** in examining relationships among variables when we believe that, in addition to the variables being *correlated* (i.e., the variables' values change similarly in both direction and magnitude), the variables are also *causally related* - that is, we believe that there is a **causal relationship** between one or more independent variables and a dependent variable.

Causality means that you believe changes in the values of the independent variables will *cause* changes to the dependent variable. In other words, it is the change in the value of A (an independent variable) that leads to a change in the value of B (the dependent variable). When causality is absent, while we might observe the values of two variables moving together (in direction and magnitude), we cannot explain the *cause* of such movements. We are often guided by theory[130]

---

[130] A theory is a set of ideas or statements aimed at explaining some observed phenomenon. Among other things, theories define how and why independent variables can cause changes to a dependent variable.

when we conceptualize causal relationships, and we use regression analysis to test and to provide empirical support for this theoretical foundation.

In regression analysis, **independent variables** (e.g., income in our example) can assume any value regardless of the values of other variables. However, the value of a **dependent variable** (e.g., consumption in our example) is determined by the values of the independent variables. Other examples of confirmed two-variable causal relationships include: lung cancer and smoking; income and years of education; gas mileage and car engine size; and, house price and house size. In each of these examples, the first variable listed is the dependent variable and the second is the independent variable. The causal influence of smoking on lung cancer, for example, has been affirmed by the *Center for Disease Control*: "Cigarette smoking is the number one risk factor for lung cancer. In the United States, cigarette smoking causes about 90% of lung cancers."[131] Hence, if we were to analyze the relationship between smoking and lung cancer, we expect to find smoking confirmed as a *cause* for lung cancer. In other words, a heavy smoker is significantly more likely to be diagnosed with lung cancer than a nonsmoker. We say *more likely to be* rather than *will be* because (1) there are other possible causes of lung cancer and (2) some heavy smokers may not be diagnosed with lung cancer.

Regression analysis is not limited to a single independent variable and we can include as many variables as guided by our understanding of what factors may explain the value of the dependent variable. For example, if we study the real estate market, we may believe that the price of a house is determined not only by

---

[131] Source: http://www.cdc.gov/cancer/lung/basic_info/risk_factors.htm

its size (square footage), but also by the number of bathrooms available, the size of the lot, whether or not there is a garage, the age of the house, the neighborhood, and other possible variables. Similarly, in studying the job market, we may believe that a person's annual salary is a function of the individual's gender, education, work field, years of work experience in this field, and work position. In these two examples, house price and annual salary are dependent variables explained by specified sets of independent variables.

Regression analysis is a powerful and versatile technique that can be applied in many contexts and answer many research questions. In this book, we provide only an introduction to this very complex subject. This chapter covers the following topics:

- Terminology and Assumptions
- Conducting Regression Analysis
- The Coefficient of Determination (R^2)
- The Standard Error of the Estimate (S_ε)
- The ANOVA Test
- The Regression Coefficients
- Improving the Model
- Qualitative Independent Variables
- Understanding the Intercept
- Confidence and Prediction Intervals

UNIT 1

Terminology and Assumptions

Suppose we are interested in estimating a person's expenditure on goods and services (consumption). We take a sample of fifty people and measure their previous month's consumption (in dollars). We find that these consumption values range from $2,819 to $5,595, with an average of $4,301.50.[132] Lacking any other information, this average consumption value is our best predictor of future consumption values.

To improve our *point estimate* of consumption, we use additional data, based on what has been learned about consumers' behaviors. As has already been suggested, we can expect to find that consumption is linked to income, such that people with higher monthly incomes tend to spend more on goods and services. Thus, we also ask our fifty respondents for their monthly (after taxes) income.[133] Figure 13-1a displays both consumption and income in a single graph, along with the computed average consumption of $4,301.50 (shown as a horizontal, dashed line).

---

[132] The data used in this example is made up for the sake of explanation, but it is based on the actual average annual expenditure per consumer unit ($54,495 in 2014). A consumer unit includes "... families, single persons living alone or sharing a household with others but who are financially independent, or two or more persons living together who share expenses..." (From *Bureau of Labor Statistics* data; Source: http://www.bls.gov/news.release/cesan.nr0.htm).

[133] As in the expenditure case, these data are made up for the sake of explanation, but based on the actual average before-tax annual income per consumer unit ($66,877 in 2014) from *Bureau of Labor Statistics* data. Source: http://www.bls.gov/news.release/cesan.nr0.htm

Figure 13-1a
Income and Consumption Data: Linear Relationship

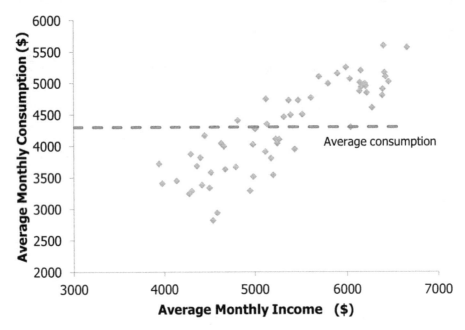

In Figure 13-1a, the vertical axis denotes consumption and the horizontal axis denotes income. Looking at the graph, there seems to be a positive relationship between income and consumption, such that a respondent with a higher income also reports higher consumption. Note that with the set of data that has been collected, the relationship between income and consumption seems *linear* - that is, it can be approximated by using a straight line. With this linear relationship, the *consumption rate* is constant. In other words, the consumption rate is the same regardless of a respondent's income. This is not always the case and non-linear relationships are possible, such as the one illustrated in Figure 13-1b.[134] Here, the collected data indicates that the consumption rate increases with income. In other words, the consumption rate of a respondent with a higher income is greater than the consumption rate of a respondent with a lower income.

---

[134] There are no values on the axes because this is a hypothetical example for illustration purpose only.

Figure 13-1b
Income and Consumption Data: Non-Linear Relationship

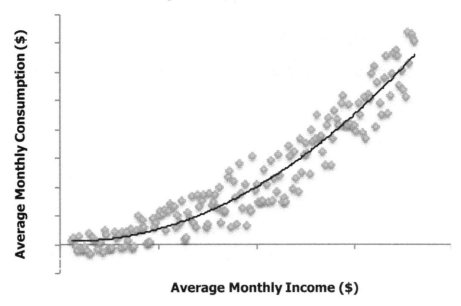

In this chapter, we restrict our focus to linear relationships, such as the relationship depicted in Figure 13-1a. We can use this linear relationship to obtain a better estimate of consumption than that obtained by using the computed average consumption value. We formulate this linear relationship between Income and Consumption using the following equation:

$$\text{Consumption} = \beta_0 + \beta_1 * \text{Income} + \varepsilon$$

where β_0 represents some base level of consumption (i.e., the amount you would consume when your monthly income is zero),[135] β_1 represents the relationship between income and consumption (i.e., the consumption rate), and ε is an error term, reflecting the reality that people with the exact same income are likely to choose different levels of consumption. Looking more closely at the meaning of β_1, β_1 represents the dollar change in consumption resulting from a one-dollar change

---

[135] We will elaborate on the interpretation of β_0 in Unit 3 of this chapter.

in income. For example, if $\beta_1=0.7$, then a one-dollar change in income would result in a +0.7 dollars (70 cents) change in consumption.

Recall that our objective is to estimate consumption. That is, we would like to know what we would expect a specific person (or a set of persons, on average) to spend on goods and services. Based on Figure 13-1a, we know that this amount likely depends on income. In order to estimate consumption based on income, we need to determine values for β_0, β_1, and ε. Regression analysis is the technique that allows us to estimate these values. More specifically, using regression analysis, we can compute estimates for β_0 and β_1. ε is a random error term that will remain unknown to us, as will be explained later in this unit.

We now direct our attention to understanding how β_0 and β_1 are estimated. Figure 13-2 displays the scatter of data points representing our collected consumption and income data, along with a superimposed line that represents the relationship between consumption and income. This upward-sloping line is the **regression line**, that here represents a linear relationship between Income (the independent variable) and Consumption (the dependent variable). This line is referred to as *upward-sloping* because higher Income values are associated with higher Consumption values. The key question you may ask is why did we choose this specific line and not others? Obviously, there are many other lines that we could pass through the cluster of data points shown in Figure 13-2. In estimating this regression line, our first requirement is that it be a straight (linear) line. This is because we are conducting **linear regression** analysis.

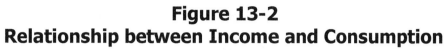
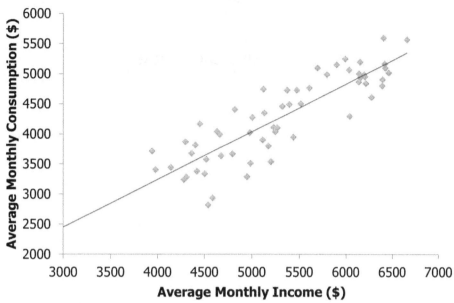

The specific regression method we apply in this chapter is called **ordinary least squares (OLS) regression**. This means that the regression line is the line that *minimizes* the sum of squared deviations between each data point and the regression line. As this concept is extremely important in OLS regression analysis, we need to explore it a bit further. Figure 13-3 depicts four vertical lines, each of which connects a data point to the regression line. Each of these lines illustrates the vertical differences between a data point and the regression line. For a given level of Income (the horizontal axis), the vertical line indicates the difference between *estimated* Consumption (the regression line) and *actual* Consumption (the data point): a *deviation from the regression line*. By squaring these values and then adding these numbers, we obtain the sum of squared deviations. Our regression line is chosen to be that line for which the sum of squared deviations is the smallest of all possible lines - hence, the regression line minimizes the squared deviations (you can also think of the regression line as the straight line that is simultaneously closest to all data points).

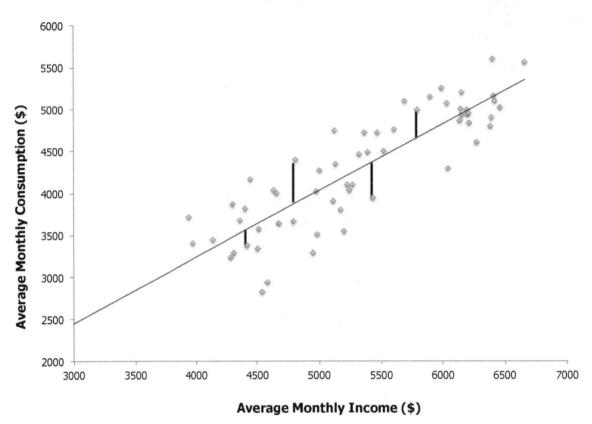

**Figure 13-3
OLS Regression Line**

Before we conclude, we will discuss the error term (ε) in more depth. The error is a random component that is not estimated by the regression line. Visually, the individual errors are represented in Figure 13-3 as the *difference* between each data point and the estimated regression line. Importantly, the regression line does not capture these errors. In our consumption example, these errors represent the differences in Consumption not being explained by Income, the independent variable – as when we observe that two people reporting the same monthly income also report that they spent different amounts of money on goods and services. (We will demonstrate later in this chapter how some of this error can be accounted for by incorporating additional independent variables into our analysis.)

These errors are assumed to be normally distributed around the regression line, with a mean of zero (0) and a constant standard deviation. That is, we do not expect to see any systematic error in the data (e.g., observing that all the data points are above or below the regression line). Further, the errors are assumed to be independent of one another; that is, one error does not affect the magnitude or direction of another error.

Unit 1 Summary

- **Regression analysis** is a statistical technique that (guided by theory) aims to estimate a **causal relationship** between a set of **independent variables** and a **dependent variable**. **Causality** implies that we believe that the values of the independent variables determine, to some extent, the value of the dependent variable.

- There are many different forms of relationships between variables. In this chapter, we focus on **linear regression**, which assumes a linear relationship between the variables.

- The specific technique we use is called **ordinary least squares (OLS) regression**, which minimizes the squared deviations between actual and estimated data values. The line representing this linear relationship is called the **regression line**.

Unit 1 Exercises

For Questions 1-6, identify the dependent variable and the independent variable (or variables) and comment on the validity of the implied causal relationship:

1. A study that looks at the effect of calorie consumption on weight gain.

2. A study that looks at students' college GPAs and how they are affected by the students' number of weekly study hours, high school GPA, and number of weekly hours of extra-curricular activities.

3. To study the return on investment from a university education, a student examines the relationship between annual tuition and the expected annual wage earned upon graduation.

4. A young pitcher wants to understand how she can improve her pitches. She collects data on the number of strikes thrown in a pitched game, the average number of practice pitches per day in the week leading up to a game, the number of hours spent working with her coach the week before the game, and the time spent warming up right before the game.

5. A start-up company is interested in improving the number of downloads of their smartphone apps. They consider three different advertisement campaigns: a *Facebook* ad, a pop-up ad in a related app, and a *Google* ad. They test the effectiveness of the campaigns for one month, collecting daily data on the number of app downloads and the number of views of each of the three ads.

6. A researcher proposes to study the effect of weight and height on a person's body mass index (BMI).

7. Consider the following dependent variables. Can you come up with independent variables that might be used in a regression model to estimate the value of each of these dependent variables?

    a. Daily number of website views

    b. Daily sales of a chocolate bar

    c. Price of personal computers

    d. Demand for the *iPhone 6s*

    e. Personal job satisfaction

UNIT 2

Conducting Regression Analysis

Suppose that we are interested in estimating the price (in dollars) of houses in Hilo, Hawaii. Based on our initial research about what is known about housing prices, we expect a house's price to be a function of the size of the house (measured in squared feet), the number of bedrooms, the number of bathrooms, and the age of the house (measured in years). We further believe that this relationship is linear.

We thus model the relationship between our dependent variable (Price) and the set of independent variables in a manner similar to that used with the income and consumption example from Unit 1:

$$\text{Price} = \beta_0 + \beta_1*\text{Size} + \beta_2*\text{Bedrooms} + \beta_3*\text{Bathrooms} + \beta_4*\text{Age} + \varepsilon$$

Or, taking the more general form used with a regression model:

$$Y = \beta_0 + \beta_1*X_1 + \beta_2*X_2 + \beta_3*X_3 + \beta_4*X_4 + \varepsilon$$

With this house prices example, X_1 is a variable measuring the size of the house, X_2 is a variable measuring the number of bedrooms, X_3 is a variable measuring the number of bathrooms, X_4 is a variable measuring the age of the house, and Y is a variable measuring the price of the house. As before, ε represents the error term.

We collect data on houses in our geographic area of interest (Hilo, Hawaii) to obtain our β estimators. These data are shown in this chapter's Excel file in the worksheet titled 'Hilo_Basic'.[136] We perform a regression analysis on this collected data set (with the help of statistical software) to obtain what is called the **_estimated regression model_**, or an equation defining the regression line. As has been the case with the other statistical techniques introduced in this book, when we move from the realm of parameters (actual values) to the realm of statistics (estimated values), we use slightly different notations. Specifically, we use $\hat{Y}$ to represent the estimated dependent variable, we replace the βs with lower case bs, and we no longer include the error term (ε), as we are unable to estimate it with the regression line. Hence, our estimated regression model for the house prices example is written as:

$$\hat{Y} = b_0 + b_1*X_1 + b_2*X_2 + b_3*X_3 + b_4*X_4$$

---

[136] Data were collected online for houses posted for sale in Hilo, Hawaii in November 2013.

We mentioned that, typically, regression analysis is conducted with the help of statistical software. What the software specifically does is find the set of **regression coefficients** (the β estimators that, respectively, represent the relationship between an independent variable and the dependent variable) that minimizes the sum of squared deviations.

While regression analysis can be conducted manually, it tends to be very labor intensive. Hence, we run our regression analysis in Excel, again using the *Analysis Toolpak*. From the 'Data' tab, select 'Data analysis' and then, from the menu that opens, select 'Regression'. You should see the input screen shown in Figure 13-4, which is then used to input the following information:

- Input:
    - Input Y Range: Select the cells A1:A42, which hold the data for the price of houses.
    - Input X Range: Select the cells B1:E42 which hold the data for the four independent variables.
    - Check the box titled 'Labels' to indicate that you have column titles in Row 1.
    - Leave unchecked the boxes for 'Constant is Zero' and 'Confidence Level'. We will explain these options later in the chapter.
- Output options:
    - Pick the options you prefer. Here, we selected to place the output table on the same worksheet.
- Residuals and Normal Probability:
    - Leave this blank for now. These options allow you to visually examine the distribution of the errors (residuals) and determine whether or not the assumptions about the error's mean and variance hold for this data set.
- Click 'OK'.

Figure 13-4
Regression Input in Excel: House Prices Example

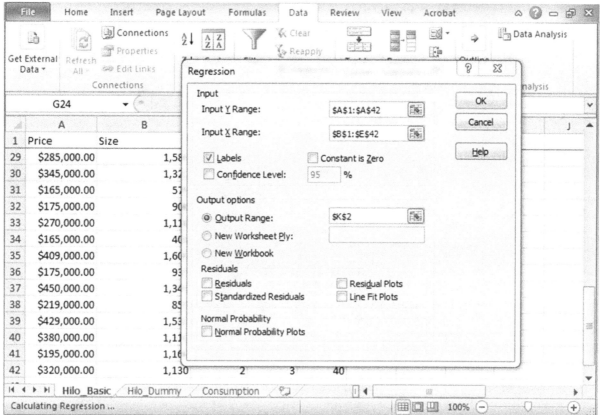

Figure 13-5 shows the resulting output table.[137] There are three parts to Figure 13-5, each of which will be described. The part of the output table likely to be of most immediate interest is the estimated regression model. The set of results at the bottom of Figure 13-5 display this estimated regression model. More specifically, the column labeled Coefficients provides the values of the computed estimate of the β's in the regression model. Hence, the estimated regression model:

$$\hat{Y} = b_0 + b_1 * X_1 + b_2 * X_2 + b_3 * X_3 + b_4 * X_4$$

---

[137] We omitted one part of the output in the bottom table that includes confidence intervals. We will add and address this part later in the chapter. We also rounded all numbers to three decimal places for presentation purposes.

can be re-written as:

$$\hat{Y} = -29414.478 + 2.682 \cdot X_1 + 56049.601 \cdot X_2 + 98811.693 \cdot X_3 - 790.94 \cdot X_4$$

**Figure 13-5
Regression Output in Excel: House Prices Example**

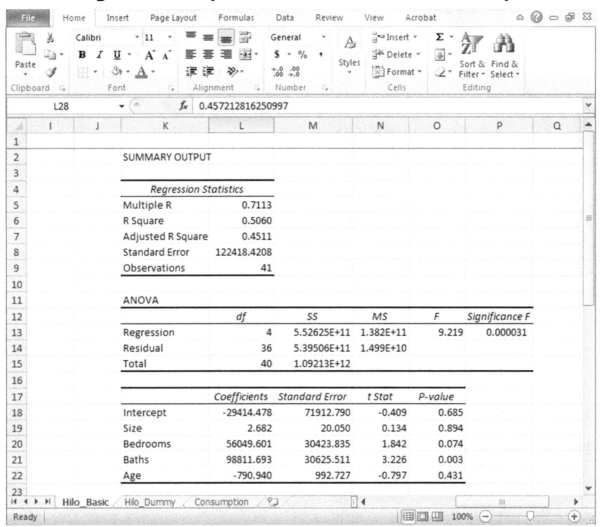

Given the estimated regression model from Figure 13-5, we now use the data provided in the 'Hilo_Basic' worksheet to substitute the X's in the above equation with actual data values, as illustrated in Table 13-1 for two of the data points. We also include in Table 13-1 the Estimated Price and Error Columns from the 'Hilo_Basic' worksheet.

Table 13-1
Actual Data: House Prices Example

Price (Y)	Size (sq. ft.) (X_1)	Bedrooms (X_2)	Baths (X_3)	Age (X_4)	Estimated Price $(\hat{Y})$	Error (ε)
$165,000	575	2	1	32	$157,728.32	(7,271.68)
$184,500	768	3	1	60	$192,149.17	7,649.17

Consider the two observations displayed in Table 13-1. Applying the regression model, we obtain:

$\hat{Y}_1 = -29414.478 + 2.682*575 + 56049.601*2 + 98811.693*1 - 790.94*32 = 157,728.32$

$\hat{Y}_2 = -29414.478 + 2.682*768 + 56049.601*3 + 98811.693*1 - 790.94*60 = 192,149.17$

The actual asking price for the house in Row 1 is $165,000, whereas our estimated price is $157,728.32. Therefore, there is a deviation (i.e., the error term) of -7,271.68 between the computed estimate and the actual price. You will see, in this chapter's Excel file, that averaging all of these errors will produce a value of zero, indicating that there is no bias in our estimate.

Of course, in this specific example we know the actual asking price and do not need to estimate it. However, we can use this same regression model to estimate any combination of values of our independent variables. For example, we would now expect that a house with 1,050 square feet, two bedrooms, two baths and 25 years in age would have an asking price of $263,350.39.[138] We will discuss more about how regression coefficients are used for prediction purposes in the next unit.

Beyond being used to compute an estimate of the dependent variable, these regression coefficients are interpreted as the change in the value of the dependent variable for a one-unit change in the independent variable. In other words, if the

---

[138] $\hat{Y} = -29414.478 + 2.682*1050 + 56049.601*2 + 98811.693*2 - 790.94*25 = $263,350.39$

value of an independent variable was to change by one (in whatever unit of measurement was applicable), then the value of the dependent variable would correspondingly change by an amount equal to the coefficient of the independent variable. For example, the coefficient of 56,049.601 for the variable Bedrooms in Figure 13-5 means that each additional bedroom adds an estimated $56,049.60 to the price of a house. Similarly, the negative coefficient of -790.94 for the Age variable means that the price of a house loses approximately $791 with each additional year of age. Remember that this $791 only represents the contribution of a house's age (that is, if you compared two identical houses in the exact same geographic area, but one of these houses was one year older than the other, then we expect the older house to be priced about $791 lower than the younger house).

Armed with this understanding of how to interpret the estimated regression model, we are ready to go over the other parts of the regression output table. These other outputs address specific tests conducted to evaluate the quality and significance of the estimated regression model.

The Coefficient of Determination (R^2)

At the top of Figure 13-5 is a set of results labeled 'Regression Statistics'. The first three values refer to three statistics that are related to one another. The first of these statistics to be discussed is the very important *R Square* statistic, formally referred to as the *coefficient of determination* and informally written as R^2. The **coefficient of determination** is a measure of the amount of variability in the dependent variable that has been explained by the independent variables. It is interpreted as a percentage. Hence, in our example, an R^2 value of 0.506 means that our regression model with four independent variables accounts for 50.6% of the variability in house prices in Hilo, Hawaii.

To understand how R^2 is computed, we need to step back and review the theory behind regression analysis. Recall that when we conducted ANOVA in Chapter 12, we looked at sums of squares as measures of variability being explained through statistical analysis (for example, with one-way ANOVA we looked at SSB, SSW and SST). We look at similar measures with regression analysis.

Before returning to our house prices example, we will use the simpler example of determining the relationship between consumption and income to explain these concepts. You may recall from this example that people spend different amounts of money on the consumption of goods and services. In the data that was collected on consumption and income, the overall variability in consumption (Y) is represented by its variance, the average squared deviations from Y's mean. In regression analysis, we specifically look at the *Total Sum of Squares (SST)*, measured by the distances between each pair of *actual-Y* and *mean-Y* data points:

$$SST = \sum (y - \bar{y})^2$$

We further know that some of this variability can be explained by the regression line. Specifically, we know that part of the variability in the amount that people spend on goods and services is due to their different income levels, represented by the upward-sloping regression line. We term this component of total variability the *Regression Sum of Squares (SSR)*, and we measure SSR by the differences between each pair of *estimated-Y* and *mean-Y* data points:

$$SSR = \sum (\hat{y} - \bar{y})^2$$

Finally, we have already mentioned that the least squares regression line is the one that minimizes the sum of the squared deviations from the regression line (or, the error term). The statistical term for this is the *Error Sum of Squares* (*SSE*), and SSE is measured by the distances between each pair of *actual-Y* and *estimated-Y* data points:

$$SSE = \sum (y - \hat{y})^2$$

Together, SSR and SSE comprise the total variability in the dependent variable, such that:

$$SST = SSR + SSE$$

Figure 13-6 should graphically help you to visualize what SST, SSR and SSE each represents. Here, we focus on the specific data point indicated by the arrow toward the bottom part of the graph. The total distance between this point and the horizontal dashed line representing average consumption would be included in the computation of SST (remember that SST is summed across all data points, whereas we are using a single data point here to illustrate these concepts). We can further split this distance into two parts: SSE, which represents the distance between the data point (an actual value of consumption) and the regression line (an estimated value of consumption); and, SSR, which represents the distance between the regression line (an estimated value of consumption) and the dashed line (the mean value of consumption). Remember, SSR represents the variation in consumption that we are able to explain using what we know about a person's income, while SSE represents the variation in consumption that we cannot explain.

Figure 13-6
Illustration of Sums of Squares

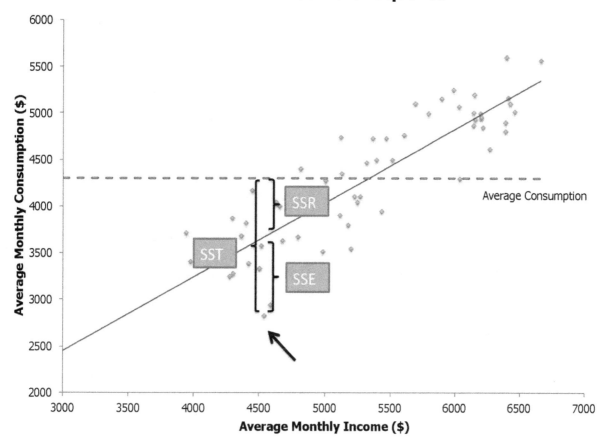

Returning to our house prices example and the coefficient of determination, look at the ANOVA table in the middle of Figure 13-5. The third column (labeled SS) of this table provides the values for SSR (first row), SSE (second row; note that *error* and *residual* are two different terms for the same concept), and SST (third row). R^2 is computed as

$$R^2 = \frac{SSR}{SST}$$

which in our example is:[139]

---

[139] The output indicates that these are very large numbers (E+11 means you need to move the decimal point 11 places to the right). We moved the decimal place for SST one place to the right to ensure that both SSR and SST have the same magnitude (now E+11), which gets canceled in the division.

$$R^2 = \frac{5.52625}{10.9213} = 0.506$$

In other words, SSR (the variability in house prices explained by the regression model) accounts for 50.6% of SST, the total variability in house prices.

A very reasonable question to ask at this point is, "Just how good is this R^2 value?" That is, are we satisfied with being able to explain 50.6% of the variability in housing prices? The answer largely depends on the dependent variable we are estimating and how much of the variable's variance is typically explained with regression analysis. In order to obtain a high R^2 value, the dependent variable being estimated must be well-defined and its potential predictors must be well understood and measurable. As this is typically the case with regression models used in estimating economic and financial variables, we often see R^2 values greater than 0.9 (meaning that more than 90% of the variance in the dependent variable is being explained). In studying human behavior, on the other hand, much lower R^2 values are typically obtained. How does one define variables such as happiness, job satisfaction or work-related stress? In thinking about happiness, what variables are likely to determine a person's happiness, what is the relative importance of each of these variables, and just how measurable is each? Because of the complexities that arise in studying human behavior, relatively low values of R^2 are typically observed, with R^2s of 0.2 or 0.3 considered as being reasonable. Remember, if you obtain a low R^2 value in estimating a dependent variable, you can always try to add additional, wisely-chosen independent variables to improve your R^2. In fact, we will do so for the house prices example data later in this chapter.

Before concluding this discussion of the coefficient of determination, the two statistics related to R^2 that are included in the Regression Statistics part of the output need to be described. First, **Multiple R** is the correlation coefficient between the *estimated-Y* values and the *actual-Y* values. It is computed as the squared root of R². **Adjusted R²** is another important statistic. Often, we add additional independent variables to a regression model in the expectation that the model's R² will increase (to explain a greater percentage of the variation in the dependent variable). If these added variables are not very good predictors of the dependent variable, then R² will not increase by very much, if at all. At the same time, as we add independent variables to a regression model, we are losing some degrees of freedom. Hence, there is a cost-benefit consideration that needs to be made when adding independent variables to a regression model. The adjusted R² statistic takes such considerations into account. We will highlight differences between R² and adjusted R² later in this chapter as we introduce different approaches used in regression analysis.

The Standard Error of the Estimate (S$_\varepsilon$)

The remaining statistic provided in the Regression Statistics results displayed at the top of Figure 13-5 is labeled Standard Error. The **standard error of the estimate** (S$_\varepsilon$) measures *how* the data points vary around the regression line (that is, do the data points tightly bunch around the regression line or are the data points more dispersed?). S$_\varepsilon$ is computed as:

$$S_\varepsilon = \sqrt{\frac{SSE}{n - k - 1}}$$

where SSE is obtained from the ANOVA table (middle row), n is the number of observations shown in the Regression Statistics table, and k is the number of independent variables (four in our example). With this house prices example, thus, S_ε is calculated as follows:

$$S_\varepsilon = \sqrt{\frac{539506000000}{41 - 4 - 1}} = 122418.4$$

The standard error of the estimate is important because the regression line reflects a series of point estimates for the dependent variable, at given levels of the independent variables. By knowing the standard error of the estimate, we can construct confidence intervals around the regression line to improve our estimate and predictions. We explain how to compute these confidence intervals in Unit 3.

The ANOVA Test

The second part of the regression output table shown in Figure 13-5 is an ANOVA table. We have already discussed the sums of squares reported in this table and we will now explain the remaining terms. Since we are conducting an ANOVA test, we need to know what hypothesis we are testing. The set of hypotheses being tested is:

H_0: $\beta_1 = \beta_2 = \beta_3 = \beta_4 = 0$

H_1: At least one of the (independent variable) coefficients differs from zero.

The null hypothesis states that all of the regression coefficients have a value of zero. (Note that we have as many β's in the null hypothesis as we have independent variables.) Recall that the relationship we are estimating is:

$$Y = \beta_0 + \beta_1 * X_1 + \beta_2 * X_2 + \beta_3 * X_3 + \beta_4 * X_4 + \varepsilon$$

If you replace β_1 through β_4 with a value of zero, then what remains is:

$$Y = \beta_0 + \varepsilon$$

Stated in words, this expression implies that Y is a function of a base value (β_0) and a random error term. Thus, if the null hypothesis is not rejected, we would conclude that none of the specified independent variables significantly influences the value of the dependent variable.

The decision whether or not to reject the null hypothesis is based on a testing procedure similar to that used with ANOVA (discussed in Chapter 12). From the ANOVA table in the middle of Figure 13-5, we are given a test statistic (F) value of 9.219 and a p-value (Significance F) of 0.000031. As per our usual decision rule, we reject the null hypothesis for any p-value that is lower than our selected level of significance, α. With a p-value of 0.000031, we would reject the null hypothesis even at the low 1% level of significance. Thus, we conclude that at least one of our independent variables significantly affects the dependent variable. To find out which of the specified independent variables does affect the dependent variable, we need to look again at the bottom part of the regression output table.

The Regression Coefficients

Statistics regarding the regression model's estimated coefficients are provided at the bottom of the output table (replicated below as Table 13-2), along with the estimated values of the coefficients. We already examined a part of this table earlier in discussing how the output should be interpreted in order to estimate the dependent variable. We now focus on the other results provided about these coefficients.

Table 13-2
Regression Coefficients: House Prices Example

	Coefficients	Standard Error	t Stat	P-value
Intercept	-29414.478	71912.790	-0.409	0.6849
Size	2.682	20.050	0.134	0.8943
Bedrooms	56049.601	30423.835	1.842	0.0737
Baths	98811.693	30625.511	3.226	0.0027
Age	-790.940	992.727	-0.797	0.4308

The first column of Table 13-2 provides the names of the independent variables. Each row provides measures and statistics for one of these independent variables. The top row refers to the Intercept, or b_0. It represents some base level of the dependent variable that holds when all the independent variables are set to values of zero. Mathematically, it is the location where the regression line meets, or intercepts, the y-axis. We will discuss how the intercept can and should be interpreted in Unit 3. Next are four similar rows, one for each of the four independent variables. The first number in each row (in the second column labeled Coefficients) provides the b values that we previously explained. Since each coefficient is an estimated value, the third column, labeled Standard Error, provides the standard error associated with each estimated coefficient. Finally, the fourth (t Stat) and fifth (P-value) columns are used in determining whether or not the relationship between an independent variable and the dependent variable is significant.

Specifically, each row displayed in Table 13-2 represents a hypothesis test of the following form:

$$H_0: \beta_i = 0$$

$$H_1: \beta_i \neq 0$$

where *i* is an index used to indicate each independent variable (in our example, *i* would take on the values of 1, 2, 3 and 4 since we have four independent variables).[140] From the discussion given above regarding the ANOVA test, we know that at least *one* (and, possibly more) of our independent variables is significant. But, the question remains, "Which independent variable or variables significantly affect the value of the house prices, our dependent variable?"

The answer to the above question can be reached using a t-test on the above hypotheses, with values for the test statistic and p-value provided in the regression output used to conduct the test. The test statistic (t Stat) is computed similar to previous hypothesis tests that we have conducted:

$$t = \frac{b - \beta}{S_b}$$

where b is the estimated regression coefficient, β is the hypothesized coefficient (zero according to our null hypothesis), and s_b is the standard error (one of the measures provided in Table 13-2 for each of the estimated coefficients). Thus, we compute the test statistic for the variable Baths as:

$$t = \frac{98811.693 - 0}{30625.511} = 3.226$$

and we could find the p-value associated with this t-value using Excel. Of course, we do not have to manually go through these steps as these values are already provided in the regression output. With a low p-value of 0.0027, we reject the null hypothesis that $\beta_{baths}=0$ and conclude that there is a significant relationship between the number of bathrooms and the asking price of a house.

---

[140] The test is also conducted for the intercept, but, as mentioned earlier, we are delaying our discussion of the intercept until Unit 3.

Browsing over the remaining results shown in Table 13-2, you can see that only one other independent variable, Bedrooms, is a significant predictor of house prices. However, with a p-value of 0.0737, this variable (representing the number of bedrooms) is only significant at the 10% level of significance.

Improving the Model

One means for improving an estimated regression model involves removing variables that are not significant. By removing variables that are highly non-significant, we may be able to improve indicators such as the adjusted R^2, the significance of the model as a whole (the F-statistic and its p-value), and possibly the significance levels of the remaining independent variables. Unlike the adjusted R^2, we expect R^2 to decrease whenever we remove any variable from the regression; however, this reduction should be minimal when removing non-significant variables since non-significant variables do not contribute much to our ability to explain the dependent variable. Specifically with this house prices example, it may be worthwhile to consider removing the variables Size and Age.

The process is quite simple, as one variable at a time is removed. After deciding on a variable to remove, we then run the regression analysis again in Excel, but now without the column of the omitted-variable's data.[141]

In deciding on which variable to remove first, we can examine the candidate variables' levels of significance. With the house prices example (see Table 13-2), it can be seen that Size is less significant than Age because Size has a lower test statistic than does Age. Thus, Size would seem to be the best variable to consider removing.

---

[141] Often, sophisticated statistical software packages automatically do these types of model adjustments for us, such that the regression model does not have to be re-specified and re-run multiple times.

Another approach to identify candidate variables for removal involves examining the *correlation matrix* for all of the independent variables used in a regression model. A **correlation matrix** is a matrix showing all pair-wise correlations in a set of variables (the correlation matrix for the house prices example is provided in Table 13-3). In order to obtain a correlation matrix in Excel, choose 'Correlation' from the 'Data analysis' menu and then select the relevant range of independent variables in your worksheet. Because correlations are symmetric (that is, the correlation of variables A and B is the same as the correlation of B and A), only the lower half of a correlation matrix is typically provided. Note that since each variable is perfectly correlated with itself, values of 1 are placed along the diagonal.

Table 13-3
Correlation Matrix: House Prices Example

	Size	*Bedrooms*	*Baths*	*Age*
Size	1			
Bedrooms	0.457212816	1		
Baths	0.406752682	0.565179961	1	
Age	0.181089714	0.199213507	-0.120438941	1

As a rule of thumb, a correlation greater than 0.7 is considered high; and, including two highly-correlated independent variables in a regression model is likely to introduce errors that need to be fixed. This problem is called **multicollinearity**. When multicollinearity exists (luckily it does not exist in our example, as none of the variables are highly correlated with other variables), the effect of one variable on the dependent variable is masked by its correlation with other variables. More formally, multicollinearity weakens the effect of the independent variables on the dependent variable by artificially inflating the standard error of these variables, thus lowering the values of test statistics.

Going back to our house prices example, the pair-wise correlation between the independent variables are within reason, though Size is more correlated with the other variables than is Age. Given that Size is also the least significant of the four independent variables, we remove Size from our input range and re-run the regression analysis.

The results from re-running the regression analysis without the Size variable are shown in Figure 13-7. Comparing these results (with three independent variables) to the previous results (with four independent variables, replicated in Figure 13-8), you will see that the R^2 value does not change (since the value is rounded to three decimal places, there is no visible change). However, the value of adjusted R^2 has improved from 0.451 to 0.466. Similarly, in the ANOVA table, the F-statistic has increased (becoming more significant), again indicating that this three-variable model is an improvement over the four-variable model. Finally, both Bedrooms and Baths have become more significant (seen by comparing the values for the test statistic and the p-value with both sets of results), indicating, again, that the decision to remove the Size variable was a good decision.

Figure 13-7
Regression Output: House Prices Example, Removing Size

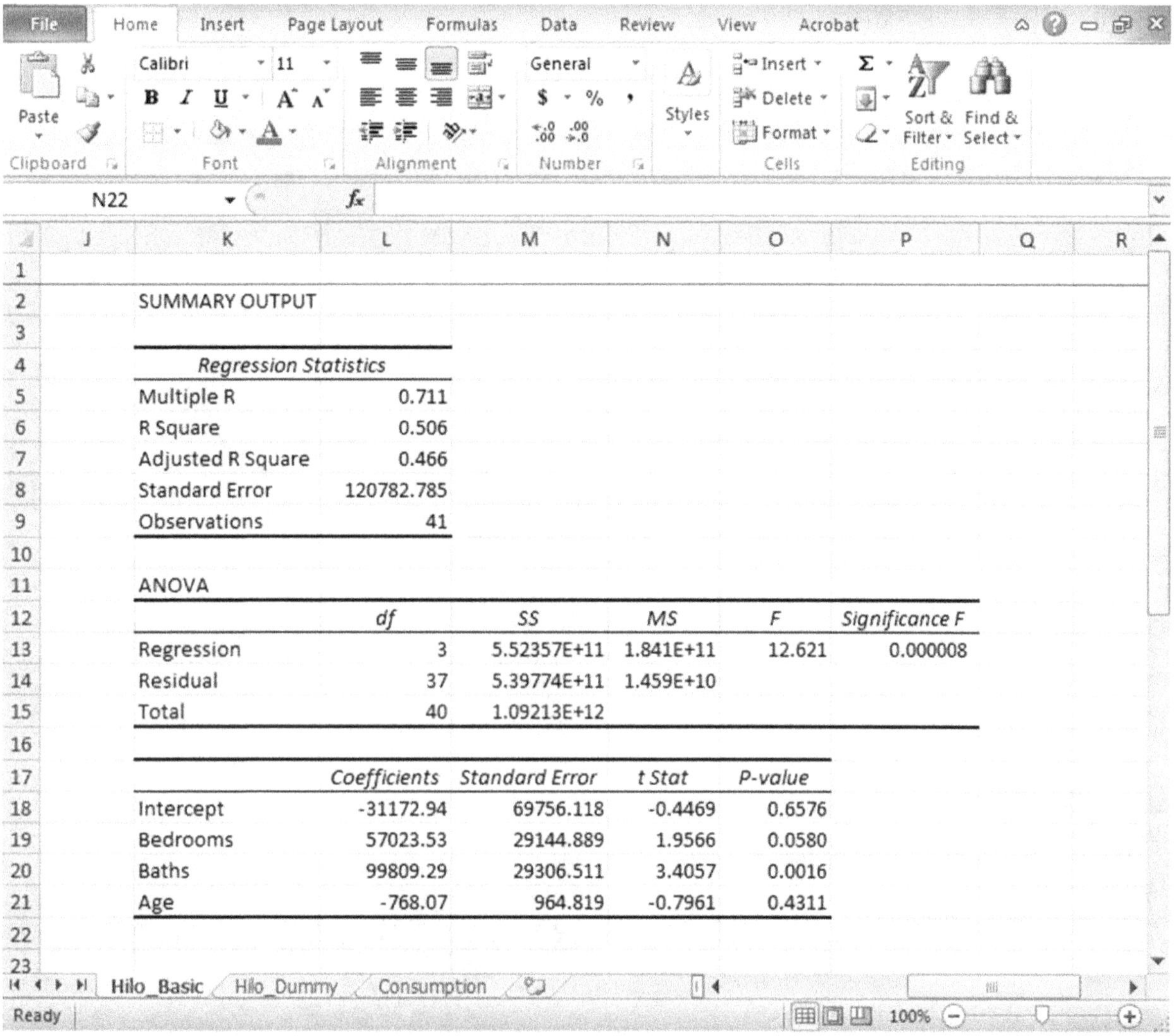

SUMMARY OUTPUT

Regression Statistics

Multiple R	0.711
R Square	0.506
Adjusted R Square	0.466
Standard Error	120782.785
Observations	41

ANOVA

	df	SS	MS	F	Significance F
Regression	3	5.52357E+11	1.841E+11	12.621	0.000008
Residual	37	5.39774E+11	1.459E+10		
Total	40	1.09213E+12			

	Coefficients	Standard Error	t Stat	P-value
Intercept	-31172.94	69756.118	-0.4469	0.6576
Bedrooms	57023.53	29144.889	1.9566	0.0580
Baths	99809.29	29306.511	3.4057	0.0016
Age	-768.07	964.819	-0.7961	0.4311

Figure 13-8
Initial Regression Output: House Price Example

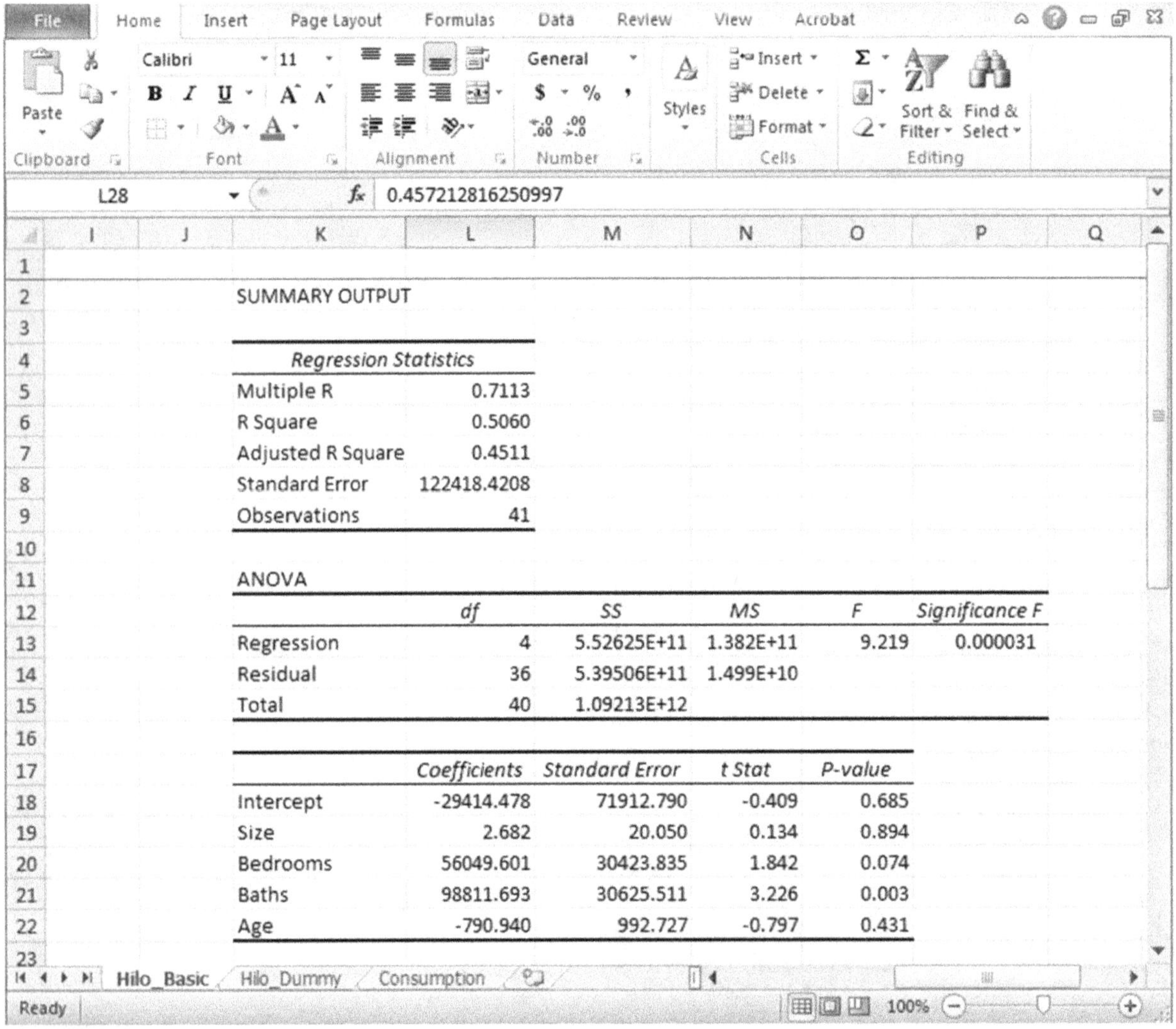

SUMMARY OUTPUT

Regression Statistics

Multiple R	0.7113
R Square	0.5060
Adjusted R Square	0.4511
Standard Error	122418.4208
Observations	41

ANOVA

	df	SS	MS	F	Significance F
Regression	4	5.52625E+11	1.382E+11	9.219	0.000031
Residual	36	5.39506E+11	1.499E+10		
Total	40	1.09213E+12			

	Coefficients	Standard Error	t Stat	P-value
Intercept	-29414.478	71912.790	-0.409	0.685
Size	2.682	20.050	0.134	0.894
Bedrooms	56049.601	30423.835	1.842	0.074
Baths	98811.693	30625.511	3.226	0.003
Age	-790.940	992.727	-0.797	0.431

We can see from Figure 13-7 that the variable Age is still not significant. Thus, we should consider removing the variable. However, decisions on removing nonsignificant independent variables from a regression model should not be based just on statistical results. If we felt strongly that Age should remain in our regression model (perhaps because of the results of analyses performed by others, or because of a personal understanding of the Hawaiian housing market, or because of theories proposed about housing prices in general), then we should keep the variable Age in the regression model. Then, when summarizing our findings, we would explain that, although there are strong arguments for Age having a significant influence on house prices, we could not find support for these arguments in the statistical results based on our specific data set.

In summary, when independent variables are found *not* to have a significant effect on a dependent variable, we should consider removing these variables in order to improve our estimated regression model. In deciding which specific variables to remove, a number of factors should be considered: the arguments that guide our thinking, the nature of the relationships that exist among the variables in our data set, and the expected improvement from removing a variable (as determined by examining the statistics produced from already-run regression analyses).

Unit 2 Summary

- Regression analysis estimates the relationship between a set of independent variables and a dependent variable. The resulting **estimated regression model** is the mathematical representation of the *regression line*. It displays the estimated *regression coefficients* for each of the independent variables.

- **Regression coefficients** (the β estimators) represent the relationship between each independent variable and the dependent variable.

- The **coefficient of determination** (R^2) is a measure of the variability in the dependent variable that can be explained by the independent variables. **Multiple R** is the correlation coefficient between the estimated-Y values and the actual-Y values and is computed as the squared root of R^2. **Adjusted R^2** is a measure of the amount of variation in the dependent variable being explained by a regression model's independent variables *after* accounting for the degrees of freedom lost by using multiple independent variables. Adjusted R^2 is the preferred statistic to examine when a regression model has more than one independent variable.

- The **standard error of the estimate** ($S\varepsilon$) measures how the data points vary around the regression line.

- The ANOVA table tests the null hypothesis that all of the coefficients of the independent variables are equal to zero:

  $H_0: \beta_1 = \beta_2 = \beta_3 = ... = 0$
  H_1: At least one of the coefficients differs from zero.

  We look at the significance of the F-statistic to decide whether or not to reject this null hypothesis. If we reject it, then we conclude that at least one of our independent variables significantly affects the value of the dependent variable.

- The significance of each estimated coefficient is assessed through a t-test, testing a null hypothesis that a coefficient is equal to zero:

  $H_0: \beta_i = 0$
  $H_1: \beta_i \neq 0$

  If we reject the null hypothesis (i.e., the p-value is sufficiently low), then we conclude that there is a significant relationship between the independent variable and the dependent variable.

- When two independent variables are highly-correlated, we say that **multicollinearity** exists. A straightforward way to identify suspected multicollinearity is to look at the **correlation matrix**, which is a matrix showing all pair-wise correlations associated with a set of variables. High correlation values (e.g., values greater than 0.7) indicate possible multicollinearity. Multicollinearity is not desired in regression analysis because it distorts the effects of the independent variables on the dependent variable.

Unit 2 Exercises

1. Do socio-economic conditions - as indicated by a country's per capita gross domestic product (GDP) - predict Internet usage?[142] Run a regression analysis to answer this question, using the data in the worksheet titled 'Internet Users' in this chapter's Excel file.

2. What affects the price of laptop computers? The worksheet titled 'Laptops' in this chapter's Excel file contains data collected from four online retailers on the price and characteristics of laptops in January 2014. Use the variables Monitor, Weight, Processor, Memory, and HD as independent variables and the variable Price as the dependent variable.

    a. Which independent variables are significant and what is each significant variable's level of significance (10%, 5% or 1%)?

---

[142] Source: http://databank.worldbank.org/

b. Create a correlation matrix to identify highly-correlated independent variables. Then, remove any such variables from the model and run the regression again. How have the results changed?

3. A researcher at the University of Denver investigated whether babies take longer to learn to crawl in cold months (when they are wrapped in many layers of clothes) than in warmer months.[143] Data were collected on 208 boys and 206 girls, or a total of 414 babies. Organizing the collected data by birth month, the researchers computed for the babies born each month: the average age at which the babies began to crawl (in months), and the average weather temperature six months after their birth month (the age at which babies typically begin to crawl). These computed data are reported in the worksheet titled 'Babies' in this chapter's Excel file. Based on these data, is there a significant causal relationship between temperature and crawling age?

4. Is the size of your brain an indicator of your mental capacity? In a study using college students as subjects, researchers recorded data on students' gender, height, weight, and three different IQ test scores.[144] They also conducted MRI scans and counted the number of pixels in a brain image as a measure of brain size. The full data set is in the worksheet titled 'Brain' in this chapter's Excel file. Estimate three linear regression equations using the MRI pixel count as a single independent variable, but vary with each respective regression model the type of IQ scores being used as the dependent variable. What did you learn from this analysis?

5. A hypothetical economist wishes to determine how a household's yearly spending on groceries (measured in hundreds of dollars) is influenced by yearly household income, household size, and the education level of the head of the household. Income is measured in thousands of dollars, Size is measured by number of household members, and Education is measured in years. The economist randomly selected 50 households and conducted a regression analysis on the collected data. The regression coefficients resulting from the regression analysis are shown in Table 13-e1.

    a. What is the effect of an additional $10,000 in yearly income on spending?

    b. Which of the independent variables is not significant at the 5% level?

    c. An individual with an income of $100,000, a five-people household, and eighteen years of education is spending $500 each month on groceries. What is the error (ε), in hundreds of dollars, for this data point?

---

[143] Source: http://lib.stat.cmu.edu/DASL/Stories/WhendoBabiesStarttoCrawl.html
[144] Source: http://lib.stat.cmu.edu/DASL/Stories/BrainSizeandIntelligence.html

Table 13-e1
Regression Coefficients Table

	Coefficients	Standard Error	t Stat	P-value
Intercept	-0.16	0.58	-0.276	0.783814
Income	0.45	0.11	4.091	0.000160
Size	0.43	0.08	5.375	0.000002
Education	0.065	0.043	1.512	0.137052

6. A researcher has collected data on the weekly numbers of questions and answers posted to a telecom company's online customer support forum.[145] These data are provided in the 'Online Community' worksheet in this chapter's Excel file. The researcher suspects that a causal relationship exists between the number of posted questions and the number of posted answers. Specifically, the researcher believes that a greater number of posted questions will generate a greater number of posted answers.

   a. Run a regression analysis to assess the researcher's claim. Do your regression results support this claim?

   b. Do you think the causal relationship can go in the other way (i.e., the more answers that are provided result in more questions being asked)? Explain your answer.

UNIT 3

Qualitative Independent Variables

Suppose that we wish to increase the amount of variation in house prices that is being explained by our regression model, i.e., increase the model's R^2. Rather than (or in addition to) removing independent variables that are not significant, we can add new independent variables that we believe have an influence on house prices. For example, in Hilo, Hawaii, it might be expected that house prices would be higher for waterfront properties. Whether or not a house is a waterfront property is a qualitative variable, since this variable reflects a simple

---

[145] Lu, Y., Singh, P., and Sun, B. 2011. "Learning from Peers on Social Media Platforms," Proceedings of the 2011 *International Conference on Information Systems*.

yes/no answer rather than being numerical in value. To incorporate qualitative variables into regression analysis, we use what are termed *dummy variables*. A **dummy variable** is a special type of variable that only has a value of one or zero. In our house prices example, the dummy variable Waterfront will receive a value of one if a property is on the waterfront and a value of zero otherwise. Note that in this specific example, *otherwise* precisely means that the house is *not* on the waterfront - there are no other alternatives.

The worksheet titled 'Hilo_Dummy' in this chapter's Excel file includes this Waterfront dummy variable. The results for performing a regression analysis on this new data set are shown in Figure 13-9. For this specific example, we have only used the two significant variables from our prior analyses: Bedrooms and Baths. We start by examining the quality of our estimated regression model. Contrasting the results shown in Figure 13-9 with those obtained earlier (refer back to Figure 13-7) for the regression model without the Waterfront dummy variable, you can see that adding the Waterfront dummy variable has increased the regression model's R^2 to 0.659. Adjusted R^2 has also increased, further indicating that the Waterfront variable does improve our ability to explain the variance in house prices. Looking at the ANOVA table, the slight increase in the F-statistic indicates that the model, as a whole, has improved. Finally, after examining the individual coefficients, we see that Baths remains significant, but Bedrooms has lost significance (though it is nearly significant at the 10% level). This sometimes happens when you add (or remove) variables from a regression model because, even with no multicollinearity present, certain variables in a model may be correlated (perhaps waterfront properties in Hilo, Hawaii tend to be larger homes, with more bedrooms).

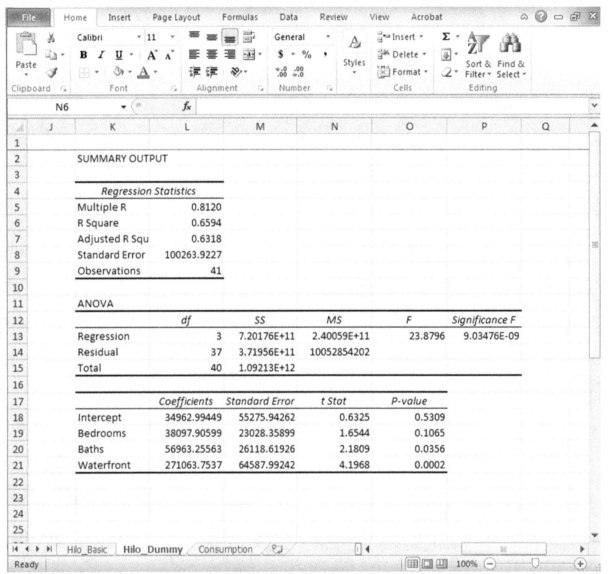

Figure 13-9
Regression Output with Dummy Variables

Recall that when examining regression coefficients, we interpret them as *the change in the dependent variable due to a one-unit change in the independent variable, holding all else constant*. In the case of dummy variables that only take on values of 1 or 0, this interpretation is slightly different. With this house prices example, the coefficient of the dummy variable represents the mean change in price between a waterfront home and a non-waterfront home. Specifically, if we examined two identical houses (here, identical in terms of the number of bedrooms and number of bathrooms), we would expect to find that the waterfront property would be listed for $271,063.75 *more* than the non-waterfront property.

Now, suppose that we add a second qualitative variable to indicate a property's geographic location. Suppose further that there are three major housing areas, which we label A, B and C. Note that this new variable can take one of three distinct values, rather than just the two distinct values (indicated by a 1 or a 0) associated with the Waterfront variable. Since we cannot input textual values for independent variables, intuitively you might suggest something like: "Why not give the three housing areas, respectively, values of 1, 2 and 3 when performing a regression analysis?" The problem with this suggestion is that numerical values will be interpreted by the regression software *as interval-scaled numbers*. For example, if we were to assign area A the value of 1 and area C the value of 3, then the regression analysis analytical procedures would interpret area C as being *two more* than area A, which doesn't really make much sense. What is needed is a way to extend the dummy variable approach beyond simple yes/no situations.

The solution is to use more than one dummy variable. Specifically, if we have a variable with n categories, we use n-1 dummy variables. Thus, in order to incorporate *three* housing areas into the house prices regression model, we would use *two* dummy variables:

D_1: takes on a value of 1 if the house is in area A and a value of 0 otherwise

D_2: takes on a value of 1 if the house is in area B and a value of 0 otherwise

We do not include a D_3 dummy variable because a house not in areas A or B, by default, is in area C.

Let us review a second, this time hypothetical, example to firm up your understanding of dummy variables and their interpretation. Suppose that we are estimating the effects of level of education and of years of experience on a person's annual salary. The dependent variable is thus Salary and the two independent variables are Experience (quantitative) and Education (categorical). We further note that a person may be exhaustively described by one of four levels of education: (1) completed grade school only, (2) completed high school, (3) completed a college degree, and (4) completed a graduate degree. In other words, Education is a qualitative (categorical) variable with *four* possible values. We represent this independent variable with *three* dummy variables:

- D_1: 1 if a person's highest education level is a *graduate* degree, otherwise 0
- D_2: 1 if a person's highest education level is a *college* degree, otherwise 0
- D_3: 1 if a person's highest education level is a *high school* degree, otherwise 0

By default, if D_1, D_2, and D_3 are each zero, then we know the person's highest education level is grade school.

Suppose that we collect a sample of data and estimate the following regression model:

Salary = 20,000 + 2,500*Experience + 15,000*D_1 + 5,000*D_2 + 1,000D_3

Now, consider four people, each with *five years of experience* but with *differing education levels.*

The first person, Joe, has only completed grammar school. With five years of experience, Joe would be expected to earn:

$$20{,}000 + 2{,}500*5 + 15{,}000*0 + 5{,}000*0 + 1{,}000*0 = \$32{,}500$$

Note that, in estimating Joe's salary with the regression model, we assigned values of zero for D_1, D_2 and D_3 because Joe only completed grammar school.

Now consider Jane. Jane has a high school diploma and will thus earn an estimated salary of:

$$20{,}000 + 2{,}500*5 + 15{,}000*0 + 5{,}000*0 + 1{,}000*1 = \$33{,}500$$

In Jane's case, we have set values of zero for D_1 and D_2 and a value of one for D_3.

The next person is Mike. As Mike has a college degree, we would expect Mike to earn a salary of:

$$20{,}000 + 2{,}500*5 + 15{,}000*0 + 5{,}000*1 + 1{,}000*0 = \$37{,}500$$

You can see we have set Mike's Education dummy variables as a value of zero for D_1, a value of one for D_2, and a value of zero for D_3.

Finally, Laura has a graduate degree. We thus expect her salary to be:

$$20{,}000 + 2{,}500*5 + 15{,}000*1 + 5{,}000*0 + 1{,}000*0 = \$47{,}500$$

As you may have expected already, we have assigned a value of one for D_1 and values of zero for D_2 and D_3.

Let us now contrast these four cases to better understand how to interpret dummy variables. Joe (our grammar school graduate) is our **base level** estimate - the level for which all three dummy variables are set to values of zero. The coefficients for the remaining three dummy variables are then interpreted with respect to this base level. Specifically, someone who has completed high school is expected to earn $1,000 *more than someone with only a grammar school education*. Someone who has completed college is expected to earn $5,000 *more than someone with grammar school education* (or $4,000 *more than someone with* *a high school diploma*). Finally, someone who has completed a graduate degree is expected to earn $15,000 *more than someone with grammar school education* (or $10,000 *more than someone with a college degree* and $14,000 *more than someone with a high school diploma*).

Understanding the Intercept

Up to this point in our treatment of regression analysis, we have acknowledged the existence of the regression intercept (β_0) but have not explicitly discussed its meaning. We will now more closely examine the role of the intercept in a regression model. We begin by using the consumption/income example introduced in Unit 1. Looking at the data provided in the worksheet titled 'Consumption' in this chapter's Excel file, the minimum value of the independent variable Income is $3,941 and the maximum value is $6,658. These two values define the *range* of our data. Figure 13-10 shows the scatter plot of our consumption and income data, along with the regression line and two vertical dashed lines representing the range of the data.

**Figure 13-10
Income and Consumption Data: The Ranges of Our Predictions**

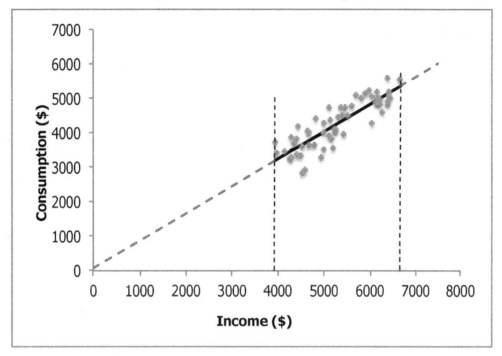

As you now understand, we use the regression line to estimate the value of the dependent variable (Consumption) for different possible values of the independent variable (Income). When these estimates are based on values of the independent variable that are within the range of our data, we are ***interpolating***, or estimating within the known range of the independent variable. These estimates are shown in Figure 3-10 by the solid line between the minimum income value of $3,941 and the maximum income value of $6,658. When we attempt to estimate values for a dependent variable based on values for the independent variable that are outside of its range (say, for incomes of $1,800 per month or $7,000 per month), we are ***extrapolating***, or making a prediction beyond our known range of the independent variable. This is illustrated through the dashed lines extending below the minimum income value of $3,941 and above the maximum income value of $6,658 in Figure 13-10.

Extrapolating is considered to be less certain than interpolating because we have no factual knowledge on the relationships among the variables in a regression model *beyond* the range of our data. Stock market data provide a nice example of why we should be cautious in the predictions we make. Figure 13-11a shows a screen shot taken from *Yahoo! Finance* of the *NASDAQ* composite index values from 1990 until March 2000. Extrapolating from this data, we would certainly expect to see gains from *NASDAQ* stock investments made after March 2000. However, as Figure 13-11b clearly shows, predictions outside the range of our data (in this case, predictions for time periods after March 2000) can be highly uncertain. As can be seen from Figure 13-11b, *NASDAQ* stocks were especially hard hit by the Spring 2000 dot.com bust.

Figure 13-11a
NASDAQ Composite Values Prior to March 2000

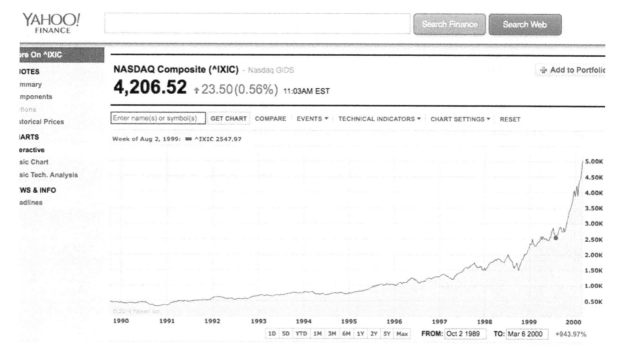

Figure 13-11b
NASDAQ Composite Values from 1990 until 2013

We now return to our discussion of a regression model's intercept. Recall that the intercept is the value taken on by the dependent variable when a regression model's independent variables are each assigned a value of zero. If a value of zero is not within the range of our independent variable(s) (as is the case for the consumption/income example), then we are essentially extrapolating the value of the intercept. Also, in many situations, a value of zero for the independent variable just does not make sense. While it is somehow possible to imagine that a person with no income would still have some level of consumption (i.e., the level needed to survive), what does the intercept mean in a regression that uses Age to estimate Time Spent Online? Literally, it means the time spent online by a person whose age is zero - which does not make any real-world sense.

So, why do we typically have an intercept term in regression equations? Actually, it is possible in Excel to remove the intercept. If you refer back to Figure 13-4 (that depicted an input screen for regression analysis), you will see that the option exists to check 'Constant is Zero' and force the regression line to go through the origin of the scatter plot chart (where the value of each variable in a regression model is set to zero), thus eliminating the estimation of the intercept. The problem with doing this is that *forcing* the regression line to go through the origin changes the *slope* of the estimated regression line. As a result, it is very likely that the resulting regression line will not satisfy the *least-squares requirement*. For this reason, we allow for an estimated intercept, but, usually, do not interpret this intercept in analyzing results.

Confidence and Prediction Intervals

Recall from Chapter 8 that a *confidence interval* is an interval constructed around a sample statistic to improve our estimate of the parameter. Since a regression analysis provides estimated coefficients for the independent variables (the b_i values) and the intercept (b_0) from a single sample, we can create confidence intervals for the true coefficients (the β values). In fact, *Lower 95%* and *Upper 95%* confidence interval values are provided with the Excel output for regression analysis (shown in the bottom right-hand corner of Figure 13-12).

Figure 13-12
Regression Output with Confidence Intervals

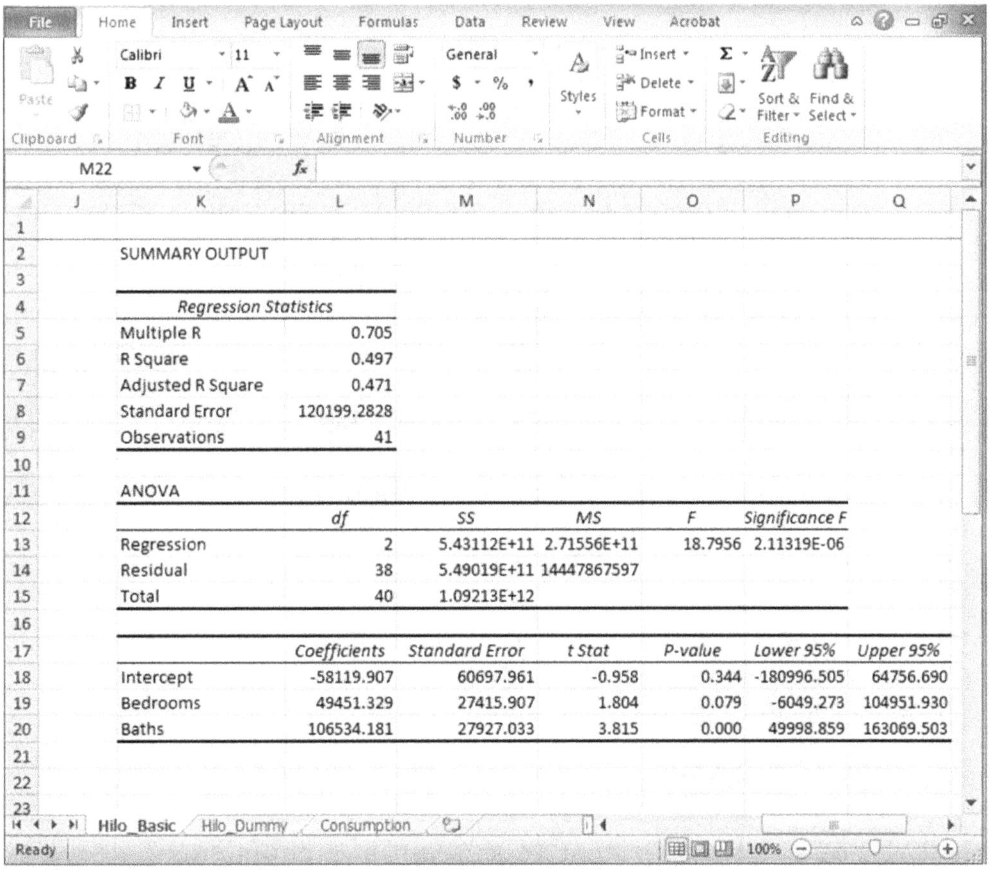

SUMMARY OUTPUT

Regression Statistics

Multiple R	0.705
R Square	0.497
Adjusted R Square	0.471
Standard Error	120199.2828
Observations	41

ANOVA

	df	SS	MS	F	Significance F
Regression	2	5.43112E+11	2.71556E+11	18.7956	2.11319E-06
Residual	38	5.49019E+11	14447867597		
Total	40	1.09213E+12			

	Coefficients	Standard Error	t Stat	P-value	Lower 95%	Upper 95%
Intercept	-58119.907	60697.961	-0.958	0.344	-180996.505	64756.690
Bedrooms	49451.329	27415.907	1.804	0.079	-6049.273	104951.930
Baths	106534.181	27927.033	3.815	0.000	49998.859	163069.503

Consider, for example, the confidence interval for the variable Baths. You can see that this confidence interval ranges from 49,998.859 to 163,069.503.

Since the value of zero is not included in this confidence interval, we know that, at the 5% level of significance (since Excel's default is the 95% confidence interval), this coefficient is significantly different than zero. Indeed, this is the same conclusion we reached by examining the p-value for Baths. Similarly, you can see that the value of zero is included in the confidence interval for Bedrooms. This is because, at the 5% level of significance (recall that this is a 95% confidence interval), we could not reject the null hypothesis that the coefficient for the variable Bedrooms was significantly different than zero. But, remember, earlier we noted that Bedrooms was significant at a 10% level of significance. Thus, if we repeated the analysis with a 90% confidence interval, we would not expect to see zero included in this interval.

We can also compute a confidence interval for the regression line and, hence, the *estimated-Y* values. Again, we are working with a single sample to estimate the regression line. If we collected a different sample of data, we would most likely compute a slightly different regression line. Our regression line is, in a sense, a point estimator of the true relationship between the independent and dependent variables. To improve the accuracy of our prediction regarding this relationship, we can compute a confidence interval. For simplicity, we restrict our discussion to the case of a single independent variable, such as in our example of income determining consumption. Regression models with multiple independent variables would require more complex computations.

In fact, there are two different regression line confidence intervals that we can compute. First, we can compute a regression line confidence interval for the estimated mean value of Y at a given value of X (x^*). For example, when examining the relationship between consumption and income, computing a regression line confidence interval enables us to estimate the average consumption of people with a monthly income of x*. As another example, consider a regression analysis of the effect of number of hours studied on students' grades. A confidence interval constructed around the regression line can provide an estimate of the average expected grade of students who put in, say, ten hours of study time. The second type of regression line confidence interval is that associated with estimating a specific predicted Y value for a given value of X (x*). Computing this second type of regression line confidence interval - termed a **prediction interval** – might prove useful, for example, when predicting the consumption of a person having a monthly income of x*. In our grades regression example, the prediction interval would provide a range prediction for the grade of a student who put in ten hours of study. Formulae for computing both of these types of regression line confidence intervals are:

$$\text{Confidence interval: } \hat{y} \pm t_{\frac{\alpha}{2}, n-2} * S_\varepsilon \sqrt{\frac{1}{n} + \frac{(x^* - \bar{X})^2}{\Sigma(x_i - \bar{X})^2}}$$

$$\text{Prediction interval: } \hat{y} \pm t_{\frac{\alpha}{2}, n-2} * S_\varepsilon \sqrt{1 + \frac{1}{n} + \frac{(x^* - \bar{X})^2}{\Sigma(x_i - \bar{X})^2}}$$

For computation purposes, note that the expression $\Sigma(x_i - \bar{X})^2$ can be re-written as $(n-1)*s_x^2$.

Figure 13-13 superimposes both of these types of regression line confidence intervals on a hypothetical set of data points. The difference between these two confidence intervals is subtle, with the confidence interval being narrower than the prediction interval. Recall that in introducing the sampling distribution of the mean of a random variable in Chapter 7, we pointed out that the sampling distribution of the mean was narrower in shape than the distribution of the actual values of the variable - because we can more accurately estimate *average* values of a variable than *individual* values of the variable. This same reasoning holds here. This also explains the slight inward-curve of the two confidence intervals, as we are better at making predictions based on the average value, as opposed to extreme values of the independent variable (the intervals are narrowest around the average value of Income in Figure 13-13). The mathematical basis for this inward-narrowing of the regression line confidence intervals is that the width of the interval depends on the expression $(x^* - \bar{X})^2$ and the value of this expression is smallest when x* is close to $\bar{X}$ in value.

Figure 13-13
Confidence and Prediction Intervals

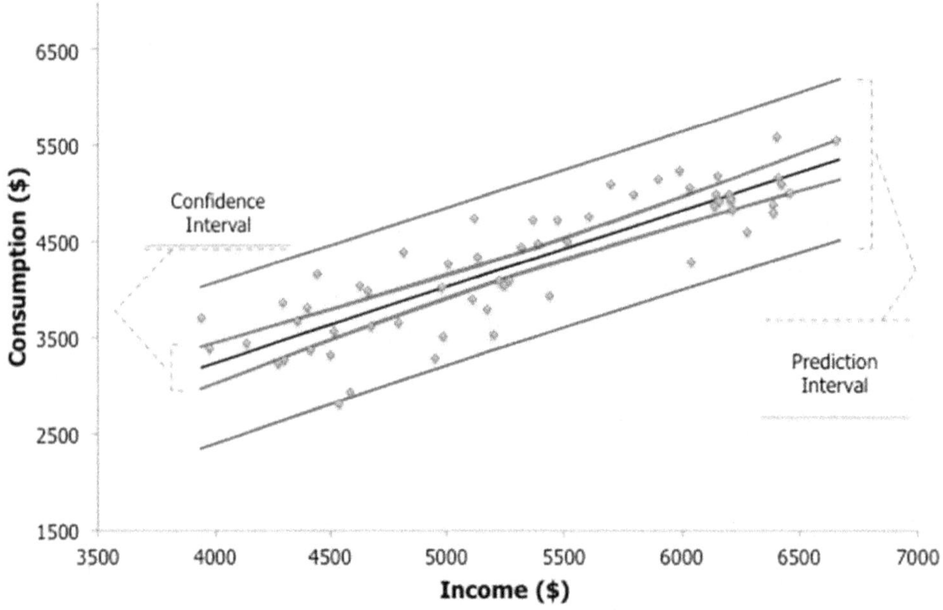

In this chapter's Excel file, the worksheet titled 'Consumption_CI' shows the calculated confidence interval (CI) and prediction interval (PI) for our consumption data. Examining this worksheet, you can see that a person earning a monthly income of $4,000 has an expected consumption of $3,244.38 and that the prediction interval provides a predicted range of consumption for this person that ranges from $2,407.47 to $4,081.28. If we wanted to know the *expected* (average) consumption of people having an income of $4,000, the estimated consumption value is still $3,244.38, but we have a narrower confidence interval, ranging from $3,033.07 to $3,455.68.

Unit 3 Summary

- Qualitative independent variables are modeled using **dummy variables**. A dummy variable can obtain one of two values: 1 or 0. For a qualitative variable that takes on n categorical values, we need to use n-1 dummy variables.

- When a qualitative variable can take on n categorical values, we interpret the set of dummy variables associated with the qualitative variable vis-à-vis the **base level**, which is the value of the qualitative variable when each of the associated dummy variables is set at zero.

- When using the estimated regression line for prediction, we need to consider the range of our independent variables. Making predictions within this range, referred to as **interpolating**, is generally more certain than making predictions outside this range, referred to as **extrapolating**.

- There are several *confidence intervals* that we can construct for our regression values. Confidence intervals for each of the estimated coefficients are provided in the Excel regression output.

- A confidence interval for the *average-Y* at a given level of X (x*) is computed as:

$$\hat{y} \pm t_{\frac{\alpha}{2}, n-2} * S_\varepsilon \sqrt{\frac{1}{n} + \frac{(x^* - \bar{X})^2}{\Sigma(x_i - \bar{X})^2}}$$

- A **prediction interval** is a confidence interval for a *specific estimated-Y* value at a given level of X (x*), and it is computed as:

$$\hat{y} \pm t_{\frac{\alpha}{2}, n-2} * S_\varepsilon \sqrt{1 + \frac{1}{n} + \frac{(x^* - \bar{X})^2}{\Sigma(x_i - \bar{X})^2}}$$

Unit 3 Exercises

1. What factors makes a video go viral? A researcher sets out to test the effect of the following factors on the number of views a *YouTube* video receives: length of the video (measured in minutes), number of likes the video received, whether the video includes kids, whether the video includes animals, what category the video falls under (the researcher examined three categories: humorous, informational, and inspirational), and, finally, the number of videos previously created by this person. Write the regression model to be estimated and clearly describe each variable in the model.

2. What factors affect a person's weight? A researcher collected data on the daily caloric intake of patients, the number of minutes that they exercise each day, the number of hours they spend watching television, and whether or not they suffer from anxiety or depression. For each person, she further recorded their age and race (i.e., White, African American, Asian, Hispanic or Other). Write the regression model to be estimated and clearly describe each variable in the model.

3. Desiring to study the effect of social life on students' academic performance, a professor asked her students to respond to the following question: "How often did you attend parties last semester?" Students were asked to choose among the following options:

   a. Never
   b. Not more than once a week
   c. Two to three times a week
   d. Four to six times a week
   e. Every day

   How would you represent these responses in the regression model?

4. Question 2 in the Unit 2 Exercises made use of the quantitative variables provided in the worksheet titled 'Laptops' in this chapter's Excel file to estimate laptop prices. Two qualitative variables, *Apple* and *Touch*, are also included in the worksheet. Redo the regression analysis by including these two qualitative variables in the regression model. What is the effect, if any, of each of these variables on the dependent variable?

5. Question 1 in the Unit 2 Exercises made use of country data on Internet usage and per capita GDP provided in the worksheet titled 'Internet Users' in this chapter's Excel file to determine the extent to which GDP influences Internet usage. Using these data along with the previous regression output, compute a 95% prediction interval for the number of Internet users for a country with a per capita GDP of $40,000.

6. Question 6.a in the Unit 2 Exercises involved the estimation of a regression equation to determine the causal influence of posted questions (to an online customer support forum) on posted answers. Using the data in the worksheet titled 'Online Community' in this chapter's Excel file and the Unit 2 Question 6.a regression analysis results, compute the 99% confidence interval for the number of answers in a week for which 150 questions were posted.

7. A skier wants to study the amount of daily snow-making (in inches) at three different ski resorts, labeled A, B and C. She collects data during 40 random days at each of the three ski resorts. In addition to the amount of man-made snow, she also collected data on the amount of natural snowfall (measured in inches) for the day prior to her measurement of man-made snow. Suppose that she analyzed these data through regression analysis and obtained the results provided in Table 13-e2.

   a. Which of the independent variables is significant?

   b. What is the difference in snow-making between Resort A and Resort C?

Table 13-e2
Regression Coefficients: Snow-Making Example

	Coefficients	Standard Error	t Stat	P-value
Intercept	22.29	3.505	6.359	1.644E-07
Natural Snow	0.6645	0.07	9.522	1.005E-11
Resort A	10.2528	2.465	4.158	0.0002
Resort B	0.4452	2.464	0.181	0.8573

END-OF-CHAPTER PRACTICE

1. The worksheet titled 'NHL' in this chapter's Excel file provides data on Points (a composite score of goals and assists) of individual NHL players in the worksheet's Column A.[146] The remaining columns provide data on potential independent variables for explaining a player's Points score. Which of these variables are significant predictors of Points?

2. The worksheet titled 'Social Media' in this chapter's Excel file provides hypothetical data on individuals' social media use, measured in minutes per day, along with data regarding four potential predictors of social media use. Run a regression model on these data and discuss your results.

3. The worksheet titled 'Movies' in this chapter's Excel file provides box office gross receipts data for twenty-five films, along with seven possible

---

[146] Source: http://www.nhl.com/ice/historicalstats.htm

predictors of a film's gross box office receipts.[147] Given these data, what are the best predictors of a film's box office success?

4. The worksheet titled 'GPA' in this chapter's Excel file provides survey data collected by this book's author from MBA students on their undergraduate GPAs, as well as other factors that may have affected the students' undergraduate GPAs. What do you conclude after running a regression analysis on these data?

5. The worksheet titled 'Swimming' in this chapter's Excel file shows the times that seventeen swimmers obtained in five events:[148] 50 yard fly, 50 yard backstroke, 50 yard breaststroke, 50 yard freestyle, and 200 yard individual medley (IM). In the IM, swimmers must swim 50 yards in each style. Which of the individual events that comprise the medley is the best predictor of the time obtained by swimmers for the medley event?

6. Return to Question 4 in the Unit 2 Exercises, where data in the worksheet titled 'Brain' from this chapter's Excel file is used in estimating an individual's IQ score using his brain size (measured as an MRI pixel count). Use the data provided in this worksheet, along with regression analysis, to determine if brain size can be determined, to some extent, by a person's height, weight or gender. What do you conclude?

7. Again using the data provided in the worksheet titled 'Brain' from this chapter's Excel file, estimate a regression model where the Full Scale IQ score is the dependent variable and MRI Pixel Count is the single independent variable. Now, create a 90% prediction interval for the Full Scale IQ for a person whose MRI Pixel Count is 850,000.

8. The worksheet titled 'Wages' in this chapter's Excel file provides data on people's yearly wages as the dependent variable and their age, gender, hours of work, and level of education as independent variables.[149]

    a. Conduct a regression analysis to determine which of these variables significantly affects a person's expected yearly wage.

    b. Holding all other variables constant, what change would you expect to see in Wage for an extra hour of work? Explain your answer.

    c. Who earns higher wages – men or women? What is the difference in the predicted wage between the two genders? Explain your answers.

    d. What wage would you expect a 40-year-old female with a high school degree who works 35 hours per week to earn? Explain your answer.

---

[147] Source: http://www.the-numbers.com/movies/records/
[148] Source: http://www.adirondackswimming.org
[149] Source: http://www.census.gov/main/www/pums.html

e. How much more money do people with a graduate degree earn compared with people with a college degree (but no graduate degree)? Explain your answer.

9. The worksheet titled 'Electric' in this chapter's Excel file provides data on households' monthly electricity cost along with: the number of people in a household, the number of bedrooms in the house, the type of house (detached single family home, attached single family home or apartment), and whether or not the household's heating is electricity-based.[150]

   a. Conduct a regression analysis to determine which of these variables significantly affect a household's electricity bill.

   b. By how much does the use of electricity-based heating increase the household's electrical bill (compared with other forms of heating)?

   c. Which type of house suffers the highest cost of electricity?

   d. What is the effect of one extra bedroom on a household's electricity bill?

10. Figure 13-e1 shows the regression output of a model using Earnings Per Share (EPS) of S&P 500 companies as the dependent variable and Sales as the independent variable.[151]

    a. What can you learn from the output about the relationship between these two variables?

    b. What is the estimated EPS for a company having a volume of Sales of $20,000?

---

[150] Source: http://www.census.gov/main/www/pums.html
[151] Source: https://www.capitaliq.com/home/what-we-offer/information-you-need/financials-valuation/compustat-financials.aspx

Figure 13-e1

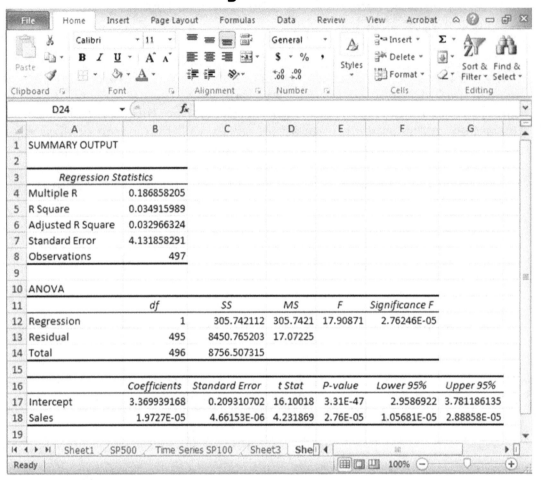

APPENDIX A: Using the Microsoft Excel *Analysis ToolPak* for Testing Hypotheses for Parameters from Two Populations

The primary purpose of this appendix is to illustrate how the Microsoft Excel *Analysis ToolPak* can be used to conduct a few of the two-populations tests that were covered in Chapter 10. (You should already be familiar with using Excel to better understand the nature of a data set and to compute the means and standard deviations used in calculating test statistics and confidence intervals.) The URL that follows provides instructions for installing the *Analysis Toolpak* into your own version of Excel:

http://office.microsoft.com/en-us/excel-help/load-the-analysis-toolpak-HP010021569.aspx

Note that if you are a *Mac* user, not all versions include the *Analysis Toolpak*. Please follow this link for instructions on obtaining statistical add-ins for *Mac*:

https://support.microsoft.com/en-us/kb/2431349

Once you have installed the *Analysis ToolPak*, you can select the menu option of 'Data Analysis' from the 'Data' tab to view the menu of statistical options available to you through the *Analysis Toolpak*. Note, in particular, that the available data analysis procedures are not limited to hypothesis tests regarding parameters from two populations. Earlier in Chapter 2, for example, we described how the *Analysis Toolpak* can be used to construct histograms. In Chapters 12 and 13, respectively, we will use the *Analysis Toolpak* to carry out two new statistical procedures, i.e., ANOVA and regression analysis, whose calculations can prove to be quite time-consuming if done by hand. In order to learn more about the *Analysis ToolPak*, a comprehensive overview of all its data analysis procedures is available at:

http://office.microsoft.com/en-us/excel-help/use-the-analysis-toolpak-to-perform-complex-data-analysis-HP010342762.aspx

The *Analysis ToolPak* contains procedures for carrying out five two-populations hypothesis tests: the test of two populations' proportions, the test of two populations' variances, and three tests of two populations' means (independent samples/equal variances, independent samples/unequal variances, and paired samples). Step-by-step guidance on using the *Analysis Toolpak* for three of these tests is described in this appendix (with the other two tests handled in essentially the same manner as the three illustrated tests). The three hypothesis tests covered in this appendix are:

- F-Test: Two-Sample for Variances (that is, test of two populations' variances)

- t-Test: Two-Sample, Assuming Unequal Variances (that is, test of two populations' means, independent samples and unequal variances)

- t-Test: Paired Two Sample for Means (that is, test of two populations' means, paired samples)

F-Test: Two Sample for Variances

We will work again with the example about the three exams, which was used in Chapter 10. Recall that the professor was interested in analyzing differences between exam averages. We will focus on two of the questions that the professor asked: "Is there a difference in the mean score between Exam II and Exam III?", and "Is there evidence of a difference in each individual student's performance on the Year III midterm exam and on the Year III final exam?" The data for the three exams are provided in the Appendix A Excel file, under the worksheet titled "Two-Populations, Exams Example". There are two tables in this worksheet. Table 1 provides 33 grades for each of the three exams (independent samples). Table 2 provides the matched grades for 33 students on the midterm exam and final exam for Year III.

We start with the first question: "Is there a difference in the mean score between Exam II and Exam III?" Recall that before we conduct the test of the two means, we need to determine if the sample variances are equal. Our hypotheses are:

$$H_0: \frac{\sigma_{II}^2}{\sigma_{III}^2} = 1$$

$$H_1: \frac{\sigma_{II}^2}{\sigma_{III}^2} \neq 1$$

We begin the analysis via Excel by clicking on the 'Data' tab and then choosing 'Data Analysis' to open the data analysis window shown in the screen shot in Figure A-1. To conduct the test, select (from the menu of tests) the 'F-Test Two-Sample for Variances' option and click 'OK'.

Figure A-1
Excel Screen Shot: Data Analysis Menu

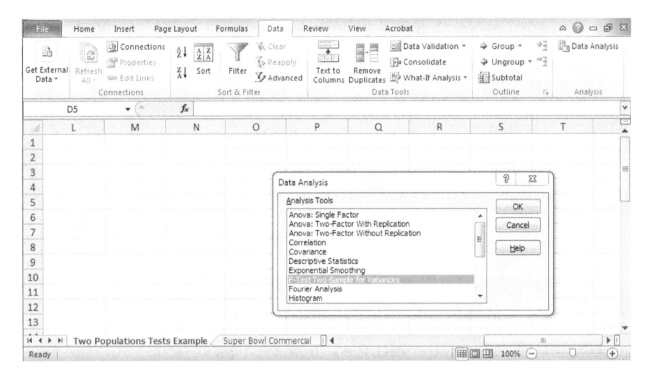

Figure A-2 displays the menu that opens next. Here, you are asked to specify several components needed in conducting the test:

- Input:
  - Variable 1 Range: Select the cells B2:B35, which hold the data for Exam II.
  - Variable 2 Range: Select the cells C2:C35, which hold the data for Exam III.
  - Check the box titled 'Labels' to indicate that you have column titles in Row 1.
  - Alpha: 0.05 is the default level. Since this is the significance level at which we wish to conduct our test, we can leave this value as is.
- Output options:
  - Pick the options you prefer. Here, we selected to place the output table in cell M1 on the same worksheet.
- Click 'OK'.

Figure A-2
Excel Screen Shot: Setting Up the F-Test

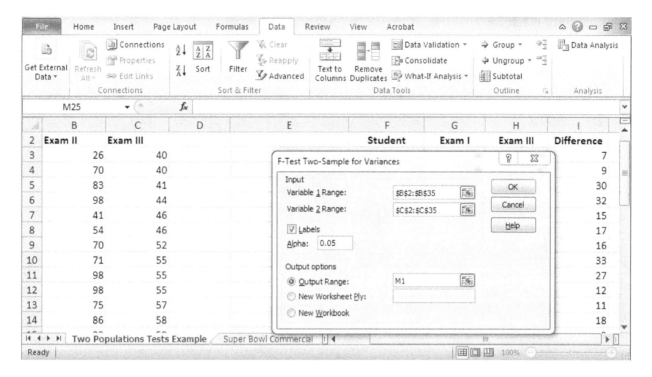

Figure A-3 provides the screen capture of the output obtained from Excel. These output results include the following elements:

- Mean: These are the means of the two exams.

- Variance: These are the variances of the two exams.

- Observations: These are the sample size for each exam.

- df: These are the degrees of freedom for each exam.

- F: This is the test statistic (variance$_{II}$ divided by variance$_{III}$).

- P(F<=f) one-tail: This is the p-value for this one-tail test.

- F Critical one-tail: This is the critical value for this one-tail test.

Figure A-3
Excel Screen Shot: F-Test Output

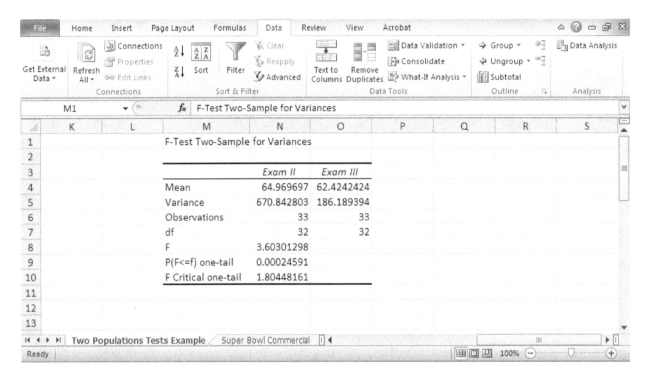

Recall that our rejection rule for this test is: reject the null hypothesis if the test statistic is greater than the critical value. Alternatively, since we are using Excel, we can also use the p-value rule: reject the null hypothesis if the p-value is less than α. Looking at the output, you will see that the test statistic's value is 3.603, which is greater than the critical value of 1.804 and, therefore, we reject the null hypothesis. The same conclusion can be reached by looking at the p-value of 0.0002, which is much smaller than the selected α of 5%. We therefore conclude that the variances are unequal; and, we proceed with the test of two populations' means, using the test that assumes unequal variances.

t-Test: Two-Sample, Assuming Unequal Variances

Before conducting the test, recall the hypotheses that were used in Chapter 10 in determining if the mean of Exam II was higher than the mean of Exam III:

$$H_0: \mu_{II}-\mu_{III} \leq 0$$

$$H_1: \mu_{II}-\mu_{III} > 0$$

To conduct the test with Excel, click again on 'Data' and then 'Data Analysis'. You should then see the input window shown in Figure A-4. Scroll down the list of tests and select the 't-Test: Two-Sample Assuming Unequal Variances' option. Click 'OK'.

Figure A-4
Excel Screen Shot: Selecting a t-Test with Unequal Variances

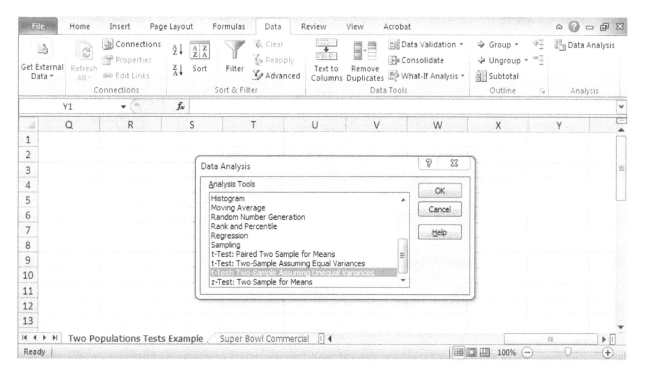

Figure A-5 displays the menu that opens next. You, again, are asked to specify several test components:

- Input:
    - Variable 1 Range: Select the cells B2:B35, which hold the data for Exam II.
    - Variable 2 Range: Select the cells C2:C35, which hold the data for Exam III.
    - Hypothesized Mean Difference: Enter 0 since our null hypothesis is that there is no difference in the means.
    - Check the box titled 'Labels' to indicate that you have column titles in Row 1.

- Alpha: 0.05 is the default level. Since we are interested in conducting this test at the 1% level of significance, we need to change the α value to 0.01.
- Output options:
  - Pick the options you prefer. Here, we selected to place the output table in cell M13 on the same worksheet.
- Click 'OK'.

Figure A-5
Excel Screen Shot: Setting Up a t-Test with Unequal Variances

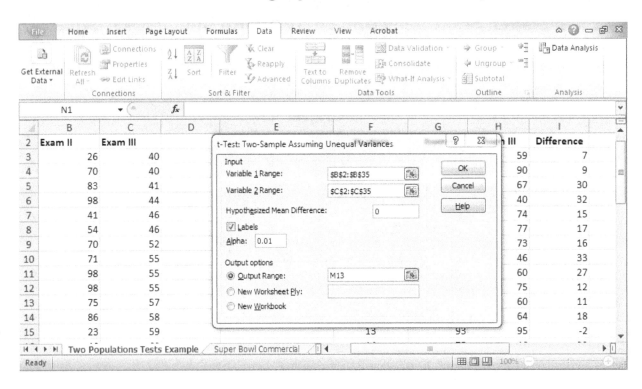

Figure A-6 provides the screen capture of the output obtained from Excel.

These output results include the following elements:

- Mean: These are the means of the two exams.

- Variance: These are the variances of the two exams.

- Observations: These are the sample size for each exam.

- Hypothesized Mean Difference: 0, as per the null hypothesis.

- df: This is the degrees of freedom for the two exams, calculated using the unequal variances formula.

- t stat: This is the test statistic for unequal variances.

- P(T<=t) one-tail: This is the p-value for a one-tail test.

- t Critical one-tail: This is the critical value for a one-tail test.

- P(T<=t) two-tail: This is the p-value for a two-tail test (you can see that it is simply twice the p-value for the one-tail test).

- t Critical two-tail: This is the critical value for a two-tail test.

Figure A-6
Excel Screen Shot: t-Test with Unequal Variances Output

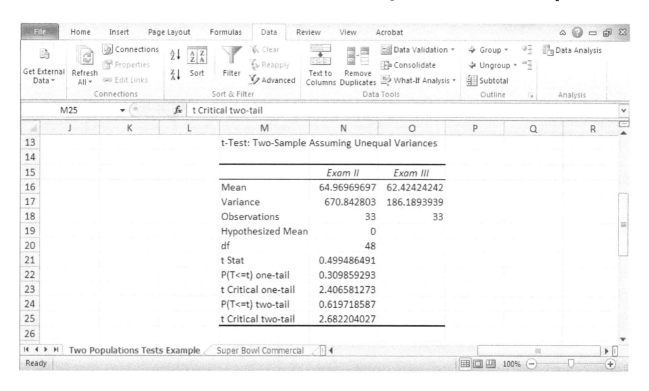

Recall that this example involves an upper-tail test. Thus, our rejection rule for this test is: reject the null hypothesis if the test statistic is greater than the critical value. Alternatively, since we are using Excel, we can also use the p-value rule: reject the null hypothesis if the p-value is less than α.

Because this is a one-tail test, we compare our test statistic of 0.499 to the one-tail critical value of 2.407. Because our test statistic is lower than the critical value, we do not reject the null hypothesis. We reach a similar conclusion by comparing the one-tail p-value of 0.309 to the desired α of 0.01, which is smaller. By not rejecting the null hypothesis, we conclude that there is insufficient evidence to conclude that the mean grade on Exam II is higher than that of Exam III.

t-Test: Paired Two Sample for Means

Another of the two-populations hypothesis tests covered in Chapter 10 involved testing the mean difference for paired samples. Specifically, we used this test to answer the professor's question about a difference in each individual student's performance on the midterm exam and on the final exam for Year III. Since the professor suspects that students will perform better on the midterm exam than on the final exam, the hypotheses were formulated as an upper-tail test:

$$H_0: \mu_d \leq 0$$

$$H_1: \mu_d > 0$$

where d represents the difference between the midterm exam grade and the final exam grade. To conduct this test using Excel, click again on 'Data' and then 'Data Analysis'. You should again see the input window shown in Figure A-7. Scroll down the list of tests, select the 't-Test: Paired Two Sample for Means' option and click 'OK'.

Figure A-7
Excel Screen Shot: Selecting a t-Test with Paired Samples

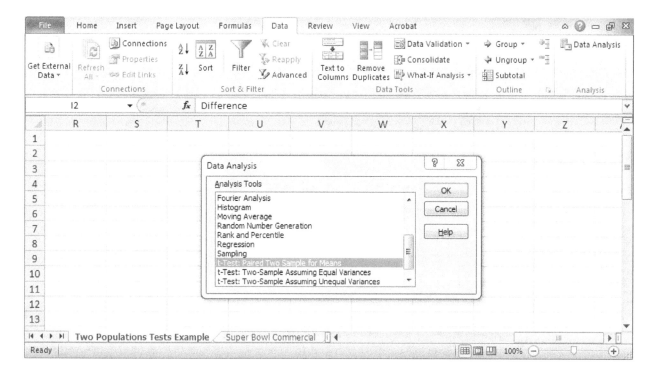

Figure A-8 displays the menu that opens next. Again, you are asked to specify several test components:

- Input:
  - Variable 1 Range: Select the cells G2:G35, which hold the data for the Year III midterm exam in the paired dataset.
  - Variable 2 Range: Select the cells H2:H35, which hold the data for the Year III final exam in the paired dataset.
  - Hypothesized Mean Difference: Type 0, since our null hypothesis is that there is no difference in the means.
  - Check the box titled 'Labels' to indicate that you have column titles in Row 1.
  - Alpha: 0.05 is the default level. Since this is the significance level at which we wish to conduct our test, we can leave this value as is.
- Output options:
  - Pick the options you prefer. Here, we selected to place the output table in cell M30 on the same worksheet.
- Click 'OK'.

Figure A-8
Excel Screen Shot: Setting Up a t-Test with Paired Samples

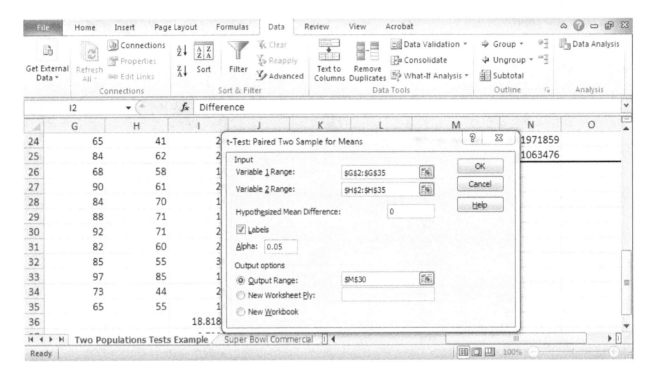

Figure A-9 provides the screen capture of the output obtained from Excel.

The output results include the following elements:

- Mean: These are the means of the two exams.

- Variance: These are the variances of the two exams.

- Observations: These are the sample sizes for each exam.

- Pearson Correlation: This is an indication of the strength of the relationship between the midterm and final grades. We introduced correlations in Chapter 5 and will discuss this statistical concept again in Chapter 13. Values closer to 1 mean that there is a strong relationship between the two variables. In this example, the value of 0.769 indicates that the grades on the two exams are strongly correlated.

- Hypothesized Mean Difference: 0, as per the null hypothesis.

- df: This is the degrees of freedom for the two exams, calculated as the number of pairs minus one (33-1=32).

- t stat: This is the test statistic for the paired sample means.

- P(T<=t) one-tail: This is the p-value for a one-tail test. (E-14 means that the number 4.8254 will begin fourteen places after the decimal place. In other words, this is a very small number.)

- t Critical one-tail: This is the critical value for a one-tail test.

- P(T<=t) two-tail: This is the p-value for a two-tail test.

- t Critical two-tail: This is the critical value for a two-tail test.

Figure A-9
Excel Screen Shot: t-Test with Paired Samples Output

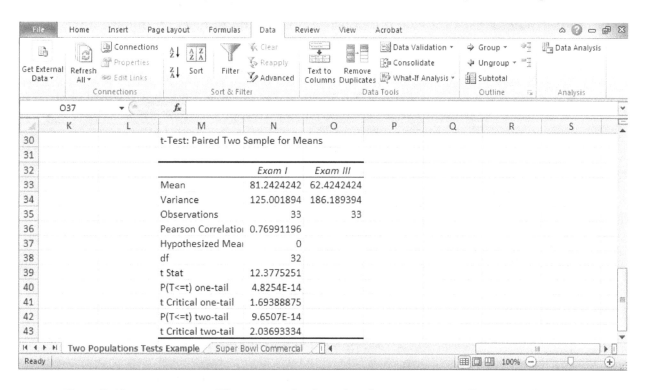

Recall that our specific example involved an upper-tail test. Our rejection rule for this test is: reject the null hypothesis if the test statistic is greater than the critical value. Alternatively, since we are using Excel, we can also use the p-value rule: reject the null hypothesis if the p-value is less than α.

Because this is a one-tail test, we compare our test statistic of 12.377 to the one-tail critical value of 1.69. Because our test statistic is greater than the critical value, we reject the null hypothesis. We reach a similar conclusion by comparing the one-tail p-value of 0.000 to the desired α of 0.05. As the p-value is less than α, we reject the null hypothesis. By rejecting the null hypothesis, we conclude that, indeed, individual students' exam grades have dropped between the Year III midterm exam and the Year III final exam, indicating that the final exam was quite likely more difficult than the midterm exam.

Exercises

The data for the following questions is provided in the Appendix A Excel file. Each worksheet provides data for one question. The beginning of each question will indicate which Excel worksheet should be used.

1. 'UFO Sighting'. The Hynek Scale is a six-fold classification of UFO sightings. It ranges from Level 1 (visual sighting of an unidentified flying object) to Level 6 (a UFO incident that causes direct injury or death - which has never been reported).[1,2] Every alleged UFO sighting since the 19th century has been assigned a Hynek rank. Given the random sample of UFO sightings from North America and Europe provided in the worksheet, is there enough evidence to conclude at the 10% significance level the mean Hynek rank is higher in North America?

2. 'Repair Costs'. In order to compare the cost of fender-bender accidents with *Mazdas* and *Toyotas*, the worksheet provides, for differing rates of speed, estimated repair costs for each of the car brands. At the 5% significance level, is there enough evidence to conclude that *Toyotas* are less expensive to repair?

3. 'Classroom vs. Online'. A teacher is testing a new method of presenting information over the Internet. In her first semester class, she just used her traditional method of teaching in the classroom; and, in the second semester, she also posted everything online. Given the two samples of exam results provided in the worksheet, is the variance of students who only obtained lessons in the classroom lower than the students who were also given access to the lessons online (at the 5% significance level)?

4. 'Stocks'. In deciding which investment is less risky, Natalie obtains stock price samples for the *Facebook* stock and for the *Microsoft* stock.[3] Can she conclude at the 1% significance level that *Facebook* is the riskier stock?

---

[1] Source: http://en.wikipedia.org/wiki/List_of_UFO_sightings
[2] Source: http://en.wikipedia.org/wiki/Close_encounter
[3] Source: http://finance.yahoo.com

5. 'Study Times'. A professor asked her students to report how many hours they invested in studying outside of classroom time. Students tracked their study time over one week. The study times are reported in the worksheet for each student (student 1 through student 29) and for each day of the week. At the 5% level of significance, is there evidence that students spend more time studying on Sunday than on other days of the week?[4]

6. 'Crime Rate'.[5] This worksheet demonstrates the usefulness of statistical software when working with large datasets. It also shows the potential complexity in many data sets that you might work with in the future. Take a few minutes to understand what the numbers in the table represent. Then, use this worksheet in answering Questions 6.a. and 6.b.

   a. Pick any two states and compare the average number of violent crimes in 2011 between these states.

   b. It can be argued that the data for specific crimes can be matched by city. Conduct a paired-samples t-test to determine whether there are more arsons than murders across all cities.

---

[4] This question requires some additional work. First, notice that the data are in rows and not in columns. Make sure that you enter the data correctly in the input window. Second, you need to compare the study hours on Sunday against those of the other days of the week combined; thus, you will need to combine all other days into a single data set.

[5] The data in this worksheet are taken from the U.S. Department of Justice, Federal Bureau of Investigation, Preliminary Annual Uniform Crime Report, January-December 2012:
http://www.fbi.gov/about-us/cjis/ucr/crime-in-the-u.s/2012/preliminary-annual-uniform-crime-report-january-december-2012

EXCEL FUNCTION INDEX[1]

FUNCTION NAME **AVERAGE**
Syntax AVERAGE(number1,number2,...)
Description Computes the arithmetic mean of the numbers indicated through the arguments.
Arguments **number1:** The first number (actual value or cell reference) of a series of numbers, or the complete range of numbers, for which you wish to compute an average.
number2: Additional numbers, cell references or ranges to be included within the average, up to a maximum of 255.
Chapter (Page Number) 3 (95)

FUNCTION NAME **BINOM.DIST**
Syntax BINOM.DIST(number_s,trials,probability_s,cumulative)
Description Computes a binomial distribution probability. Use BINOM.DIST in problems with a fixed number of trials, when trial outcomes are either a success or a failure, when trials are independent, and when the probability of success is constant for all trials.
Arguments **number_s:** The number of successes observed.
trials: The number of trials.
probability_s: The probability of success.
cumulative: Takes a value of either 0 or 1. A value of 0 will compute the probability of observing the specified number of successes, whereas a value of 1 will return the cumulative probability for the specified number of successes.
Chapter (Page Number) 5 (185)

FUNCTION NAME **CHISQ.DIST**
Syntax CHISQ.DIST(x,deg_freedom,cumulative)
Description Computes the lower-tail probability of the chi-squared (χ^2) distribution. To compute the upper-tail probability, you would type:
=1-CHIDIST(x,deg_freedom,cumulative)
Arguments **x:** A chi-square value (e.g., the test statistic).
deg_freedom: The number of degrees of freedom associated with the test.
cumulative: Takes a value of either 0 or 1, but is generally set as 1 for this course as the focus is on continuous distributions rather than discrete distributions.
Chapter (Page Number) 9 (340)

---

[1] All of the functions are from the Office 2016 version

FUNCTION NAME   **CHISQ.DIST.RT**
Syntax   CHISQ.DIST.RT(x,deg_freedom)
Description   Computes the upper-tail probability of the chi-squared (χ^2) distribution.
Arguments   **x:** A chi-square value (e.g., the test statistic).
deg_freedom: The number of degrees of freedom associated with the test.
Chapter (Page Number)   9 (340)

FUNCTION NAME   **CHISQ.INV**
Syntax   CHISQ.INV(probability,deg_freedom)
Description   Computes the inverse of the lower-tail probability of the chi-squared (χ^2) distribution. In other words, given a specified significance level, the function returns the chi-square value defining the upper-bound of the left-tail probability (e.g., the critical value).
Arguments   **probability:** The significance level for the chi-square value to be obtained.
deg_freedom: The number of degrees of freedom associated with a test.
Chapter (Page Number)   9 (340)

FUNCTION NAME   **CHISQ.INV.RT**
Syntax   CHISQ.INV.RT(probability,deg_freedom)
Description   Computes the inverse of the upper-tail probability of the chi-squared (χ^2) distribution. In other words, given a specified significance level, the function returns the chi-square value defining the lower-bound of the right-tail probability (e.g., the critical value).
Arguments   **probability:** The significance level for the chi-square value to be obtained.
deg_freedom: The number of degrees of freedom associated with a test.
Chapter (Page Number)   9 (340)

FUNCTION NAME   **COUNTIF**
Syntax   COUNTIF(range,criteria)
Description   Counts the number of cells within a range that meet a single criterion that you specify.
Arguments   **range:** The range of spreadsheet cells containing the data set values to be counted.
criteria: The value to be matched in counting. This value can be a number, an expression or a text string (a series of characters).
Chapter (Page Number)   2 (44)

FUNCTION NAME	**F.DIST**
Syntax	FDIST(x,deg_freedom1,deg_freedom2)
Description	Computes the lower-tail probability for the F-distribution (as when performing a hypothesis test on the difference in the variance for two populations).
Arguments	**x:** An F value (e.g., the test statistic).
	deg_freedom1: The number of degrees of freedom for the numerator.
	deg_freedom2: The number of degrees of freedom for the denominator.
Chapter (Page Number)	10 (*not specifically mentioned within the text*)

FUNCTION NAME	**F.DIST.RT**
Syntax	F.DIST.RT(x,deg_freedom1,deg_freedom2)
Description	Computes the upper-tail probability for the F-distribution (as when performing a hypothesis test on the difference in the variance for two populations).
Arguments	**x:** An F value (e.g., the test statistic).
	deg_freedom1: The number of degrees of freedom for the numerator.
	deg_freedom2: The number of degrees of freedom for the denominator.
Chapter (Page Number)	10 (375)

FUNCTION NAME	**F.INV**
Syntax	F.INV(probability,deg_freedom1,deg_freedom2)
Description	Computes the inverse of the lower-tail probability for the F-distribution. In other words, given a specified significance level, the function returns the F value defining the upper-bound of the left-tail probability.
Arguments	**probability:** The significance level for the F value to be obtained.
	deg_freedom1: The number of degrees of freedom for the numerator.
	deg_freedom2: The number of degrees of freedom for the denominator.
Chapter (Page Number)	10 (374)

FUNCTION NAME	**F.INV.RT**
Syntax	F.INV.RT(probability,deg_freedom1,deg_freedom2)
Description	Computes the inverse of the upper-tail probability for the F-distribution. In other words, given a specified significance level, the function returns the F value defining the lower-bound of the right-tail probability.
Arguments	**probability:** The significance level for the F value to be obtained.
	deg_freedom1: The number of degrees of freedom for the numerator.
	deg_freedom2: The number of degrees of freedom for the denominator.
Chapter (Page Number)	10 (374)

FUNCTION NAME   **FREQUENCY**
Syntax   FREQUENCY(data_array,bins_array)
Description   Computes how often values occur within specified classes for a data set and then returns these frequency counts as a column of numbers. To use this function, you select a range of cells (the same number of cells as the number of classes) for the function's output prior to typing the formula and then you simultaneously press the 'CTRL', 'SHIFT' and 'ENTER' keys to obtain these output values. The function cannot be typed into a single cell and then copied across a desired range.
Arguments   **data_array:** The range of the data set for which class frequencies are to be counted.
    bins_array: A range of data containing the upper-class bounds defining the classes for which frequencies are to be counted.
Chapter (Page Number)   2 (52)

FUNCTION NAME   **GEOMEAN**
Syntax   GEOMEAN(number1,number2,…)
Description   Computes the geometric mean of the numbers indicated through the arguments.
Arguments   **number1:** The first number (actual value or cell reference) of a series of numbers, or the complete range of numbers, for which you wish to compute a geometric mean.
    number2: Additional numbers, cell references or ranges to be included within the geometric mean, up to a maximum of 255.
Chapter (Page Number)   3 (95)

FUNCTION NAME   **MEDIAN**
Syntax   MEDIAN(number1,number2,…)
Description   Returns the median, or the value in the middle of the set of data values indicated through the arguments.
Arguments   **number1:** The first number (actual value or cell reference) of a series of numbers, or the complete range of numbers, for which you wish to compute a geometric mean.
    number2: Additional numbers, cell references or ranges to be included within the geometric mean, up to a maximum of 255.
Chapter (Page Number)   3 (99)

FUNCTION NAME   **MODE.MULT**
Syntax   MODE.MULT(number1,number2,…)
Description   Returns the most frequently occurring value(s) in the set of data values indicated through the arguments.
Arguments   **number1:** The first number (actual value or cell reference) of a series of numbers, or the complete range of numbers, for which you wish to compute a geometric mean.
    number2: Additional numbers, cell references or ranges to be included within the geometric mean, up to a maximum of 255.
Chapter (Page Number)   3 (*100 – refer to Footnote #25*)

FUNCTION NAME	**MODE.SNGL**
Syntax	MODE.SNGL(number1,number2,...)
Description	Returns the most frequently occurring value in the set of data values indicated through the arguments.
Arguments	**number1:** The first number (actual value or cell reference) of a series of numbers, or the complete range of numbers, for which you wish to compute a geometric mean.
	number2: Additional numbers, cell references or ranges to be included within the geometric mean, up to a maximum of 255.
Chapter (Page Number)	3 (100)

FUNCTION NAME	**NORM.DIST**
Syntax	NORM.DIST(x,mean,standard_dev,cumulative)
Description	Computes the area under the normal distribution curve up to a specified value. In other words, the function returns the cumulative probability that a value less than or equal to the specified value will be obtained.
Arguments	**x:** The specified value.
	mean: The arithmetic mean of the distribution.
	standard_dev: The standard deviation of the distribution.
	cumulative: Takes a value of either 0 or 1, but is generally set as 1 for this course as the focus is on continuous distributions rather than discrete distributions.
Chapter (Page Number)	6 (214)

FUNCTION NAME	**NORM.INV**
Syntax	NORM.INV(probability)
Description	Computes the inverse of the normal distribution. In other words, given a specified probability, the function returns the Z value defining the upper-bound of the lower-tail area under the normal distribution curve.
Arguments	**probability:** The specified probability.
Chapter (Page Number)	6 (227)

FUNCTION NAME	**NORM.S.DIST**
Syntax	NORM.S.DIST(z,cumulative)
Description	Computes the area under the standard normal distribution curve up to a specified Z value. The standard normal cumulative distribution function has a mean of 0 and a standard deviation of 1. In other words, the function returns the cumulative probability that a value less than or equal to the specified value will be obtained.
Arguments	**z:** The specified Z value.
	cumulative: Takes a value of either 0 or 1, but is generally set as 1 for this course as the focus is on continuous distributions rather than discrete distributions.
Chapter (Page Number)	6 (219)

FUNCTION NAME	**NORM.S.INV**
Syntax	NORM.S.INV(probability)
Description	Computes the inverse of the standard normal distribution. In other words, given a specified probability, the function returns the Z value defining the upper-bound of the lower-tail area under the standard normal distribution curve.
Arguments	**probability:** The specified probability.
Chapter (Page Number)	6 (225)

FUNCTION NAME	**PERCENTILE.EXC**
Syntax	PERCENTILE.EXC(array,k)
Description	Computes the kth percentile for the values in a data set.
Arguments	**array:** The range of the data set for which the kth percentile is to be computed.
	k: The percentile to be computed.
Chapter (Page number)	3 (122)

FUNCTION NAME	**POISSON.DIST**
Syntax	POISSON.DIST(x,mean,cumulative)
Description	Computes a Poisson distribution probability.
Arguments	**x:** The number of successes observed.
	mean: The expected (mean) number of successes.
	cumulative: Takes a value of either 0 or 1. A value of 0 will compute the probability of observing the specified number of successes, whereas a value of 1 will return the cumulative probability for the specified number of successes.
Chapter (Page Number)	5 (193)

FUNCTION NAME	**RAND**
Syntax	RAND()
Description	Computes an evenly distributed random real number greater than or equal to 0 and less than 1. A new random real number is returned every time the worksheet is calculated.
Arguments	This function has no arguments.
Chapter (Page Number)	1 (15)

FUNCTION NAME	**RANDBETWEEN**
Syntax	RANDBETWEEN(bottom,top)
Description	Computes a random integer number between the numbers you specify. A new random integer number is returned every time the worksheet is calculated.
Arguments	**bottom:** The smallest integer RANDBETWEEN will return.
	top: The largest integer RANDBETWEEN will return.
Chapter (Page Number)	7 (262)

FUNCTION NAME	**STDEV.P**
Syntax	STDEV.P(number1,number2,...)
Description	Estimates the standard deviation of the data population indicated through the arguments.
Arguments	**number1:** The first number (actual value or cell reference) of a series of numbers, or the complete range of numbers, for which you wish to compute a standard deviation.
	number2: Additional numbers, cell references or ranges to be included within the standard deviation, up to a maximum of 255.
Chapter (Page number)	3 (112)

FUNCTION NAME	**STDEV.S**
Syntax	STDEV.S(number1,number2,...)
Description	Estimates the standard deviation of the data sample indicated through the arguments.
Arguments	**number1:** The first number (actual value or cell reference) of a series of numbers, or the complete range of numbers, for which you wish to compute a standard deviation.
	number2: Additional numbers, cell references or ranges to be included within the standard deviation, up to a maximum of 255.
Chapter (Page number)	3 (112)

FUNCTION NAME	**SUMPRODUCT**
Syntax	SUMPRODUCT(array1,array2,array3,...)
Description	Multiplies corresponding components of a series of data sets (with each data set placed within a range of Excel cells) and then returns the sum of those products. There can be up to 255 data sets, and each data set must have the same number of observations. While there is no direct function to calculate the *weighted mean* of a data set, the SUMPRODUCT function can be used to multiply the values of a data set (array1) by their respective weights (array2) and, by doing so, obtain the numerator for the weighted mean formula.
Arguments	**array1:** The range of one of the data sets whose components are to be multiplied.
	array2: The range of another data set whose components are to be multiplied.
	array3: The range of another data set whose components are to be multiplied.
Chapter (Page Number)	3 (95)

FUNCTION NAME	**T.DIST**
Syntax	T.DIST(x,deg_freedom,cumulative)
Description	Computes the left-tailed t-distribution (the area under the t-distribution curve in the lower-tail rejection region) given a t value (generally, the test statistic) and the degrees of freedom associated with the sample of data.
Arguments	**x:** The specified t value.
	deg_freedom: An integer indicating the number of degrees of freedom associated with the sample of data.
	cumulative: Takes a value of either 0 or 1, but is generally set as 1 for this course as the focus is on continuous distributions rather than discrete distributions.
Chapter (Page Number)	8 (288)

FUNCTION NAME	**T.INV**
Syntax	T.INV(probability,deg_freedom)
Description	Computes the inverse of the t distribution. In other words, given a specified probability, the function returns the t value defining the upper-bound of the lower-tail rejection region. Since the t-distribution is symmetric, use a positive value for the returned t value for an upper-tail test.
Arguments	**probability:** The specified probability.
	deg_freedom: The number of degrees of freedom associated with the sample of data.
Chapter (Page Number)	8 (288)

FUNCTION NAME	**VAR.P**
Syntax	VAR.P(number1,number2,...)
Description	Calculates the variance of the data population indicated through the arguments.
Arguments	**number1:** The first number (actual value or cell reference) of a series of numbers, or the complete range of numbers, for which you wish to compute a variance.
	number2: Additional numbers, cell references or ranges to be included within the variance, up to a maximum of 255.
Chapter (Page number)	3 (112)

FUNCTION NAME	**VAR.S**
Syntax	VAR.S(number1,number2,...)
Description	Estimates the variance of the data sample indicated through the arguments.
Arguments	**number1:** The first number (actual value or cell reference) of a series of numbers, or the complete range of numbers, for which you wish to compute a variance.
	number2: Additional numbers, cell references or ranges to be included within the variance, up to a maximum of 255.
Chapter (Page number)	3 (112)

EQUATION INDEX

CHAPTER 2

- **Relative Frequency of a Class within a Frequency Table**

$$rf_i = \frac{f_i}{N}$$

Where i is the ith class in the frequency table, f_i is the number of observations in class i, and N is the total number of observations.

CHAPTER 3

- **Arithmetic Mean of a Sample**

$$\bar{x} = \frac{\sum_{i=1}^{n} x_i}{n}$$

Where n is the number of observations in a sample, x_i is the value of each observation in the sample, and Σ denotes a summation operation.

- **Geometric Mean of a Series of Growth Rates**

$$\text{Geometric Mean} = \left(\prod_{t=1}^{n} (1 + r_t) \right)^{1/n} - 1$$

Where r_t is the tth growth rate in a series of growth rates, n is the number of rates in the series, and Π denotes a multiplication operation.

- **Weighted Average of a Sample**

$$\text{Weighted Average} = \frac{\sum_{i=1}^{n} w_i x_i}{\sum_{i=1}^{n} w_i}$$

Where n is the number of observations, w_i is the weight assigned to each observation, x_i is the value of each observation, and Σ denotes a summation operation.

- **Location of the Median in a Sorted Sample**

  L = 0.5*(n+1)

  Where n is the number of observations in the sample.

- **Population Variance**

  $$\sigma^2 = \frac{\sum_{i=1}^{N}(x_i - \mu)^2}{N}$$

  Where x_i is the value of each observation, μ is the population mean, N is the population size, and Σ denotes a summation operation.

- **Sample Variance**

  $$s^2 = \frac{\sum_{i=1}^{n}(x_i - \overline{X})^2}{n-1}$$

  Where x_i is the value of each observation, $\overline{X}$ is the sample mean, n is the sample size, and Σ denotes a summation operation.

- **Sample Variance (Shortcut)**

  $$s^2 = \frac{1}{n-1}\left[\sum_{i=1}^{n}x_i^2 - \frac{\left(\sum_{i=1}^{n}x_i\right)^2}{n}\right]$$

  Where x_i is the value of each observation, n is the sample size, and Σ denotes a summation operation.

- **Population Standard Deviation**

  $$\sigma = \sqrt{\sigma^2}$$

  Where σ^2 is the population variance and $\sqrt{}$ denotes a square root operation.

- **Sample Standard Deviation**

    $$s = \sqrt{s^2}$$

    Where s² is the sample variance and $\sqrt{}$ denotes a square root operation.

- **Coefficient of Variation**

    $$CV = \frac{\sigma}{\mu} * 100$$

    Where σ is the population standard deviation and μ is the population mean.

- **Location of a Percentile P in a Sorted Sample**

    $$L_p = (n + 1) * \frac{P}{100}$$

    Where P is the sought percentile and n is the sample size.

CHAPTER 4

- **Probability of Elementary Event A Occurring**

    P(A) = (number of times event A occurs) / (total number of trials)

- **Probability of a Complex Event E Occurring**

    $P(E) = P(E_1) + P(E_2) + P(E_3) + ... + P(E_n)$

    Where n is the number of elementary events comprising E.

- **Probability of the Complement of Event A**

    $P(A^c) = 1 - P(A)$

- **Probability of the Union of Events A and B**

    P(A or B) = P(A) + P(B) − P(A and B)

- **Probability of Event A Occurring Given that Event B Has Occurred**

$$P(A|B) = \frac{P(A \text{ and } B)}{P(B)}$$

- **Probability of Events A and B Occurring Together (Non-Independent Events)**

$$P(A \text{ and } B) = P(B)*P(A|B) = P(A)*P(B|A)$$

- **Probability of Events A and B Occurring Together (Independent Events)**

$$P(A \text{ and } B) = P(A)*P(B)$$

- **Bayes' Theorem**

$$P(B|A) = \frac{P(B \text{ and } A)}{P(A)} = \frac{P(B \text{ and } A)}{P(B \text{ and } A) + P(B^C \text{ and } A)}$$

CHAPTER 5

- **Mean of a Probability Distribution (or Expected Value)**

$$E(X) = \mu = \sum_x xP(x)$$

Where x represents each value in the distribution, P(x) is the probability that a specific value will occur, and Σ denotes a summation operation.

- **Variance of a Probability Distribution**

$$V(X) = \sigma^2 = \sum_x (x-\mu)^2 P(x)$$

Where x represents each value in the distribution, P(x) is the probability that a specific value will occur, μ represents the mean of the probability distribution, and Σ denotes a summation operation.

- **Covariance of Two Variables (X and Y)**

$$COV(X,Y) = \sigma_{xy} = \sum_{x}\sum_{y}(x-\mu_X)(y-\mu_Y)P(x,y)$$

Where x represents each value of X, μ_X represents the mean of X, y represents each value of Y, μ_Y represents the mean of Y, P(x,y) represents the joint probability of the specific x and y values occurring, and Σ denotes a summation operation.

- **Coefficient of Correlation**

$$COR(X,Y) = \rho = \frac{COV(X,Y)}{\sqrt{V(X)}\sqrt{V(Y)}} = \frac{\sigma_{XY}}{\sigma_X\sigma_Y}$$

Where COV(X,Y) represents the covariance of X and Y, V(X) represents the variance of X, V(Y) represents the variance of Y, and $\sqrt{}$ denotes a square root operation.

- **Binomial Formula (Probability of Finding *x* Successes in *n* Trials)**

$$P(x) = \frac{n!}{x!\,(n-x)!}p^x(1-p)^{n-x}$$

Where x is the number of successes, n is the number of trials, p is the probability of x occurring, and ! denotes a factorial operation.

- **Mean of a Binomial Distribution**

$$\mu = np$$

Where p is the probability of a success and n is the number of trials.

- **Variance of a Binomial Distribution**

$$\sigma^2 = np(1-p)$$

Where p is the probability of a success and n is the number of trials.

- **Standard Deviation of a Binomial Distribution**

$$\sigma = \sqrt{np(1-p)}$$

Where p is the probability of a success, n is the number of trials, and $\sqrt{}$ denotes a square root operation.

- **Probability of x (Poisson Distribution)**

$$P(x) = e^{-\lambda} \frac{\lambda^x}{x!}$$

Where x represents the number of successes, e is the base of the natural logarithm, λ (lambda) represents the Poisson parameter (which denotes the mean and the variance of the Poisson distribution), and ! denotes a factorial operation.

CHAPTER 6

- **Probability of a Range of Values (Uniform Distribution)**

$$P(x_1 \leq X \leq x_2) = (x_2 - x_1) * \frac{1}{(b-a)}$$

Where x_1 and x_2 define the interval of interest and a and b, respectively, are the lower and upper values of the random variable.

- **Z Score**

$$Z = \frac{X - \mu}{\sigma}$$

Where X represents a value of a normally-distributed random variable, μ is the mean of the random variable, and σ is the standard deviation of the random variable.

CHAPTER 7

- **Standard Error of the Mean**

$$\text{standard error of the mean} = \frac{\sigma}{\sqrt{n}}$$

Where σ represents the standard deviation of the population, n represents the sample size, and $\sqrt{}$ denotes a square root operation.

CHAPTER 8

- **Test Statistic Used with the Critical Value Approach to Hypothesis Testing**

$$\text{test statistic} = t = \frac{\bar{X} - \mu}{\frac{s}{\sqrt{n}}}$$

Where $\bar{X}$ represents the sample mean, μ represents the hypothesized mean, s represents the sample standard deviation, n represents the sample size, and $\sqrt{}$ denotes a square root operation.

- **Confidence Interval for the Population Mean (Population Variance Known)**

$$\bar{X} \pm Z_{\alpha/2} \frac{\sigma}{\sqrt{n}}$$

Where $\bar{X}$ represents the sample mean, $\frac{\sigma}{\sqrt{n}}$ represents the standard error of the mean, and $Z_{\alpha/2}$ represents the number of standard errors we need to add to and subtract from the sample mean in order to obtain an interval estimate with a (1-α) confidence level.

- **Confidence Interval for the Population Mean (Population Variance Unknown)**

$$\bar{X} \pm t_{\alpha/2} \frac{s}{\sqrt{n}}$$

Where $\bar{X}$ represents the sample mean, $\frac{s}{\sqrt{n}}$ represents the standard error of the mean, and $t_{\alpha/2}$ represents the number of standard errors we need to add to and subtract from the sample mean in order to obtain an interval estimate with a (1-α) confidence level.

CHAPTER 9

- **Sample Proportion**

$$\hat{p} = \frac{x}{n}$$

Where x represents the number of observations holding the attribute of interest (or, the number of successes) and n represents the sample size.

EI-7

- **Standard Error of the Proportion**

    standard error of the proportion = $\sqrt{\dfrac{p(1-p)}{n}}$

    Where p represents the population proportion, n represents the sample size, and $\sqrt{}$ denotes a square root operation.

- **Test Statistic Used with Proportion Hypothesis Testing**

    test statistic = $Z = \dfrac{(\hat{p}-p)}{\sqrt{\dfrac{p(1-p)}{n}}}$

    Where $\hat{p}$ represents the sample proportion, p represents the hypothesized proportion, and $\sqrt{\dfrac{p(1-p)}{n}}$ represents the standard error of the proportion.

- **Confidence Interval for a Single Population's Proportion**

    $\hat{p} \pm Z_{\alpha/2} \sqrt{\dfrac{\hat{p}(1-\hat{p})}{n}}$

    Where $\hat{p}$ represents the sample proportion, n represents the sample size, and $Z_{\alpha/2}$ represents the number of standard errors we need to add to and subtract from the sample proportion in order to obtain an interval estimate with a (1-α) confidence level.

- **Test Statistic Used for a Single Population's Variance: Hypothesis Testing**

    test statistic = $\chi^2 = \dfrac{(n-1)s^2}{\sigma^2}$

    Where s^2 represents the sample variance, σ^2 represents the hypothesized variance, and n represents the sample size.

- **Confidence Interval for a Single Population's Variance**

  Lower Confidence Limit = $LCL = \dfrac{(n-1)s^2}{\chi^2_{\alpha/2}}$

  Upper Confidence Limit = $UCL = \dfrac{(n-1)s^2}{\chi^2_{1-\alpha/2}}$

  Where s^2 represents the sample variance, n represents the sample size, $\chi^2_{\alpha/2}$ represents the number of standard errors we need to subtract from the sample variance in order to obtain an interval estimate with a (1-α) confidence level, and $\chi^2_{1-\alpha/2}$ represents the number of standard errors we need to add to the sample variance in order to obtain an interval estimate with a (1-α) confidence level.

CHAPTER 10

- **Test Statistic Used with Two Populations' Proportions Hypothesis Testing**

  test statistic = $Z = \dfrac{(\hat{p}_1-\hat{p}_2)-(p_1-p_2)}{\sqrt{\dfrac{\hat{p}_1(1-\hat{p}_1)}{n_1}+\dfrac{\hat{p}_2(1-\hat{p}_2)}{n_2}}}$

  Where $(\hat{p}_1 - \hat{p}_2)$ represents the sample proportions difference, (p_1-p_2) represents the hypothesized proportions difference, and $\sqrt{\dfrac{\hat{p}_1(1-\hat{p}_1)}{n_1}+\dfrac{\hat{p}_2(1-\hat{p}_2)}{n_2}}$ represents the shared standard error.

- **Pooled Proportion (for Two Populations)**

  $\hat{p} = \dfrac{x_1 + x_2}{n_1 + n_2}$

  Where x_1 represents the number of successes in sample 1, x_2 represents the number of successes in sample 2, n_1 represents the size of sample 1, and n_2 represents the size of sample 2.

- **Test Statistic Used with Two Populations' Proportions Hypothesis Testing (Pooled)**

test statistic = $Z = \dfrac{(\hat{p}_1 - \hat{p}_2) - (p_1 - p_2)}{\sqrt{\hat{p}(1-\hat{p})(\frac{1}{n_1}+\frac{1}{n_2})}}$

Where $(\hat{p}_1 - \hat{p}_2)$ represents the sample proportions difference, $(p_1 - p_2)$ represents the hypothesized proportions difference, and $\sqrt{\hat{p}(1-\hat{p})(\frac{1}{n_1}+\frac{1}{n_2})}$ represents the pooled standard error.

- **Confidence Interval for the Difference between Two Populations' Proportions**

$$(\hat{p}_1 - \hat{p}_2) \pm z_{\alpha/2} \sqrt{\dfrac{\hat{p}_1(1-\hat{p}_1)}{n_1} + \dfrac{\hat{p}_2(1-\hat{p}_2)}{n_2}}$$

Where $(\hat{p}_1 - \hat{p}_2)$ represents the sample proportions difference, $\sqrt{\dfrac{\hat{p}_1(1-\hat{p}_1)}{n_1} + \dfrac{\hat{p}_2(1-\hat{p}_2)}{n_2}}$ represents the shared standard error, and $Z_{\alpha/2}$ represents the number of standard errors we need to add to and subtract from the sample proportion in order to obtain an interval estimate with a $(1-\alpha)$ confidence level.

- **Test Statistic Used with Two Populations' Variances: Hypothesis Testing**

test statistic = $F = \dfrac{s_1^2}{s_2^2}$

Where s_1^2 represents the sample variance of sample 1 and s_2^2 represents the sample variance of sample 2.

- **Confidence Interval for the Ratio of Two Populations' Variances**

Lower Confidence Limit = $LCL = \left(\dfrac{s_1^2}{s_2^2}\right) \dfrac{1}{F_{\alpha/2, df_1, df_2}}$

Upper Confidence Limit = $UCL = \left(\dfrac{s_1^2}{s_2^2}\right) F_{\alpha/2, df_2, df_1}$

Where s_1^2 represents the sample variance of sample 1, s_2^2 represents the sample variance of sample 2, and $\dfrac{1}{F_{\alpha/2, df_1, df_2}}$ and $F_{\alpha/2, df_2, df_1}$ represent, respectively, appropriate values from the F-distribution to obtain an interval estimate with a $(1-\alpha)$ confidence level.

- **Test Statistic Used with Two Populations' Means Hypothesis Testing (Equal Variances and Independent Samples)**

    test statistic $= t = \dfrac{(\bar{X}_1 - \bar{X}_2) - (\mu_1 - \mu_2)}{\sqrt{s_p^2 (\frac{1}{n_1} + \frac{1}{n_2})}}$

    Where $(\bar{X}_1 - \bar{X}_2)$ represents the observed difference between the two means, $(\mu_1 - \mu_2)$ represents the hypothesized difference between the two means, S_p^2 is the pooled variance, n_1 is the size of sample 1, n_2 is the size of sample 3, and $\sqrt{}$ denotes a square root operation.

- **Pooled Variance**

    $$s_p^2 = \dfrac{(n_1 - 1)s_1^2 + (n_2 - 1)s_2^2}{n_1 + n_2 - 2}$$

    Where s_1^2 represents the variance of sample 1, s_2^2 represents the variance of sample 2, n_1 represents the size of sample 1, and n_2 represents the size of sample 2.

- **Test Statistic Used with Two Populations' Means Hypothesis Testing (Unequal Variances and Independent Samples)**

    test statistic $= t = \dfrac{(\bar{X}_1 - \bar{X}_2) - (\mu_1 - \mu_2)}{\sqrt{(\frac{s_1^2}{n_1} + \frac{s_2^2}{n_2})}}$

    Where $(\bar{X}_1 - \bar{X}_2)$ is the observed difference between the two means, $(\mu_1 - \mu_2)$ is the hypothesized difference between the two means, S_1^2 is the variance of sample 1, S_2^2 is the variance of sample 2, and $\sqrt{}$ denotes a square root operation.

- **Degrees of Freedom with Two Populations' Means Hypothesis Testing (Unequal Variances and Independent Samples)**

    $$df = \dfrac{\left(\frac{s_1^2}{n_1} + \frac{s_2^2}{n_2}\right)^2}{\dfrac{\left(\frac{s_1^2}{n_1}\right)^2}{n_1 - 1} + \dfrac{\left(\frac{s_2^2}{n_2}\right)^2}{n_2 - 1}}$$

    Where s_1^2 represents the variance of sample 1, s_2^2 represents the variance of sample 2, n_1 represents the size of sample 1, and n_2 represents the size of sample 2.

- **Test Statistic Used with Two Paired Populations' Means Hypothesis Testing**

  Test statistic = $t = \dfrac{\bar{d} - \mu_d}{s_d/\sqrt{n}}$

  Where d is the difference ($d = x_1 - x_2$) between each set of the paired observations, $\bar{d}$ is the mean difference across all paired observations ($\bar{d} = \dfrac{\sum_{i=1}^{n} d_i}{n}$), μ_d is the hypothesized mean difference, s_d is the sample standard deviation of the differences, n is the number of pairs in the sampled data, and $\sqrt{}$ denotes a square root operation.

- **Confidence Interval for the Difference between Two Populations' Means (Equal Variances and Independent Samples)**

  $$(\bar{X}_1 - \bar{X}_2) \pm t_{\alpha/2, df} \sqrt{s_p^2 \left(\dfrac{1}{n_1} + \dfrac{1}{n_2}\right)}$$

  $$df = n_1 + n_2 - 2$$

  Where $(\bar{X}_1 - \bar{X}_2)$ represents the observed difference between the two means, S_p^2 represents the pooled variance, n_1 represents the size of sample 1, n_2 represents the size of sample 2, df represents the number of degrees of freedom, $t_{\alpha/2, df}$ represents the number of standard errors we need to add to and subtract from the sample proportion in order to obtain an interval estimate with a (1-α) confidence level, and $\sqrt{}$ denotes a square root operation.

- **Confidence Interval for the Difference between Two Populations' Means (Unequal Variances and Independent Samples)**

  $$(\bar{X}_1 - \bar{X}_2) \pm t_{\alpha/2, df} \sqrt{\left(\dfrac{s_1^2}{n_1} + \dfrac{s_2^2}{n_2}\right)}$$

  $$df = \dfrac{\left(\dfrac{s_1^2}{n_1} + \dfrac{s_2^2}{n_2}\right)^2}{\dfrac{\left(\dfrac{s_1^2}{n_1}\right)^2}{n_1 - 1} + \dfrac{\left(\dfrac{s_2^2}{n_2}\right)^2}{n_2 - 1}}$$

  Where $(\bar{X}_1 - \bar{X}_2)$ represents the observed difference between the two means, S_1^2 represents the variance of sample 1, S_2^2 represents the variance of sample 2, n_1 represents the size of sample 1, n_2 represents the size of sample 2, df represents the number of degrees of freedom, $t_{\alpha/2, df}$ represents the number of standard errors we need to add to and subtract from the sample proportion in order to obtain an interval estimate with a (1-α) confidence level, and $\sqrt{}$ denotes a square root operation.

- **Confidence Interval for the Difference between Two Paired Populations' Means**

$$\bar{d} \pm t_{\alpha/2, df} \frac{s_d}{\sqrt{n}}$$

df = n-1

Where d is the difference (d = $x_1 - x_2$) between each set of the paired observations, $\bar{d}$ is the mean difference across all paired observations ($\bar{d} = \frac{\sum_{i=1}^{n} d_i}{n}$), s_d is the sample standard deviation of the differences, n is the number of pairs in the sampled data, df represents the number of degrees of freedom, $t_{\alpha/2, df}$ represents the number of standard errors we need to add to and subtract from the sample proportion in order to obtain an interval estimate with a (1-α) confidence level, and $\sqrt{}$ denotes a square root operation.

CHAPTER 11

- **Test Statistic Used with the Goodness of Fit Test**

$$\text{test statistic} = \chi^2 = \sum_{i=1}^{k} \frac{(o_i - e_i)^2}{e_i}$$

Where o_i represents the observed cell frequency for category i, e_i represents the expected cell frequency for category i, k represents the number of categories, and Σ denotes a summation operation.

- **Computing Expected Cell Frequencies in Contingency Tables**

$$e_{ij} = \frac{(i^{th} \text{ row total}) * (j^{th} \text{ column total})}{\text{Total sample size}}$$

- **Test Statistic Used with Test of Independence**

$$\text{test statistic} = \chi^2 = \sum_{i=1}^{r} \sum_{j=1}^{c} \frac{(o_{ij} - e_{ij})^2}{e_{ij}}$$

Where r is the number of rows, c is the number of columns, o_{ij} is the observed cell frequency, e_{ij} is the expected cell frequency, and Σ denotes a summation operation.

- **Degrees of Freedom Used with Test of Independence**

  $df = (r-1)*(c-1)$

  Where r is the number of rows and c is the number of columns.

CHAPTER 12

- **Sum of Squares between Groups: One-Way ANOVA**

$$SSB = \sum_{j=1}^{k} n_j(\bar{X}_j - \bar{\bar{X}})^2$$

Where k represents the number of groups, $\bar{X}_j$ represents the mean of group j, $\bar{\bar{X}}$ represents the grand mean, n_j is the size of group j, and Σ denotes a summation operation.

- **Sum of Squares within Groups: One-Way ANOVA**

$$SSW = \sum_{j=1}^{k}\sum_{i=1}^{n_j}(x_{ij} - \bar{X}_j)^2 = \sum_{j=1}^{k}(n_j - 1)s_j^2$$

$$s_j^2 = \frac{\sum_{i=1}^{n}(x_i - \bar{X})^2}{n-1}$$

Where k represents the number of groups, $\bar{X}_j$ represents the mean of group j, n_j is the size of group j, s_j^2 represents the variance of group j, and Σ denotes a summation operation.

- **Sum of Squares Total: One-Way ANOVA**

  SST = SSB + SSW

- **Mean Squares Between: One-Way ANOVA**

$$MSB = \frac{SSB}{k-1}$$

Where k represents the number of groups.

- **Mean Square Within: One-Way ANOVA**

$$MSW = \frac{SSW}{n-k}$$

Where k represents the number of groups and n represents the number of observations.

- **Test Statistic for One-Way ANOVA**

$$F = \frac{MSB}{MSW}$$

- **Mean Square Blocking: Randomized Block ANOVA**

$$MSBL = \frac{SSBL}{b-1}$$

Where b represents the number of blocks.

- **Mean Square Factor (Between): Randomized Block ANOVA**

$$MSB = \frac{SSB}{k-1}$$

Where k represents the number of groups.

- **Mean Square Error (Within): Randomized Block ANOVA**

$$MSW = \frac{SSW}{n-k-b+1}$$

Where n represents the number of observations, k represents the number of groups, and b represents the number of blocks.

- **Blocking Test Statistic: Randomized Block ANOVA**

$$F = \frac{MSBL}{MSW}$$

- **Factor Test Statistic: Randomized Block ANOVA**

$$F = \frac{MSB}{MSW}$$

- **Mean Square Factor A: Two-Way ANOVA**

$$MSA = \frac{SSA}{a-1}$$

Where a represents the number of levels of Factor A.

- **Mean Square Factor B: Two-Way ANOVA**

$$MSB = \frac{SSb}{b-1}$$

Where b represents the number of levels of Factor B.

- **Mean Square Interaction: Two-Way ANOVA**

$$MSAB = \frac{SSAB}{(a-1)*(b-1)}$$

Where a represents the number of levels of Factor A and b represents the number of levels of Factor B.

- **Mean Square Error (Within): Two-Way ANOVA**

$$MSW = \frac{SSW}{n-(a*b)}$$

Where n represents the number of observations, a represents the number of levels of Factor A, and b represents the number of levels of Factor B.

- **Factor A Test Statistic: Two-Way ANOVA**

$$F = \frac{MSA}{MSW}$$

- **Factor B Test Statistic: Two-Way ANOVA**

$$F = \frac{MSB}{MSW}$$

- **Interaction Test Statistic: Two-Way ANOVA**

$$F = \frac{MSAB}{MSW}$$

CHAPTER 13

- **General Form of a Regression Model**

$$Y = \beta_0 + \beta_1{*}X_1 + \beta_2{*}X_2 + \beta_3{*}X_3 + \beta_4{*}X_4 + \varepsilon$$

Where Y represents a dependent variable, X_i represents an independent variable whose value influences the value of Y, β_i represents the estimated regression coefficient for X_i, and β_0 represents the estimated intercept term, and ε represents the error term.

- **Regression Sum of Squares**

$$SSR = \sum (\hat{y} - \bar{y})^2$$

Where $\hat{y}$ represents the each estimated value of Y, $\bar{y}$ represents the mean value of Y, and Σ denotes a summation operation.

- **Error Sum of Squares**

$$SSE = \sum (y - \hat{y})^2$$

Where $\hat{y}$ represents the each estimated value of Y, y represents the actual values of Y, and Σ denotes a summation operation.

- **Total Sum of Squares**

$$SST = \sum (y - \bar{y})^2 = SSR + SSE$$

Where $\bar{y}$ represents the mean value of Y, y represents the actual values of Y, and Σ denotes a summation operation.

- **Coefficient of Determination (R²)The**

$$R^2 = \frac{SSR}{SST}$$

- **The Standard Error of the Estimate (Sε)**

$$S_\varepsilon = \sqrt{\frac{SSE}{n-k-1}}$$

Where n represents the number of observations and k represents the number of independent variables.

- **Regression Coefficient Test Statistic**

$$\text{test statistic} = t = \frac{b-\beta}{S_b}$$

Where b represents the estimated regression coefficient, β represents the hypothesized coefficient (zero according to the null hypothesis), and s_b represents the standard error of the regression coefficient.

- **Regression Line Confidence Interval (One Independent Variable)**

$$\hat{y} \pm t_{\frac{\alpha}{2}, n-2} * S_\varepsilon \sqrt{\frac{1}{n} + \frac{(x^* - \bar{X})^2}{\sum(x_i - \bar{X})^2}}$$

Where $\hat{y}$ represents the estimated value of Y, x^* represents a given value of X, $\bar{X}$ represents the mean value of X, S_ε represents the standard error, n represents the number of observations, $t_{\frac{\alpha}{2}, n-2}$ represents the number of standard errors we need to add to and subtract from the sample proportion in order to obtain an interval estimate with a (1-α) confidence level, and $\sqrt{}$ denotes a square root operation.

- **Regression Line Prediction Interval (One Independent Variable)**

$$\hat{y} \pm t_{\frac{\alpha}{2}, n-2} * S_\varepsilon \sqrt{1 + \frac{1}{n} + \frac{(x^* - \bar{X})^2}{\Sigma(x_i - \bar{X})^2}}$$

Where $\hat{y}$ represents the estimated value of Y, x^* represents a given value of X, $\bar{X}$ represents the mean value of X, S_ε represents the standard error, n represents the number of observations, $t_{\frac{\alpha}{2}, n-2}$ represents the number of standard errors we need to add to and subtract from the sample proportion in order to obtain an interval estimate with a (1-α) confidence level, and $\sqrt{}$ denotes a square root operation.

KEY TERM INDEX

KEY TERM	CHAPTER (UNIT SUMMARY)	PAGE
Addition Rule	4 (2)	140
Adjusted R^2	13 (2)	514
Alternative Hypothesis	7 (1)	237
Analysis of Variance (ANOVA)	12 (1)	440
Analytics Techniques	1 (4)	31
ANOVA: Two Factor with Replication	12 (3)	477
Arithmetic Mean	3 (1)	91
Bar Chart	2 (4)	80
Base Level	13 (3)	532
Bayes' Theorem	4 (3)	150
Big Data	1 (4)	30
Bimodal Distribution	3 (1)	101
Binomial Experiment	5 (2)	180
Binomial Random Variable	5 (2)	181
Bivariate Distribution	5 (1)	170
Blocking (ANOVA)	12 (3)	461
Box Plot	3 (3)	119
Business Analytics	1 (Introduction)	1
Causal Relationship	13 (Introduction)	493
Causality	13 (Introduction)	493
Census	1 (2)	12
Central Limit Theorem	7 (2)	248
Chi-squared (χ^2) Distribution	6 (1)	210
Class Width Interval (frequency distribution)	2 (1)	50
Classes (frequency distribution)	2 (1)	46
Classical Probabilities	4 (1)	131
Cluster Sample	1 (2)	17
Coefficient of Correlation	5 (1)	173
Coefficient of Determination	13 (2)	509
Coefficient of Variation	3 (2)	113
Column Chart	2 (2)	59
Complement of an Event	4 (2)	136
Conditional Probability	4 (2)	143
Confidence Interval	8 (2)	307
Confidence Level	8 (2)	307
Consistent (point estimate)	8 (2)	306
Context (data)	3 (Introduction)	88
Contingency Table	4 (2); 11 (2)	139; 428
Continuous Random Variable	6 (1)	204
Continuous Data	1 (1)	5
Correlation Matrix	13 (2)	520
Covariance	5 (1)	172
Critical Value (hypothesis testing)	8 (1)	295
Cumulative Relative Frequency	2 (3)	75
Data	1 (1)	3
Data Visualization	2 (Introduction)	35
Data Wrangling	1 (4)	31
Degrees of Freedom	6 (1)	209

KEY TERM	CHAPTER (UNIT SUMMARY)	PAGE
Dependent Variable	12 (1); 13 (Introduction)	440; 494
Descriptive Statistics	1 (Introduction)	1
Discrete Data	1 (1)	5
Distribution	2 (2)	62
Dummy Variable	13 (3)	528
Elementary Event (probability)	4 (1)	129
Empirical Rule	3 (2)	112
Estimated Regression Model	13 (2)	504
Events (probability)	4 (1)	129
Exhaustive (class)	2 (1)	47
Exhaustive (hypotheses)	7 (1)	239
Expected Value	5 (1)	169
Experiments	1 (3)	26
Extrapolating	13 (3)	534
F-Distribution	6 (1); 10(2)	211; 371
F-Statistic	12 (2)	449
Factor (ANOVA)	12 (1)	440
Frequency	2 (1)	40
Frequency Distribution Table	2 (2)	62
Frequency Table	2 (1)	41
Generalizable (survey)	1 (3)	25
Geometric Mean	3 (1)	93
Goodness of Fit	11 (1)	416
Goodness of Fit Test for Normality	11 (1)	419
Grand Mean	12 (2)	445
Histogram	2 (2)	60
Hypothesis	7 (1)	236
Independent (events)	4 (2)	138
Independent Samples	10 (3)	381
Independent Variables	11 (2); 13 (Introduction)	428; 494
Inferential Statistics	1 (Introduction)	1
Interaction (ANOVA)	12 (3)	474
Interaction Plot	12 (3)	482
Interpolating	13 (3)	534
Interquartile Range (IQR)	3 (3)	118
Intersection of Events	4 (2)	136
Interval Estimator	8 (2)	307
Interval Scale	1 (1)	8
Interviews	1 (3)	27
Inverse Probability Calculation	6 (2)	225
Joint Probabilities	4 (2)	140
Levels (ANOVA)	12 (1)	440
Level of Significance	7 (3)	269
Line Graph	2 (4)	83
Linear Regression	13 (1)	499
Lower Bound (class)	2 (1)	50
Lower Confidence Limit (LCL)	8 (2)	310
Lower-Tail Test	7 (1)	238
Marginal Probabilities	4 (2)	140
Matched Samples	10 (3)	391
Mean	3 (1)	90

KEY TERM	CHAPTER (UNIT SUMMARY)	PAGE
Mean Difference	10 (3)	392
Median	3 (1)	97
Measure of Relative Standing	3 (3)	116
Measures of Centrality	3 (Introduction)	89
Measures of Variation	3 (Introduction)	89
Modal Class	3 (1)	100
Mode	3 (1)	99
Multicollinearity	13 (2)	520
Multiple R	13 (2)	514
Multiplication Rule (probability)	4 (2)	145
Mutually Exclusive (class)	2 (1)	47
Mutually Exclusive (probability)	4 (2)	137
Mutually Exclusive (hypotheses)	7 (1)	239
Nominal Scale	1 (1)	6
Non-response Bias	1 (3)	25
Normal Distribution	6 (1)	207
Null Hypothesis	7 (1)	236
Observations	1 (3)	26
Observer Bias	1 (3)	27
Ogive	2 (3)	76
One-Way ANOVA	12 (2)	445
Ordinal Scale	1 (1)	7
Ordinary Least Squares (OLS) Regression	13 (1)	500
Outliers	3 (3)	122
P^{th} Percentile	3 (3)	117
P-Value	6 (2)	224
Paired Samples	10 (3)	391
Pie Chart	2 (3)	75
Point Estimator	8 (2)	306
Poisson Experiment	5 (3)	191
Poisson Random Variable	5 (3)	191
Pooled Proportion	10 (1)	360
Population	1 (2)	12
Population Parameter	1 (2)	20
Prediction Interval	13 (3)	540
Primary Data	1 (3)	24
Probability	4 (1)	129
Probability Density Function	6 (1)	205
Probability Distribution	5 (1)	166
Probability Tree	4 (3)	154
Qualitative Data	1 (1)	5
Quantitative Data	1 (1)	5
Quartiles	3 (3)	118
Random Sampling	1 (2)	14
Random Variable	5 (1)	164
Randomized Block ANOVA	12 (3)	462
Range (data values)	2 (1)	50
Ratio Scale	1 (1)	8
Regression Analysis	13 (Introduction)	493
Regression Coefficients	13 (2)	505
Regression Line	13 (1)	499

KEY TERM	CHAPTER (UNIT SUMMARY)	PAGE
Relative Frequency (class)	2 (3); 4 (1)	73; 132
Rejection Region	8 (1)	296
Replications (ANOVA)	12 (3)	475
Research Hypothesis	7 (1)	236
Response Rate	1 (3)	24
Rule of Five	11 (1)	423
Sample	1 (2)	13
Sample Space	4 (1)	129
Sampling	1 (2)	13
Sampling Bias	1 (2)	14
Sampling Distribution	7 (2)	244
Sampling Distribution of the Mean	7 (2)	244
Sampling Distribution of Proportions	9 (1)	324
Sampling Error	1 (2)	20
Sample Statistic	1 (2)	20
Scales of Measurement	1 (1)	6
Scatter Plot	2 (4)	82
Secondary Data	1 (3)	24
Simple Random Sample	1 (2)	15
Skewness	2 (2)	62
Standard Deviation	3 (2)	108
Standard Error of the Estimate	13 (2)	514
Standard Error of the Mean	7 (2)	248
Standard Normal Distribution	6 (2)	215
Statistics	1 (Introduction)	1
Stem-and-Leaf	2 (2)	66
Stratified Sample	1 (2)	16
Sturges' Formula	2 (1)	48
Subjective Probability	4 (1)	132
Sum of Squares (ANOVA)	12 (2)	447
Sum of Squares between Groups (SSB)	12 (2)	448
Sum of Squares within Groups (SSW)	12 (2)	448
Surveys	1 (3)	24
Symmetric (distribution)	6 (1)	208
Symmetrical (distribution)	2 (2)	62
t-Distribution	6 (1)	209
Test of Independence	11 (2)	427
Test Statistic (hypothesis testing)	7 (3)	268
Time Series Data	2 (4)	83
Two-Factor ANOVA without Replication	12 (3)	462
Two-Factors ANOVA	12 (3)	473
Two-Tail Test	7 (1)	238
Two-Way ANOVA	12 (3)	458
Type I Error	7 (3)	276
Type II Error	7 (3)	276
Unbiased (point estimate)	8 (2)	306
Uniform Distribution	6 (1)	206
Unimodal (distribution)	3 (1)	101
Union of Events (probability)	4 (2)	136
Univariate Distribution	5 (1)	170
Upper Bound (class)	2 (1)	50

KEY TERM	CHAPTER (UNIT SUMMARY)	PAGE
Upper Confidence Limit (UCL)	8 (2)	310
Upper-Tail Test	7 (1)	238
Variable	1 (1)	3
Variance	3 (2)	105
Variation	1 (2)	19
Variety	1 (4)	30
Velocity	1 (4)	30
Venn Diagram	4 (2)	135
Volume	1 (4)	30
Weighted Mean	3 (1)	94
Z Score	6 (2)	215

CPSIA information can be obtained
at www.ICGtesting.com
Printed in the USA
BVHW091504220119
538218BV00007B/40/P